Invasion Biology

Invasion Biology

Mark A. Davis

OXFORD
UNIVERSITY PRESS

Great Clarendon Street, Oxford OX2 6DP

Oxford University Press is a department of the University of Oxford.
It furthers the University's objective of excellence in research, scholarship,
and education by publishing worldwide in

Oxford New York

Auckland Cape Town Dar es Salaam Hong Kong Karachi
Kuala Lumpur Madrid Melbourne Mexico City Nairobi
New Delhi Shanghai Taipei Toronto

With offices in

Argentina Austria Brazil Chile Czech Republic France Greece
Guatemala Hungary Italy Japan Poland Portugal Singapore
South Korea Switzerland Thailand Turkey Ukraine Vietnam

Oxford is a registered trade mark of Oxford University Press
in the UK and in certain other countries

Published in the United States
by Oxford University Press Inc., New York

First published 2009

British Library Cataloguing in Publication Data
Data available

Library of Congress Cataloging in Publication Data
Data available

Typeset by Newgen Imaging Systems (P) Ltd., Chennai, India
Printed in Great Britain
on acid-free paper by
CPI Antony Rowe, Chippenham, Wiltshire

ISBN 978–0–19–921875–2 978–0–19–921876–9 (Pbk.)

10 9 8 7 6 5 4 3 2 1

To my wife, Jean;
my children: Emily, Zack, Ben, and Hazel;
my mother; and, in memory of my father.

Preface

Trying to write a book on invasion biology is a daunting task. First, one's book inevitably will be compared to Mark Williamson's 1996 classic on the topic. Second, articles and books on the subject of biological invasions are being published at a furious pace. Writing this book was a lot like trying to shovel a sidewalk in the middle of a Minnesota blizzard. As soon as I cleared a path, a new gust of wind would sweep in new drifts that needed my attention. Third, while much progress has been made since Williamson's book was published, uncertainty and some lively contentiousness still characterize much of this exciting and rapidly growing field. Thus, it would be impossible for any author to please all the readers. I am quite confident that all readers will find something in this book to contest. Since the cumulative wisdom and instincts of an entire research community will normally trump that of any individual member, I am less concerned that I got everything right than I am that the book promotes focused reflection and dialogue.

During the twentieth century, Elton provided the field its primary operating paradigm. By the end of the first decade of the twenty-first century, the field of invasion biology is in a state of transition, somewhere between its Eltonian-based roots and a reframed and robust discipline of the twenty-first century. I hope that this book can help the field redefine itself during this period of self-transformation. My intended primary audience for this book includes students, researchers, land managers, and science writers, although I believe that policy makers and general citizens would find certain portions of the book interesting and informative as well. As a teacher, I have always valued thoughtful and critical discussions. It is my hope that this book will fuel many such discussions, both inside and outside the classroom.

To try to meet the needs and interests of different readers, and to make the book as user-friendly as possible, I have included a taxonomic index and a geographic index, in addition to a general subject index.

Acknowledgements

During the writing of this book, I contacted many of my colleagues with questions about their work or inquiries regarding particular organisms or ecological systems. Without exception, colleagues responded promptly and generously provided me with the information, articles, and manuscripts I requested. Thank you to all helped me in this way. I am enormously grateful to my colleagues who were kind enough to read initial drafts of chapters and provide me invaluable feedback, insights, and suggestions: Peter Alpert, Laura Calabrese, James Carlton, Aleja Carvajal, Matthew Chew, Scott Collins, Frank Egerton, Paul Gobster, Randall Hughes, Clive Jones, Sally Koerner, John Lambrinos, Jeff McNeeley, Etsuko Nonaka, Petr Pyšek, Marcel Rejmánek, David Richardson, Robert Ricklefs, Dov Sax, Jay Stachowicz, and Mark Williamson. I am indebted to Stan Wagon, who produced the invasion pressure graphs shown in Chapters 6 and 8. I also want to thank Macalester College for its support during this project. I particularly want to express my appreciation to the countless Macalester students with whom, over the years, I have discussed and developed many of the ideas in this book. Their fresh ideas and perspectives often prompted me to think in new ways. I am very grateful to the National Science Foundation, which has generously funded me over the years, including grants that supported my work on invasive species. Thank you to Editor Ian Sherman for inviting me to write a book on invasion biology to Associate Editor Helen Eaton for her prompt and invaluable assistance during the writing stage, and Jeannie Labno and Carol Bestley for their help during production process. Finally, I want to thank my wife, Jean, and my four children—Hazel, Ben, Zack, and Emily, and my many friends and colleagues for their support and interest shown in this project.

Contents

CHAPTER 1

Introduction

In any venture, it is a good idea to pause periodically and reflect on what one is doing and where one is heading. As it turns out, this might be a particularly opportune time to reflect on what we have been doing in invasion biology. This book might have been titled, *Invasion biology fifty years later*. It was completed in 2008, marking the fiftieth anniversary of the 1958 publication of Elton's well-known book, *The ecology of invasions by animals and plants*. Coincidentally, 2008 also marked the twenty-fifth anniversary of the creation of the SCOPE (Scientific Committee on Problems of the Environment, a committee formed by the International Council of Scientific Unions) scientific advisory committee, which was charged with the responsibility of focusing scientific attention on invasive species. In addition, more than a decade has passed since the 1996 publication of Williamson's book, *Biological invasions*. As good as Williamson's book was, it has become dated in the wake of the immense amount of research conducted since its publication, much of which was inspired and guided by Williamson's book. With the exception of climate change, biological invasions probably have received more attention in recent years, both from ecologists and the public-at-large, than any other ecological topic. In the past ten years, the number of papers, books, workshops, and symposia addressing biological invasions extends into the thousands.

There have been several outstanding edited volumes on the topic of invasion biology published in recent years (e.g. Ruiz and Carlton 2003a, Inderjit 2005, Mooney *et al.* 2005, Sax *et al.* 2005a, Cadotte *et al.* 2006, Gherardi 2007, Nentwig 2007). In addition, Lockwood *et al.* (2007) wrote an excellent introductory textbook, *Invasion ecology*, that primarily targeted new students, and Cox (2004) provided a detailed review of biological invasions from a focused evolutionary perspective in his book, *Alien species and evolution*. There have also been many recent thoughtful review papers on aspects of invasion biology (a small sampling includes Duncan *et al.* 2003, Dietz and Steinlein 2004, Carlton and Ruiz 2005, Alpert 2006, Henderson *et al.* 2006, Richardson and Pyšek 2006, Sax *et al.* 2007, Theoharides and Dukes 2007). However, there has been no single-authored text, written to provide a consistent and unifying perspective to the field, since Williamson's book. I hope this book can at least partially fill this role.

The reader may have noticed that the title of this book consists of the same two words used by Williamson in his 1996 volume. However, I have reversed their order. The change was intentional and done to emphasize that a primary subject of this book is the discipline of invasion biology itself. My intent was not to present an exhaustive report on the spread and impact of invasive species. Countless books, articles, and reports have provided such details, with new ones coming out all the time. While I have tried to review much of the current research and discussions regarding invasions, neither was my goal to provide a comprehensive accounting of all invasion literature. This would have resulted in a tome, and I wanted to produce something more focused and readable. Moreover, given the large number of articles being published in the field every year, this book would rapidly become outdated if its primary function was to serve as a bibliographic repository. Rather, in writing this book, I have tried to review and reflect on the approaches, findings, controversies, and conclusions that have defined invasion biology in recent years. My hope is that these assessments will help the field decide whether it should continue along its current tack, or whether it might want to consider adjusting some of the rigging in order to make some mid-course corrections.

I certainly have no pretense of writing a final word on invasion biology. As scientists we can only write middle words. I have noticed that some of my colleagues in the arts seem concerned that their work may eventually be forgotten after they die. I do not think many scientists experience this concern. Whereas most artistic work involves individual creation, our work is ultimately communal. Moreover, scientific knowledge is fundamentally cumulative, in a way that most art is not. As a scientific community, we are constructing a fabric of knowledge, with individual scientists contributing their own threads and weave. We expect our contributions to be modified and superseded, and even possibly rejected, in the future. Our pride and sense of purpose usually does not come from individual creations but from the privilege of contributing to the extraordinary collaborative human endeavor known as the natural sciences.

Thus, this book should be viewed simply as 'a word' on invasion biology, and even then, just a word on certain aspects of the discipline. The field of invasion biology, with its many dimensions, is much too large to be completely covered in a single book. Each of us has our own areas of interest and expertise and undoubtedly many readers will wish more attention had been given to certain areas and less to others. Likewise, although more than 1000 works are cited in this book, they represent only a small portion of the invasion literature, and many readers will likely think the book should have recognized other sources, possibly including some of their own.

There is an enormous amount of first-rate research being conducted around the world and not all of it receives the attention it deserves. While emphasizing the important contributions to the discipline made by recognized leaders in the field, I have tried to cite the work of lesser known invasion biologists as well, many of these individuals being part of the new generation of invasion scientists. In addition, I have done my best to avoid a pronounced geographical or taxonomic bias in the book, although this is a challenge given that so much of the field has been dominated by plant invasion research invasions related to North America (Pyšek *et al.* 2006, 2008).

While there is a clear historical dimension to this book, I wanted to focus principally on the most recent perspectives, those that are currently guiding the field. For the most part, I have tried to characterize the state of the field of invasion biology at the end of the first decade of the twenty-first century. Thus, more than 75% of the works cited in this book were published since 2000, with nearly half published since 2005. Excellent detailed treatments of the field prior to 2000 exist for those interested in the state of the field at that time, e.g. Shigesada and Kawasaki (1997) and Williamson's 1996 book, along with the several edited volumes on invasion biology produced by SCOPE. If readers believe this book could have been improved with the recognition of other specific publications, research efforts, or perspectives, they are invited to communicate their ideas to the author. Perhaps in a few years there will be a subsequent edition that will be able to incorporate these suggestions.

Terminology used in this book

The terminology used in the field of invasion biology has been the subject of much discussion and dispute, since at least the mid-1800s. At that time in England, botanist Hewett Cottrell Watson (1847, 1859), distressed over the the lack of a standardized terminology to describe plant species with different geographic histories, proposed the following vocabulary scheme (presented in Watson's words, as reported by Chew 2006).

Native: Apparently an aboriginal British species; there being little or no reason for supposing it to have been introduced by human agency.

Denizen: At present maintaining its habitats, as if a native, without the aid of humans, yet liable to some suspicion of having been originally introduced by human agency, whether by design or by accident.

Colonist: A weed of cultivated land, by road-sides or about houses, and seldom found except in places where the ground has been adapted for its production and continuance by the operations of humans; with tendency to appear on the shores, landslips, road-sides, rubbish heaps, and dunghills.

Incognita: Reported as British, but requiring confirmation as such. Some of these have been reported through mistakes of the species... others may have been really seen in the character of temporary stragglers from gardens... others cannot now be found in the localities published for them... though it is not improbable that some of these may yet be found again. A few may have existed for a time, and become extinct.

Hibernian, or Sarnian: Native, or apparently so, in Ireland, or in the Channel Isles, though not found in Britain proper.

More than a century and a half later, the field is still trying to create a standardize terminology. If anything, the variety of terms has multiplied. Colautti and MacIsaac (2004) listed more than thirty terms that have been used in the invasion literature to describe species that have recently dispersed into an area.

There have been a number of recent attempts to standardize the field's terminology (Davis and Thompson 2000, 2001, Richardson *et al.* 2000a, Daehler 2001a, Rejmánek *et al.* 2002a, Colautti and MacIsaac 2004, Occhipinti-Ambrogi and Galil 2004, Larson 2005). Although previously involved in the effort to standardize the field's vocabulary (Davis and Thompson 2000, 2001, 2002), I believe, at this point, our time is better spent on other things. Based on his review of the field's use of terminology, Carlton (2002) concluded that terminology varied among countries and scientists, and that there was no indication that the field would be able to achieve uniformity in language in the near future. Hodges (2008) reviewed and assessed efforts by ecologists to define terminology and argued that delimited definitions can actually constrain a field's development and recommended that ecologists stop trying to impose a particular terminology scheme on the field.

I do think the lack of a clearly articulated and widely utilized terminology in invasion biology has its costs. For example, invasion ecologists, managers, and policy makers can easily end up miscommunicating with one another by not realizing that they mean different things while using the same words, or the same thing with different words. Pondering the general issue of scientific terminology more than three hundred years ago, Antoine Lavoisier came to a similar conclusion, stating:

> We cannot improve the language of any science without at the same time improving the science itself; neither can we, on the other hand, improve a science without improving the language or nomenclature which belongs to it (cited in Goldenfeld 2007).

Pickett *et al.* (2007) concurred, concluding that 'a failure to develop good definitions may constitute a major impediment to progress in a discipline.' However, I agree with Hodges that language is very much a living entity and that it is extremely difficult to impose one's will on it, hence my current pragmatic view that our time now is probably better spent on other things. Certainly, if we have been unable to agree upon a common invasion lexicon in 160 years, perhaps we at least deserve a break. However, this does mean that it is vital that any authors writing in the invasion field make clear their terminology. Thus, I would like to explain at the outset my rationale for proceeding with the approach that I have used in this book.

I have never liked the term 'invasion' and think the field would have been much better off had it never been adopted, along with its accompanying military metaphors. Although the usage of military language may help to attract a group of highly motivated supporters, this same language may help foment a strongly confrontational approach, making it much more difficult to negotiate and resolve conflicts (Schroeder 2000). Despite my misgivings over this term, I recognize the absurdity of trying to write a book on invasion biology without using the word invasion. Thus, I have used the term throughout the book, although I have consciously avoided use of the terms invade or invader, outside of quotes.

Some have used the term invasion to refer to any process of colonization and establishment beyond a former range (Reise *et al.* 2006). I actually prefer this usage, since it applies to all species. In the future, I hope that researchers will take this more inclusive approach. However, at the current time, I am afraid that taking this approach would be more confusing than helpful. Thus, for purposes of this book, I have restricted the use of the word invasion to those range expansions in which the transport of the organisms to a new region was mediated by humans.

I believe an ideal approach to the study of the global redistribution of species would be one that simply distinguished species on the basis of how long they have been a resident in a region. European plant ecologists have partially taken this approach, distinguishing native species from archaeophytes, which were introduced with the spread of humans and agriculture up to the year 1500, and neophytes, which were introduced after 1500 (strictly speaking 1492), when Europe began to engage in regular biotic exchange with the western hemisphere. Chew (2006) took a similar approach and used the term neobiota. From a strictly scientific perspective, I think neobiota is a better label than non-native. Nevertheless, given that I am likely to upset the reader in other ways, I have chosen not to cause additional annoyance by using the term neobiota. I have not liked the use of terms such as aliens, exotics, and invaders, primarily owing to the pejorative implications of the terms in general usage. For the same reason, I do not like the term xenodiversity ('strange' diversity), used by Leppäkoski and Olenin (2000a) to refer to the diversity patterns of recently introduced species, in contrast to native biodiversity. Thus, in most cases I have opted for native and non-native. The terms native and non-native, and indigenous and non-indigenous, also can have normative implications, but in general their connotations are more benign. I should note that it is perfectly possible to write in this field without using any of the above words (e.g. Davis 2003). In the 2003 article, I referred to long-term resident species, recently arrived species, new species, and introduced species. A benefit of the first three terms in particular is that one avoids the strictly dichotomous paradigm typically imposed by the field's terminology (McNeely 2005).

There is a long history of using the term 'introduced species' for non-native species, and in this book, I have used the term introduced species interchangeably with non-native species. Consistent with the usages of the term by the 1992 Convention on Biological Diversity, the term introduced does not imply that the species was introduced intentionally, but is meant to apply to all species transported by humans, whether intentional or not. The synonymous usage of the terms 'non-native' and 'introduced' in this book is consistent with the recommendation by Richardson *et al.* (2000a) that non-native species be considered those whose presence is due to the intentional or accidental introductions by humans.

One of the points of contention has been whether the term invasive should incorporate impact, or whether it should solely describe the tendency of a species to spread rapidly (Pyšek and Richardson 2006). There is a large overlap of species that meet both criteria. That is, a great many of the non-native species that have spread rapidly, also are producing a sizeable impact, whether health, economic, or ecological. However, there are also many non-native species that have spread widely that are not regarded as being problematic or as having made a large ecological impact (Ricciardi and Cohen 2007). In the management and policy field, the term invasive is generally used to refer to species that have an undesirable impact, whether that be economic, health, or ecological. International initiatives, such as the Global Invasive Species Program and the Invasive Species Specialist Group, and national programs such as the US National Invasive Species Council, define and use the term invasive species to refer to species causing harm. The 1992 international Convention on Biological Diversity likewise incorporated harm in its conception of the word invasive defining invasive alien species as 'alien species whose introduction and/or spread threaten biological diversity.' In their writings, many ecologists have also incorporated impact into their understanding of the word invasive (Kiritani and Yamamura 2003, Sax *et al.* 2005b). Pyšek and Richardson (2006) urge against incorporating impact because it brings human values into the discussion. However, the reality is that definitions of the word invasive fall along a continuum with respect to spread and impact (Ruiz and Carlton 2003b). I believe that human values cannot be excluded from the discussion and it is better to be up-front with them. Thus, except where otherwise indicated, I have used the term invasive throughout the book to refer to a species, or population, that is rapidly spreading in a particular area and producing undesirable impacts, both of which, spread and impact, are assessed from the human perspective, recognizing that individuals may differ on what constitutes harm and undesirable impact.

Another point of contention has been whether the term invasive should be applied only to non-native species, as some have suggested (Pyšek and Richardson 2006). If one is involved in trying to restore a prairie in Minnesota, it seems reasonable to regard native sumac, *Rhus typhina* and *R. glabra*, as invasive since they are rhizomatous and rapidly spreading woody species, and neither is desired in the restored landscape. In a review of the impacts on the vegetation of changing fire regimes in the American southwest, Brooks and Pyke (2001) took a similar approach and used the term invasive to describe creosote bush, *Larrea tridentata*, and mesquite, *Prosopis* spp., native shrubs that have spread widely in recent decades due to fire suppression. During the past several decades, many wetlands near Lake Superior have become populated by an invasive form of *Phragmites australis*, and genetic analyses have determined that the invasive form is native in origin (Lynch and Saltonsall 2002). Irrespective of the geography of origin, the invasive *Phragmites* variety has just recently emerged in this landscape (or lakescape), has spread rapidly, has produced undesirable effects, and is the target of control and management efforts. Not being able to apply the term invasive to this variety simply because it is of native origin does not seem to make good ecological sense. This view notwithstanding, in order to be consistent with my use of the term invasion, I decided to confine the use of the term invasive in this book to non-native species.

The difference between the word invader, which I have refrained from using in this book, and the word invasive, which I have used abundantly, is that the former forces a species into a nominal (and in this case, normative) category, while the latter simply describes a behavior, which under the right conditions, practically any species is going to be able to exhibit invasive behavior. Also, as Colautti and MacIsaac (2004) pointed out, what we call invasive species are really invasive populations, since very few species are invasive everywhere they are found. To be clear, when I refer to non-native species, I am referring to all recently introduced species, invasive or not. If I just want to refer to the recently introduced invasive species, I refer to them as non-native invasive species.

A look to the past

Although the primary objective of this book is to describe and assess the current state of the field, with a look to the future, if there is one thing that we have learned as invasion biologists it is that one cannot fully make sense of the present without some understanding of what has taken place in the past. More specifically, maintaining a strong historical perspective as a researcher is very important since it reduces the likelihood of recycling ideas and losing real scientific momentum (Graham and Dayton 2002, Pianka and Horn 2005). In addition, knowing what preceded us can help us identify key questions and determine what important data and understandings are missing (Beisner and Cuddington 2005). Or, to put it more bluntly, 'to be ignorant of history is to be ignorant about current scientific developments' (Cuddington and Beisner 2005). Thus, I would like to begin this book, which is intended primarily as a look forward, with a look to the past.

Interest in what have come to be known as biological invasions did not begin with Elton (1958), of course. One would imagine that as far back as several thousand years ago, careful and interested observers of nature would have noticed the establishment and spread of new species brought to their locale by travelers. Chew (2006) provided a detailed and thorough accounting of the interest in non-native species during the 200 years prior to Elton. According to his research, written accounts of non-native species (or neobiota in Chew's words) began appearing in Western writings in the mid-1700s. One of the first such accounts was by a student of Linnaeus, Pehr Kalm. Interestingly, Kalm traveled to North America seeking new plants that might be brought back to Sweden and commercially grown for the country's economic benefit (Chew 2006). When in North America, Kalm noticed and recorded in his journals the names of European species of plants, and sometimes insects, that he frequently encountered. Explorer and naturalist Alexander Humboldt, whose life extend over the latter three decades of the eighteenth century through the middle of the nineteenth century, was cognizant of the worldwide redistribution of flora and fauna that was taking place. For example, he noted the spread of American *Opuntia* cactus throughout Europe, the Middle East, and northern Africa by the mid-nineteenth century (Humboldt 1850).

More formal and explicit treatment of native and non-native species did not begin, and Chew argued, probably could not begin, until the field of biogeography emerged in the 1800s. Led most notably by Alfred Russell Wallace and Charles Darwin, but actually consisting of hundreds of other botanists and zoologists intent on describing the distribution of the world's biota, this field provided much of the conceptual groundwork and data sets that led to the rise of the modern science of ecology in the early years of the twentieth century. Of particular consequence was Wallace's division of the earth into six distinct bio-regions, which thereby identified home regions for species. Not surprisingly, then, around this same time the European plant biogeographers, or phytographers, developed the specialty area of adventive floristics, which focused on the distributions of non-native plant species. Although most of the earliest commentaries on non-native species were by botanists, by the mid-1800s, zoologists were also beginning to note and comment on the redistribution of animal species due to human activity. In 1858, ornithologist Philip Sclater wrote:

> We do not find that the Nightingale extends its range farther to the west one year than another, nor that birds looked upon as occasional visitors to this country, grow more or less frequent. If the contrary be the case, it may always be accounted for by some external cause, generally referable to the agency of man, and not to any change in Nature's unvarying laws of distribution.

Around this same time, as described by Cadotte (2006), North American agricultural scientists began to comment on the presence and negative impacts of non-native species (Fitch 1861), and by the end of the century such commentaries became more common (Forbes 1883, 1886, 1887, 1898, Howard 1893, 1897a), with some discussions of non-native species and their negative impacts taking place in general scientific venues (Howard 1897b). Based on his research of the writings of Forbes, Cadotte (2006) observed that it was surprising that Forbes studied and wrote about introduction and agricultural impacts of non-native insects, while at the same time generally endorsing a balance of nature paradigm for the study of ecology. Cadotte pointed out that this dissociation portended similar dynamics that were described as characterizing the field of invasion biology in the latter part of the twentieth century (Davis *et al.* 2001).

By the end of the nineteenth century, European scientists were beginning to comment on the arrival and distributions of non-native species (Drude 1896, Warming 1909—translation of Warming's 1895 original text). As Cadotte (2006) pointed out, many aspects of Drude's writings have a modern flavor. Although not using contemporary terminology, Drude emphasized a variety of factors influencing successful colonization and establishment of new species, including the role of geography, propagule pressure, biotic resistance, and chance. In the first few decades of the twentieth century, ecologists continued to recognize and write about non-native species. In his monograph on the distribution of desert plants, Spalding (1909, cited in Cadotte 2006) emphasized that invasion was an integral and constant process, also stressing that the invasion process was the same for native and non-native species, the only difference being that the latter were transported by humans. New Zealand scientist George Thomson (1922) wrote a 600+ page book on the topic, *The naturalisation of animals & plants in New Zealand*, in which he documented and described more than 1000 non-native species that had been introduced into the country. A thoughtful scientist, Thomson recognized that the decline in some native species was likely due to multiple causes, non-native species being just one of them. He wrote:

> It must not be supposed that it is the introduced animals alone which have produced [the retreat of the natives], even though rats, cats, rabbits, stoats and weasels, as well perhaps as some kinds of introduced birds, have penetrated beyond the settled districts. It is largely the direct disturbance of their haunts and breeding places, and the interference with their food supply, which has caused this destruction and diminution of the native fauna... many insects which were common in the bush fifty years ago must have been displaced and largely disappeared. I cannot appeal to figures, but the surface burning of open land which prevailed, especially in the South Island, and

the wanton destruction and burning of forest which has marked so much of the North Island clearing, must have destroyed an astonishing amount of native insect life, and made room for introduced forms. The clearing of the surface for cultivation and grazing, the draining of swamps, and the sowing down of wide areas in European pasture plants, have all contributed to this wholesale destruction and displacement of indigenous species.

According to Chew (2006), some of the earliest known examples of the nativism paradigm appearing in scientific writings occurred during this time. (By nativism paradigm, Chew meant a way of thinking that regarded species as inherently more desirable than non-native species.) Douglas Campbell (1926), long time chair of the Stanford Botany Department, wrote:

The extraordinary and rapid change in the vegetation of a large area, due to man's activities is especially apparent in the United States, which a century ago was to a great extent untouched by man.

And later in the same book:

With the facilities for transportation developed during the past century, migration has reached a stage absolutely unheard of in previous history, and the influx of millions of men into previously unoccupied regions is reflected in immense changes in the vegetation of nearly all parts of the world—far greater than in any previous period of the world's history. Forests have been swept away until the world is menaced with a timber famine, and their place has been taken by crops of all kinds, which are entirely alien to the country and completely alter the appearance of the landscape.

Pauly (2008), described a dramatic shift in the United States toward non-native species at this time. For many years, there had been a strong desire to bring new species into the United States. In fact, part of the stated mission of the US Department of Agriculture (USDA) when it was created in 1862 by Abraham Lincoln was 'to procure, propagate and distribute... new and valuable seeds and plants.' However, by the early twentieth century, the desire to create a more cosmopolitan country began to be replaced with 'enthusiasms for the native, and fears of the alien' as the American public realized that importations of desired plant species also brought in agricultural pests (Pauly 2008). The Plant Quarantine Act was passed in 1912, enabling the USDA to declare quarantines, and in 1928 the Plant Quarantine and Control Administration was established.

In the 1930s and 40s, some ecologists began to publish papers based on an 'indigene vs alien' paradigm. Allan (1936) focused on the indigene and alien in the New Zealand flora, while Egler (1942) focused on the indigene and alien in Hawaiian vegetation. Both papers were published in the journal *Ecology*. Significantly, both Allen and Egler recognized the importance of landscape alterations by humans, e.g. clearing and grazing, in facilitating the establishment and spread of non-native species. These papers were followed by similar publications in the late 1940s and 1950s (Baker 1948, Stewart and Hull 1949, Huffaker 1951). In 1955, an international symposium was held in Princeton, New Jersey, with the title, *Man's role in changing the face of the earth*. In the conference proceedings, Marston Bates (1956), a tropical biologist and entomologist, authored a chapter titled, 'Man as an agent in the spread of organisms', which Elton acknowledged in his 1958 book.

As the reader may know, Elton's 1958 book was based on a series of radio broadcasts, which he had presented to the general public under the banner 'Balance and barrier'. Perhaps because of the book's broader audience, Elton used colorful language and metaphor to their fullest. In particular, he embraced a militaristic characterization of the introduction and spread of non-native species. In the first paragraph of the book, Elton wrote:

It is not just nuclear bombs and war that threatens us. There are other sorts of explosions, and this book is about ecological explosions.

Elton continued:

[there are] two rather different kinds of outbreaks in populations: those that occur because a foreign species successfully invades another country, and those that happen in native or long-established populations. This book is chiefly about the first kind, the invaders.

In 2001, Phil Grime, Ken Thompson, and I suggested that by conceiving and presenting the introductions and spread of non-native species as ecologically distinct, Elton unwittingly helped promote the eventual dissociation of invasion biology from the rest of ecology, and that by presenting these 'ecological explosions' as distinct ecological phenomena, fostered the belief that invasion biology requires both a distinct conceptual framework and research approach (Davis *et al.* 2001). Cadotte (2006) made a good case that roots of this dissociation extended back earlier in the century, although not all earlier ecologists operated this way, e.g. Volney Spalding (1909) described above. Chew (2006) came to a different conclusion on this point. He argued that a more accurate characterization is that Elton did not so much inaugurate the dissociation as much as he provided the modern field of invasion biology, which emerged approximately twenty-five years following Elton's book, a convenient patriarch, whose antagonistic views toward non-native species could be used as a kind of authoritarian support for similar views that had been arrived at largely independently from Elton. In fact, as pointed out by Richardson and Pyšek (2007), in his 1958 book, Elton brought together themes and subdisciplines, which, at the time, were not well-connected, e.g. biogeography, conservation biology, epidemiology, human history, and population ecology. Thus, whether Elton's book actually initiated the dissociation, or was simply effectively used to promote a particular perspective that was already viewing invasions as a unique ecological phenomenon, Elton's book ended up playing a significant role in the development of invasion ecology in the latter two decades of the twentieth century.

It is interesting, and perhaps a bit surprising, that Elton's book had only a very modest immediate impact on the rapidly developing field of ecology. Hutchinson (1959) briefly acknowledged it in the published version of his famous 1958 address to the American Society of Naturalists titled, *Homage to Santa Rosalia, or Why are there so many kinds of animals?*, describing it as 'a fascinating work largely devoted to the fate of species accidentally or purposefully introduced by man.' According to Chew (2006), E. O. Wilson read Elton's book shortly after it was published and recommended it to Rachel Carson, who was in the process of writing *Silent spring.* After reading Elton's book, Carson responded to Wilson, 'I found [*Invasions*] enormously stimulating. It cuts through all the foggy discussions of insect pests and their control like a keen north wind.' Interestingly, as described by Chew, Carson's British publisher asked Elton to write a preface to the British edition of Carson's book, but Elton declined.

Six years following the publication of Elton's book, the International Union of Biological Scientists (IUBS) held their first Biological Sciences Symposium, which took place in Asilomar, California. Organized by Conrad H. Waddington and Ledyard Stebbins, President and Vice-President of the IUBS at that time, the symposium was intended to attract the top geneticists, ecologists, taxonomists, and applied scientists working in the area of pest control and to discuss 'the kinds of evolutionary change which take place when organisms are introduced into new territories' (Waddington 1965). This was an international initiative. The resulting proceedings, which were published as a book that has become a classic, *The genetics of colonizing species* (Baker and Stebbins 1965), were authored by twenty-seven authors representing eleven countries. The symposium successfully attracted the top evolutionary biologists of the time, including Mayr, Stebbins, Dobzhansky, Wilson, Carson, Lewontin, and Waddington.

The difference between the language used in the proceedings from the 1964 Asilomar conference and that used by Elton in his 1958 book is striking. Most of the 1964 authors never used the words alien, exotic, invader, or invasion. In their place was terminology such as colonizers, founding populations, introduced, non-native, new arrivals, migration, spread, and geographically widespread. The one perplexing exception involved the contribution by plant ecologist, John Harper. Titled *Establishment, aggression, and cohabitation by weedy species*, Harper's (1965) chapter included some of the same evocative language that Elton had used. In fact, the opening of Harper's chapter was so similar to some of Elton's characterizations that one can hardly imagine that Harper was not consciously adopting Elton's perspective. Harper began his chapter:

The movements of man and his goods have resulted in a bombardment of areas of land and sea by alien species, both by chance and by the deliberate introduction of cultivated plants of the farm and garden.

I referred to Harper's article as perplexing because the Eltonian approach that he took in the article was in complete contrast with his characterization of non-native species in his 1977 book. Commenting on the appearance and spread of *Opuntia* cactus in Australia, Harper (1977) wrote:

The spread of *Opuntia* in Australia is a rare, large-scale event, but of the same fundamental nature as the common, small-scale shifting occupancy of sites by a species in its native vegetation.'

Elton's book and the Asilomar meeting represented two very different paths to studying non-native species. Elton emphasized the conservation and environmental impacts of non-native species, and certainly took a normative approach, while the 1964 conference participants viewed non-native species more in a value-neutral sense, and believed that their study could inform more general ecological principles and theory (Davis 2006). Curiously, just like Elton's book, the 1964 conference also did not prompt much new interest in non-native species in the field of ecology. In the 1960s and 70s, Elton's urgent conservation message was not being assimilated into the field of ecology, nor did the evolutionary perspective of colonizing species that distinguished the 1964 conference seem to have any widespread effect on ecological research. At least neither elicited much new interest in taking on non-native species as a focused area of ecological research (Fig. 1.1). This is not to say that no research on non-native species was taking place in the decade following Elton's book or the Asilomar conference. Some research was ongoing, particularly in Germany and some of the Eastern European countries, by researchers such as Sukopp (1962), Hejný and Lhotská (1964), Kohler and Sukopp (1964), Holub and Jirásek (1967), Jehlík and Slavík (1968), Faliński (1968, 1969), and Kornás (1968).

The 1970s brought the founding of Earth Day and more widespread interest in the environment. SCOPE published the first of its series titles, now numbering more than sixty, and conservation biology began to emerge as a distinct discipline. The 1970s were the first time that articles on non-native species began to appear with some

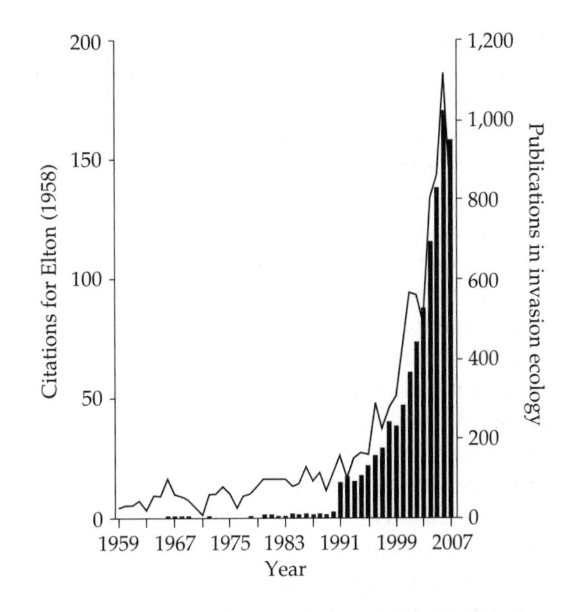

Fig. 1.1 The number of biological invasion publications since Elton published *The ecology of invasions by animals and plants* in 1958 (columns). Also shown are the number of publications that cited Elton's 1958 book during this time period (linegraph). Redrawn and printed, with permission, from Ricciardi and MacIsaac (2008), copyright Nature Publishing Group.

frequency in mainstream ecological journals, the focus primarily being a conservation one, with the emphasis on impacts (Christie 1972, Moyle 1973, Owre 1973, Baker 1974, Courtenay and Robins 1975, Burdon and Chilvers 1977, Embree 1979). However, there was little disciplinary infrastructure created to support more focused discussions on non-native species, and invasion biology had not yet emerged as a well-defined specialty area of research. But that was soon to change.

In 1980, the Third International Conference on Mediterranean Ecosystems was held in Stellenbosch, South Africa. At the meeting, there was considerable interest in biological invasions, and this interest led to a proposal to the SCOPE general assembly in Ottawa in 1982. In turn, this proposal resulted in the creation in 1983 of a scientific advisory committee that was to encourage and facilitate focus on the impacts of biological invasions on natural ecosystems. During this same time, several important papers and books on biological invasions were published (Brown and Marshall 1981, Mack 1981,

Simberloff 1981, Moulton and Pimm 1983, King 1984, MacDonald and Jarman 1984, Groves and Burdon 1986, MacDonald *et al.* 1986, Mooney and Drake 1986, Kornberg and Williamson 1987). The 1980s ended with the publication of a SCOPE synthesis publication titled *Biological invasions: a global perspective* (Drake *et al.* 1989). A reading of these publications indicates that during this time the field of invasion biology was aligning itself more with the Eltonian normative and conservation approach to non-native species (Davis 2006). This trend was most evident in the writings of North American ecologists (Davis 2006).

The publications and interest at the end of the 1980s seemed to have reached a critical mass, because the field of invasion ecology emerged and grew at an astounding rate in the 1990s, exhibiting a trajectory remarkably similar to the population growth rates of many non-native invasive species (Fig. 1.1). Annual publication rates, which not many years earlier numbered in the dozens, or less, soon numbered in the hundreds, and eventually exceeded one thousand (Fig. 1.1). The journal *Diversity and Distributions* was founded in 1998 (succeeding the journal *Biodiversity Letters*) with a focus on biological invasions and biodiversity. David Richardson has been the journal's editor-in-chief since its inception. A year later, the journal *Biological Invasions* (Kluwer) was founded, with James T. Carlton as its first editor-in-chief, a position, as of 2008, occupied by James A. Drake. The 1990s ended with a new global initiative addressing non-native invasive species, the Global Invasive Species Programme (GISP). GISP was created to help prevent, manage, and control the negative impacts of non-native invasive species, as is clear from its stated mission, 'to conserve biodiversity and sustain human livelihoods by minimizing the spread and impact of invasive alien species.'

Invasion biology of the twenty-first century

During the 1990s, the emphases in publications, both scientific and popular, indicated that the field continued to be dominated by the Eltonian conservation paradigm. However, by the end of the century, some scholars, both within and outside the field of ecology, were beginning to challenge the dichotomous and normative perspectives that had come to dominate invasion biology (Eser 1998, Gould 1998, Sagoff 1999, Davis *et al.* 2001, Slobodkin 2001). At the same time, some ecologists began to dispute the traditional notion that introduced species pose an inevitable threat to local and regional biodiversity (Rosenzweig 2001, Davis 2003, Sax and Gaines 2003, Gurevitch and Padilla 2004). In response to concerns that invasion biology had dissociated itself from the rest of ecology (Davis *et al.* 2001), a conscious effort was made by some ecologists to resurrect the more natural philosophy approach that had characterized the Asilomar conference, i.e. studying non-native species and invasions as a way to inform larger ecological principles (Sax *et al.* 2005a, Cadotte *et al.* 2006).

A strength of science is that it is self-correcting. Over-simplifications and enthusiastic, but sometimes misguided, emphases that often characterize a field when it is young, can be tempered and revised in the face of new data and new perspectives. Maturation, whether it involves a person or a discipline, normally involves a growing recognition of complexity and ambiguity. During the first decade of the twenty-first century, I believe that the field of invasion biology is undergoing considerable change as it develops into a more mature discipline, one that is more nuanced and less intellectually isolated. While the field may have dissociated itself from related subdisciplines for a period of time, current activity in the field suggests the opposite is now taking place. Callaway and Maron (2006) describe the field as helping to 'catalyze a healthy fusion between fields and subdisciplines that have historically operated in isolation.' In addition, the current field has at its disposal, data and technology not available to earlier investigators. As described by Richardson and Pyšek (2007), these include advances in computer power, new statistical methods and modeling approaches, the development of geographic information systems (GIS), advances in molecular biology, new field and data collection technology, and comprehensive data bases. It is this invasion biology, the invasion biology of the twenty-first century, that is my primary subject for the remainder of this book. Specifically, I have tried to characterize the nature of our current understandings and disagreements regarding biological invasions, including their processes, impacts, and management. In addition, I have tried to delineate what I believe are some steps we can take to develop and nurture an even more vital and robust field.

The next ten chapters are grouped into three sections. Chapters 2 through 5 focus on the invasion process. My primary objective in these chapters is to describe the field's current understanding of the patterns of dispersal, establishment, and spread of non-native species and the state of our knowledge regarding the underlying mechanisms that produce these patterns. Chapter 6 presents a proposed integrated approach to understanding and predicting invasions. The primary scientific research literature constituted the principal source of information for these five chapters, which constitute Part I. Chapters 7 and 8, Part II, have a more applied focus, addressing impacts and management. The range of sources for these two chapters tended to be broader. While many sources still came from the scientific research literature, others came from the management and social science fields. In Chapters, 9, 10, and 11, Part III, I offer some personal reflections on the field of invasion biology and its future.

Depending on one's interests, some readers may be more motivated to read certain sections more than others. Some may be mostly interested in Part I, which focuses on the science of invasions. This group might include those primarily interested in reading a review of recent literature and a summary of the current state of our understanding of biological invasions. Certainly, many newcomers to the field of invasion biology may be part of this group. At the same time, I hope that my documentation, analysis, and synthesis of current ideas, publications, and research agendas, which are currently defining the field of invasion biology, will prove to be of interest for the field's longer term residents as well. Others may be more interested in reading about the health, economic, and ecological impacts of non-native species, and the challenges faced, and the progress made, by those trying to manage non-native species and their impacts in Part II. Still others may enjoy taking some time to reflect on the way that our paradigms and inclinations may shape the way we think as invasion biologists, and as scientists in general, in Part III. My motivation in writing Part III stems from my belief that the investigation of the nature of science should not be left up to the philosophers and historians of science, but that it should be 'carried on by scientists as they themselves work on scientific problems' (Simpson *et al.* 1961). I am most hopeful that the book finds itself to the hands of students who are still developing their paradigms and perspectives, and that these students find value in all three portions of the book.

PART I

The invasion process

CHAPTER 2

Dispersal

Stages of the invasion process

The invasion process has been described as a series of stages (Carlton 1985, Williamson and Fitter 1996, Richardson *et al.* 2000a, Kolar and Lodge 2001). Some have characterized three stages: arrival, establishment, and spread (Williamson 1996, Freckleton *et al.* 2006) or establishment, spread, and integration (Marchetti *et al.* 2004); some have proposed four stages: arrival, establishment, spread, and adjustment (Ricklefs 2005, Reise *et al.* 2006). Some schemes have identified as many as six stages: introduction, establishment, naturalization, dispersal, population distribution, and invasive spread (Henderson *et al.* 2006).

There is definitely value in focusing on particular portions of the invasion process. However, the practice of identifying a series of distinct stages may have some drawbacks. A stage-based approach normally presents the stages as sequential, with an implication, even if not intended, that the invasion process is a kind of moving front that moves across the land or seascape. The stage-based characterization of the invasion process suggests that first there is the dispersal stage; when that is over, and if it succeeds, then there is the establishment phase; when the establishment phase is over, and if it is successful, then there is the naturalization stage; when the naturalization stage is over, and if it succeeds, then there is the spread phase. Of course, those who have presented the stage-based approach to invasions did not mean to suggest that invasions proceed in such an extremely discrete fashion; they have emphasized that the process is really more continuous (Richardson *et al.* 2000a, Daehler 2006). Nevertheless, the stage-based models do tend to suggest a temporal shift in the ecological sphere of influence in the invasion process, such that a particular invasion might be termed to be in the establishment phase, while another might be termed to be in the spread phase. In reality, activities in prior stages do not stop with the inauguration of a subsequent stage. An invasion in the spread phase is actually also usually experiencing continuing dispersal episodes and establishment periods as well.

I am not recommending that we discard the stage-based model approach. Without question, there are practical benefits to using stage-based models to characterize invasions, since this approach can help identify potentially effective management strategies. However, models that depict invasions as a series of distinct steps and processes may distort the invasion process, in which case such models may be hampering our scientific understanding of invasions. An alternative approach is to think of the process as an ongoing series of cyclical iterations. In this approach (Fig. 2.1), there are only two fundamental processes—dispersal and establishment, both which operate at the individual organism level. In this approach, persistence and spread are viewed as emergent properties at the population and metapopulation levels, both arising from the two individual-based processes of dispersal and establishment.

As illustrated in Fig. 2.1, many, probably most, propagules from an external dispersal pool (meaning propagules originating from outside the site under consideration) never reach a particular environment. They may successfully disperse to other sites or die during transport. Of those that are successfully dispersed to the target environment, most probably never establish. (Establishment is defined here as living long enough to be able to reproduce.) In most cases this will mean that established individuals have been able to access resources in their

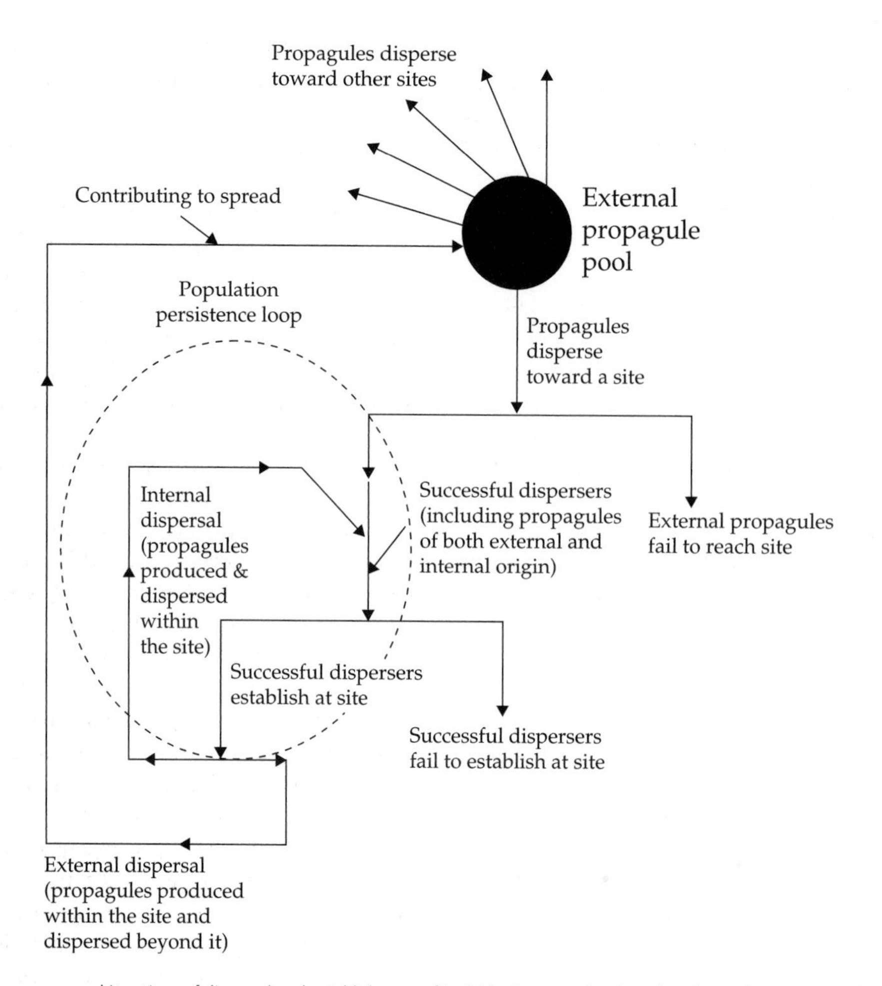

Fig. 2.1 The sequence and iterations of dispersal and establishment of individual propagules that give rise to the emergent phenomena we call persistence and spread.

new home, e.g. as opposed to surviving and reproducing solely based on resources brought with them from their original environment.

Once an established individual produces propagules, these propagules then face the same fundamental challenges that the propagules that first arrived in the environment experienced. That is, they need to find a safe and suitable site in which to establish themselves. The continued iteration of this internal loop is what we describe as persistence. Some use the term naturalization to describe this process or stage (Richardson *et al.* 2000a, Pyšek and Richardson 2006). There are actually three ways a population could persist. The population could persist solely on the basis of internally produced propagules, meaning that there was no further supplement of propagules from the external pool following the initial dispersal event. Such a population could be described as completely self-sustaining. Even if the population itself was producing no propagules that remained on site, the population could persist due to regular dispersal from the external propagule pool. This would describe the extreme case of a sink population. Or the population may persist as the result of both types of propagules, ones produced on site by already established residents and ones arriving from the external pool.

Of course, some of the propagules produced by the newly established individuals may be dispersed beyond the local environment, in which case they become part of the external dispersal pool for other environments and thereby part of the next iteration of dispersal at the metapopulation level. If the external dispersal from the first established population contributes to the founding of a new population, then the species is said to have spread. Thus, in this model, neither persistence nor spread should be considered as distinctly separate and subsequent stages. They are merely emergent manifestations of dispersal and establishment taking place at the individual level. Population persistence is simply what we call the ongoing accumulation of establishment successes of individuals produced within a site. In turn, species spread is what we call the ongoing accumulation of establishment successes at the population level. Thus, metapopulation persistence (also an emergent entity and property) is fundamentally the integrated result of individual dispersal and establishment episodes occurring throughout the region. If enough of the individual dispersal and establishment episodes are successful, populations persist, the species spreads, and the metapopulation persists.

I think that the model described above will facilitate our efforts to understand the invasion process. However, as also emphasized, other invasion models may be better at meeting other objectives. For example, in trying to develop strategies to prevent introductions, it is very useful to be able to identify specific stages of the dispersal process, each of which may occur at a particular time or place, hence requiring a different intervention strategy.

Dispersal and propagule pressure

Some organisms have experienced very long-distance dispersal events for millions of years. The low degree of regional divergence in bacterial and fungal communities throughout the world is believed to be due to the long-distance dispersal abilities of these small organisms (Hillebrand *et al.* 2001, Drakare *et al.* 2006). Recent research has shown that some pathogens can be transported for thousands of kilometers in the air (Brown and Hovmøller 2002). African dust storms can catapult pollen and spores to high altitudes, where they can be dispersed long distances, e.g. across the Atlantic Ocean and into Europe and Asia (Kellogg and Griffin 2006; Fig. 2.2). Wyatt and Carlton (2002) emphasized the long-distance dispersal capabilities of very small organisms, suggesting that they are much more likely to owe their distribution to natural processes than are larger organisms, which have been more dependent on humans for most of their long-distance dispersal during the past 500 years. While such transport may be an ancient process, there is concern that new plant and animal pathogens might be dispersed in this manner (Kellogg and Griffin 2006).

Some larger organisms have also exhibited natural dispersal histories that could be considered global in extent, or at least covering very long distances. Some plant species have been found to disperse so successfully throughout the arctic regions that researchers have suggested that unlimited dispersal models should be considered when trying to predict range shifts due to climate (Alsos *et al.* 2007). In this case, it is believed that plant propagules are transported by both wind and ice (Alos *et al.* 2007). Based on data from molecular phylogeny studies, Renner (2004) reported that at least 110 plant genera contain species inhabiting both sides of the tropical Atlantic Ocean, indicating that trans-Atlantic dispersal via wind and currents occurred in the past. Some land animals are capable of trans-oceanic movements as well. Cattle egrets, *Bubulcus ibis*, are believed to have dispersed to the Western Hemisphere from Africa, without any human assistance, sometime during the past 150 years. These examples notwithstanding, for most organisms, global dispersal represents a new phenomenon.

Along with local and regional environmental conditions (Francis and Currie 2003, Freestone and Inouye 2006, Storch *et al.* 2006), dispersal dynamics play a central role in determining the species composition of communities (Hubbell 2001, Freestone and Inouye 2006, Storch *et al.* 2006). Not surprisingly, then, invasion biologists have recognized the importance of dispersal, or propagule pressure, in the invasion process (Williamson 1996, Kolar and Lodge 2001, Lockwood *et al.*, 2005 Rejmánek *et al.* 2005a, Colautti *et al.* 2006). In the invasion

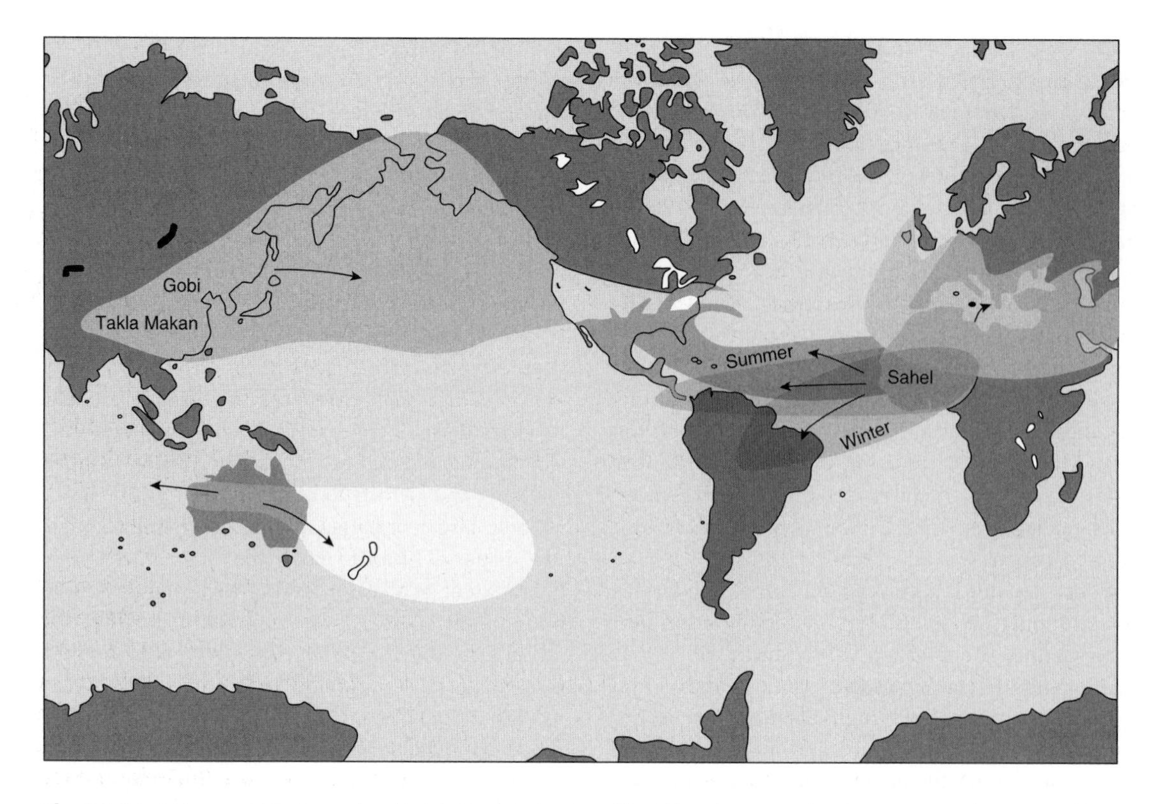

Fig. 2.2 Principal ranges of the two major global dust transport systems. The African dust system has a strong seasonal component. From about May–November, trade winds carry Saharan dust to the Caribbean and USA. From December–April, the African dust-flow is shifted to South America, where air-plants in the Amazon rainforest derive nutrients from the dust. Throughout the year, pulses of dust from northern Africa cross into the Mediterranean and Europe, impacting air quality. The Asian dust system exports dust primarily during March–May. These dust events can incorporate emissions from factories in China, Korea, and Japan, carrying a 'brown smog' across the Pacific to the west coast of North America. Occasionally, extremely large Asian dust events can travel across the entire USA and then impact Europe, making an almost complete circuit of the globe. Although not an intercontinental dust source, Australian deserts produce large dust storms that can reach New Zealand and halfway to South America. Redrawn and printed, with permission, from Kellogg and Griffin (2006), copyright Elsevier Limited, and Garrison *et al.* (2003), copyright American Institute of Biological Sciences.

literature, propagule pressure involves two aspects of dispersal: the number of individuals arriving at a site in a dispersal event, and the number of dispersal events (Lockwood *et al.* 2005). The importance of propagule pressure has been demonstrated by experimental data from studies in which propagules have been introduced into environments by the researcher (Foster and Tilman 2003, Ehrlén *et al.* 2006, Zeiter *et al.* 2006). Differences in propagule pressure are frequently invoked as a likely cause for observed differences in the invasion history of different environments (Levine 2000). For example, freshwater reservoirs situated low in the landscape have been found to be more likely inhabited by non-native and invasive zooplankton invasion than those higher in the watershed, a fact that has been partly attributed to the likely increase in propagule pressure experienced by these reservoirs, being recipients of zooplankton dispersing from upstream systems (Havel and Medley 2006).

In a comparison of introduced flora on isolated and less isolated islands, Daehler (2006) found that while factors associated with isolation (possibly including competition and/or presence or absence of enemies) contributed to the geographic patterns of the floras, propagule pressure (the number of intentional introductions) played a large role in accounting for the data, accounting for approximately half of the variation. Křivánek *et al.* (2006) conducted a study of the distributions of 28 non-native and invasive tree species in the Czech Republic, which had been intentionally introduced and planted as part of forestry practices. Their analysis underscored the importance of propagule pressure, showing that two factors explained much of the variation in distribution: time since introduction, and the number of areas in which the non-native species was planted (Fig. 2.3). Similar studies have also found that the abundance of non-native and invasive species tends to negatively correlate with the distance from dispersal loci (Bossenbroek *et al.* 2001, Rouget and Richardson 2003). Ozinga *et al.* 2005 analyzed species composition data for more than 22,000 vegetation plots in The Netherlands and concluded that the composition of individual plots was profoundly influenced by dispersal limitation. Riparian plant communities have been found to be highly prone to invasion; one of the driving factors is believed to be that the rivers act as efficient dispersal corridors for the plant propagules (Thébaud and Debussche 1991, Pyšek and Prach 1993).

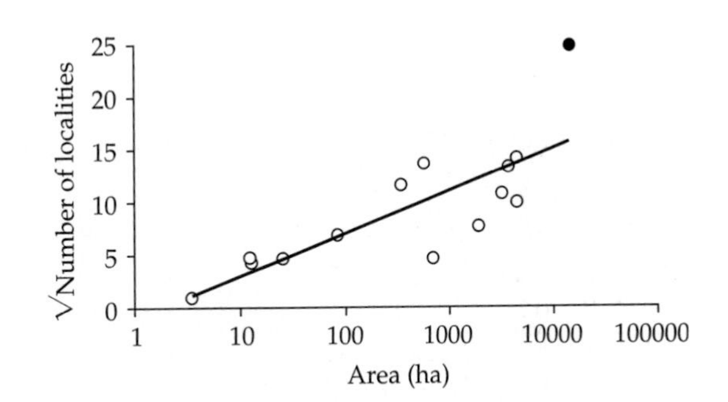

Fig. 2.3 Relationship between the number of localities of alien tree species reported from natural and seminatural habitats in the Czech Republic and the planting area. The black point is *Robinia pseudoaccacia*. Redrawn and printed, with permission, from Křivánek *et al.* (2006), copyright Blackwell Publishing.

Although Europe has been more a donor than a recipient of terrestrial plant species (Lonsdale 1999), this is not the case with marine macrophytes. The Mediterranean Sea alone now harbors more non-native macrophytes than any other marine region of the world (Williams and Smith 2007), and its numbers have been doubling every twenty years since early in the twentieth century (Fig. 2.4). In particular, the introductions have been facilitated by frequent transport of Pacific oysters (*Crassostrea gigas*) associated with the strong aquaculture industry along the northern coast, which has provided an effective dispersal vector for the marine plants (Verlaque 2001). A comprehensive analysis of introductions of vertebrates in Europe and North America (Fig. 2.5) found that the number of introductions of fish, mammals, and birds introduced was strongly associated with human immigration rates, consistent with a propagule pressure explanation for the introduction trends shown (Jeschke and Strayer 2005). A recent comprehensive analysis of the annual spread of the influenza A (H3N2) virus showed that the virus originates each year in East and Southeast Asia (E–SE Asia) and that spread to other regions in the world is a very predictable process, one associated with patterns of human travel (Russell *et al.* 2008). Specifically, the virus is initially transported from Southeast Asia to North America, Europe, and Australia, owing to the strong trade and travel connections between E–SE Asia and these regions. South America, which experiences less frequent contact with E–SE Asia, then receives the virus from Europe and North America (Russell *et al.* 2008; Fig. 2.6). Africa, which is comparatively less connected to the continents through trade and travel, does not experience the annual flu outbreak to the same extent as most other regions.

Not all studies have found propagule pressure to be of significant importance. In a comprehensive review of non-native plant species in the Czech Republic, in which land use and human population density were used as surrogates for propagule pressure, Chytrý *et al.* (2008) concluded that propagule pressure was of low importance, compared to habitat characteristics. Working at a smaller spatial scale in a temperate forest, Gilbert and Lechowicz (2005) found that distance to areas of human disturbance (e.g. trails, picnic areas, reserve boundaries), was not associated with the richness of non-native plants, suggesting that dispersal and propagule pressure is not limiting the establishment of these species. Analyzing zooplankton

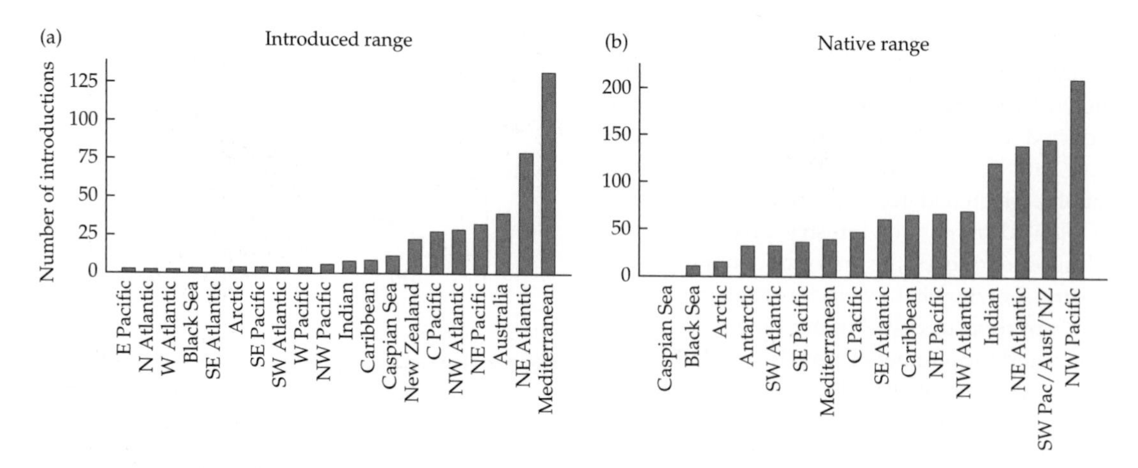

Fig. 2.4 The (a) introduced and (b) native range of seaweed introductions worldwide. Redrawn and printed, with permission, from Williams and Smith (2007), copyright Annual Review of Ecology, Evolution and Systematics, by Annual Reviews (www.annualreviews.org).

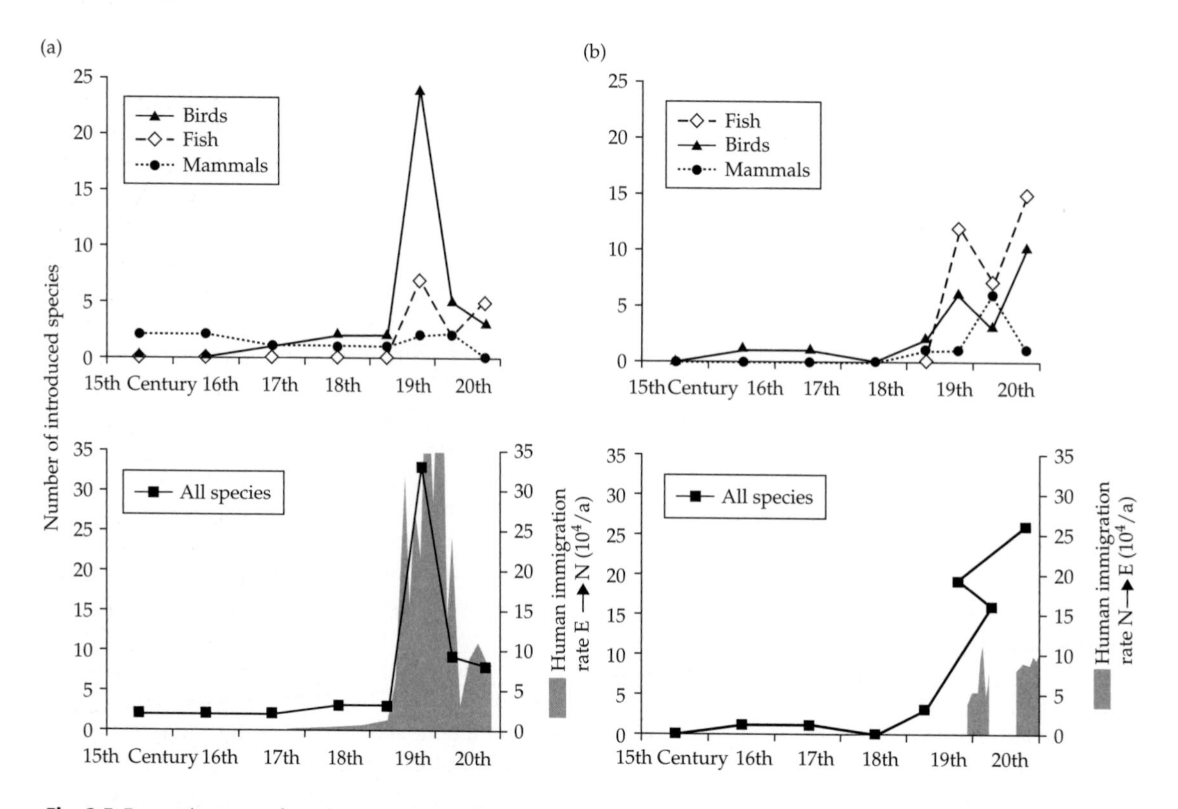

Fig. 2.5 Temporal patterns of vertebrate introductions from Europe to North America (a) and vice versa (b). Up to the eighteenth century, the numbers of introductions are given per century because of the rarity of these events. For the nineteenth and twentieth centuries, the numbers of introductions are given per half-century. Human immigration rate data are from the literature (see Jeschke and Strayer 2005); missing data are indicated by question marks. Redrawn and printed, with permission, from Jeschke and Strayer (2005), copyright National Academy of Sciences, USA.

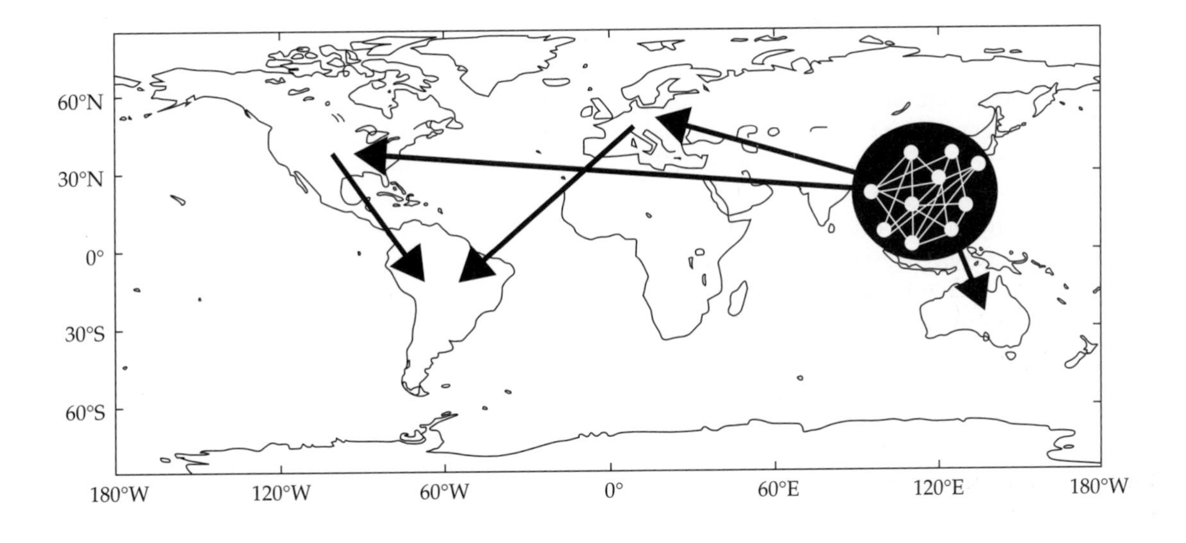

Fig. 2.6 Schematic of the global circulation of seasonal influenza A (H3N2) viruses. The structure of the network within E–SE Asia is unknown. Redrawn and printed, with permission, from Russell *et al.* (2008), copyright American Association for the Advancement of Science.

samples from ballast water taken from a variety of types of ships with different departure and destination ports, Verling *et al.* (2005) found that the number of ship arrivals to a port is normally a poor predictor of propagule pressure. Rather they found that propagule supply was the result of a complex interaction of a number of factors, including the type of ship, the source regions of the ballast water, and the survival rates of zooplankton during transport. Based on their findings, Verling *et al.* warned against using ship arrivals as a surrogate for propagule pressure of ballast-transported organisms and instead emphasized the need to incorporate several aspects of vector operation into predictive models of propagule supply. Even when propagule pressure is important, quality as well as quantity of propagules needs to be taken into account when assessing the impact of propagule pressure on the likelihood of an invasion. For example, organisms may arrive but be physiologically in such poor condition that establishment and or reproduction is precluded (Carlton and Ruiz 2005).

The dispersal process

While not every species that has dispersed to a new region in the past several centuries is the result of human activity, the vast majority have been transported by humans, both intentionally and by accident (Ruiz and Carlton 2003c, Vermeij 2005, Pauchard and Shea 2006, Keller and Lodge 2007). No doubt, ever since humans left Africa roughly one hundred thousand years ago (Templeton 2002), people have been acting as dispersal vectors for other organisms. Evidence indicates that humans were traversing distances of hundreds of kilometers between islands in the Southeast Pacific Ocean as long ago as 40,000–50,000 years (Balter 2007), movements that eventually resulted in the introduction of other organisms as well (e.g. pigs, taro, yams, rats, and lizards), during the past several thousand years (McNeely 2005). Recent carbon dating of chicken bones at an archaeological site in Chile indicated that the bones predated the arrival of Europeans, and DNA analysis showed the bones to be of Polynesian origin (Storey *et al.* 2007). One of the most common temperate-water shipworms in the world, *Teredo navalis*, was apparently dispersed so long ago by human seafarers that its origin has not yet been determined (Hoppe 2002). These demonstrate that long-distance human transport of other organisms, while much more common now, is not entirely a recent phenomenon.

International commerce is believed to be the primary driver of species introductions in the world today (Ruiz and Carlton 2003c), although other activities have contributed substantially as well. For example, the global movements of military forces and the transport of supplies to maintain them have introduced species, sometimes intentionally, for centuries (McNeely 2005). The importance of international commerce as a current vector for many non-native species is illustrated by the fact that the number of introductions in a region is often positively correlated with the volume of trade in the region (Levine and D'Antonio 2003, Semmens *et al.* 2004). An increase in international trade increases the number of introductions in two ways. First, increasing transport episodes increases the number of different species introduced, and second, repeated introductions of the same species increases propagule pressure, and hence the likelihood of establishment (Williamson 1996, Lockwood *et al.* 2005). For example, the enormous economic growth by China during the past 25 years has resulted in a dramatic increase in imported goods, e.g. an approximate six-fold increase in Shanghai of goods arriving by rail, boat, and air (Ding *et al.* 2008). During this time, the number of harmful introduced plants and animals intercepted at China's borders has increased by more than ten-fold (Ding *et al.* 2008). The anticipated continued expansion of economic development in China is expected to result in high rates of biological invasions in the future (Weber and Li 2008).

Many species have been introduced unintentionally through international commerce. For example, insects can hitchhike on nursery stock, cut flowers, fruits and vegetables, grain, and wood (Kiritani and Yamamura 2003). Similarly, reptiles and amphibians can enter a new region in cargo containers or by hiding on imported plants (Kraus 2003); the latter also a common pathway of unintentional introductions for snails and slugs (Cowie and Robinson 2003). The transport of oysters for aquaculture purposes has provided an important dispersal vector for many other marine invertebrates all over the world (Elton 1958), and in some regions of Europe, oyster imports may be the primary dispersal vector for introduced marine species (Wolff and Reise 2002). The fact that trade is a primary driver of species introductions means that those countries and regions of the world more involved in international trade are exposed to increased propagule pressure (Pyšek and Richardson 2006). Dasmann (1988) used the term 'biosphere people' to refer to these societies, i.e. ones that regularly imported material and resources from other parts of the earth, distinguishing them from 'ecosystem people', who rely much more on local resources.

In the coastal waters of the northeast Pacific Ocean, over 100 introduced invertebrate species have been identified, the majority native to the northern hemisphere and unintentionally introduced (Wonham and Carlton 2005, Carlton, personal communication). The predominant introduction pathways in this environment included Pacific and Atlantic oyster shipments (bringing in additional invertebrate 'hitchhikers'), ballast water transport, and fouling, with approximately half of the introduced species relying mostly on a single introduction pathway (Wonham and Carlton 2005; Fig. 2.7). Wonham and Carlton concluded that fouling has declined in importance as a pathway of introduction, while ballast water release has increased. During the past fifty years, an especially large number of non-native species have been introduced and have established in brackish-water seas, such as the Baltic Sea, the Caspian Sea, and the Black Sea. This may partly be due to the fact that most ports worldwide are located at river mouths, areas where the loaded ballast water often contains euryhaline species, species that can tolerate a wide range in salinity (Paavola *et al.* 2005).

Worldwide, it has been estimated that more than 10,000 different species are being transported in the ballast water of ocean vessels during any 24-hour period (Carlton 1999). Taxa identified in ballast water include viruses, bacteria, protists, fungi, algae, crustaceans, mollusks, other invertebrates representing many additional phyla, and fish (Ruiz *et al.* 2000, Wonham *et al.* 2000, Gollasch *et al.* 2002). Hülsmann and Galil (2002) estimated that more than 250 protist taxa are commonly present in ballast water. Besides potentially having impacts in the environment where the ballast water is released, many are important members

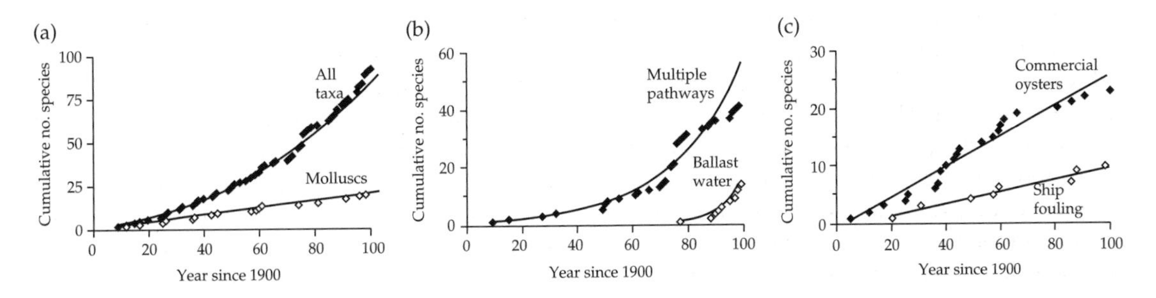

Fig. 2.7 Cumulative number of established non-native marine and estuarine species in the Northeast Pacific, by date of first record, for different taxa and invasion pathways. (a) All invertebrates and algae together (solid diamonds), and mollusks only (open diamonds); (b) multiple invasion pathways (likely via one or more of ballast water, ship fouling, and commercial oysters; solid diamonds) and ballast water only (open diamonds); (c) commercial oyster (solid diamonds) and ship fouling (open diamonds) invasion. Redrawn and printed, with permission, from Wonham and Carlton (2005), copyright Springer.

of the marine food web and thus their presence increases the likelihood that other marine hitch-hikers will survive the trip, such as filter feeders and biofilm-grazers.

A variety of dispersal vectors introduce aquatic species into freshwater environments. In the Dutch Rhine Delta region, an estimated 10% of the introduced species come from ocean vessels, while 65% have been the result of escapes from horticulture and aquaria, and 25% dispersed into the freshwater environments through the extensive canal system in the region. Canals, which have dramatically increased the connectivity between rivers and other freshwater habitats (van der Velde *et al.* 2002), have been serving as an important dispersal conduit for many freshwater non-native species in Europe. For example, canals have facilitated the spread of the zebra mussel, *Dreissena polymorpha*, into Central and Northern Europe during the eighteenth and nineteenth centuries (Jazdzewski and Konopacka 2002). Since, *D. polymorpha*'s range in Europe was much more widespread during pre- or interglacial periods (Jazdzewski and Konopacka 2002), its re-establishment in areas where it had once existed can be viewed as a human-assisted, albeit accidental, return of a native species. Besides serving as efficient dispersal corridors for riparian plants, river habitat can provide suitable conditions for some upland plant species, thereby enabling the latter to disperse through otherwise inhospitable upland habitat (Pyšek and Prach 1995), and sometimes serving as a departure point for subsequent upland spread (Pyšek *et al.* 2007).

While many introductions are accidental, many species of plants, fish, birds, mammals, and invertebrates (such as insects and snails) have been intentionally introduced into new regions of the world. Thomas Jefferson was an early advocate of introducing species, particularly plants. He once wrote, 'the greatest service which can be rendered any country is to add a useful plant to its culture' (Ford 1892–99). According to Jewett (2005), while Jefferson was an envoy to France, he sent seeds of various grasses, acorns of the cork oak, olive plants, and innumerable fruits and vegetable seeds to agricultural societies, farmers, and botanists in the United States, and that while in Italy, he smuggled out Italian rice. Whereas plants have usually been introduced for horticultural, agricultural, and forestry reasons (Pyšek *et al.* 2002, Mack 2003, Křivánek *et al.* 2006), fish have been introduced primarily for game, aquariums, and aquaculture (Fuller 2003, Keller and Lodge 2007), birds for aesthetics (Nummi 2002) and for the pet trade (Temple 1992), mammals for game and the fur trade (Nummi 2002), insects, and to a lesser extent mammals, primarily as a part of biological control efforts (Simberloff and Stiling 1996). Figure 2.5 shows the temporal patterns of intentional vertebrate introductions between Europe and North America over the past several centuries. In some cases, e.g. fish in North America, many of the introductions are not of foreign origin but consist of introductions from one part of a country to one where the species had not previously existed (Fuller 2003).

Even scientific research has been identified as a cause for species introductions. African clawed frogs, *Xenopus laevis*, are the research animal of choice for many developmental biologists. In addition, from 1934 until the 1950s, *X. laevis* was imported and used as a pregnancy test (the urine of pregnant women stimulates the production of the frog's eggs) (Marris 2008). However, *X. laevis* can host the chytrid fungus, *Batrachochytrium dendrobatidis*, which causes the disease chytridiomycosis, a major source of mortality for frogs worldwide (Weldon *et al.* 2004, Marris 2008; Fig. 2.8). The initial spread of the disease, which is believed to have originated in South Africa, was most likely the result of global commercial trade of *X. laevis* (Weldon *et al.* 2004).

For some species, animal rights activists play a role in introducing non-native species. For example, in Europe, activists have broken into mink farms rearing North American mink and released the mink into the wild (GISP database, http://www.issg.org/database/). Today, even video-arcade games are instigating the export of species from one part of the world to another. In Japan, an arcade game involving battles between stag beetles has provoked strong interest in non-Japanese beetles, resulting in the importation of more than one million beetles, including some rare and endangered species. Besides threatening the populations in the countries of their origin, there is concern that some of the imported beetles will escape into Japan's environments creating potentially undesirable impacts (Holden 2007).

As part of their effort to define a new field of vector science, Carlton and Ruiz (2005) presented a useful conceptual framework to characterize the various aspects associated with dispersal and the introduction process. The purpose of this framework was to clarify discussions involving dispersal vectors and to facilitate the development of effective strategies to reduce the successful introductions

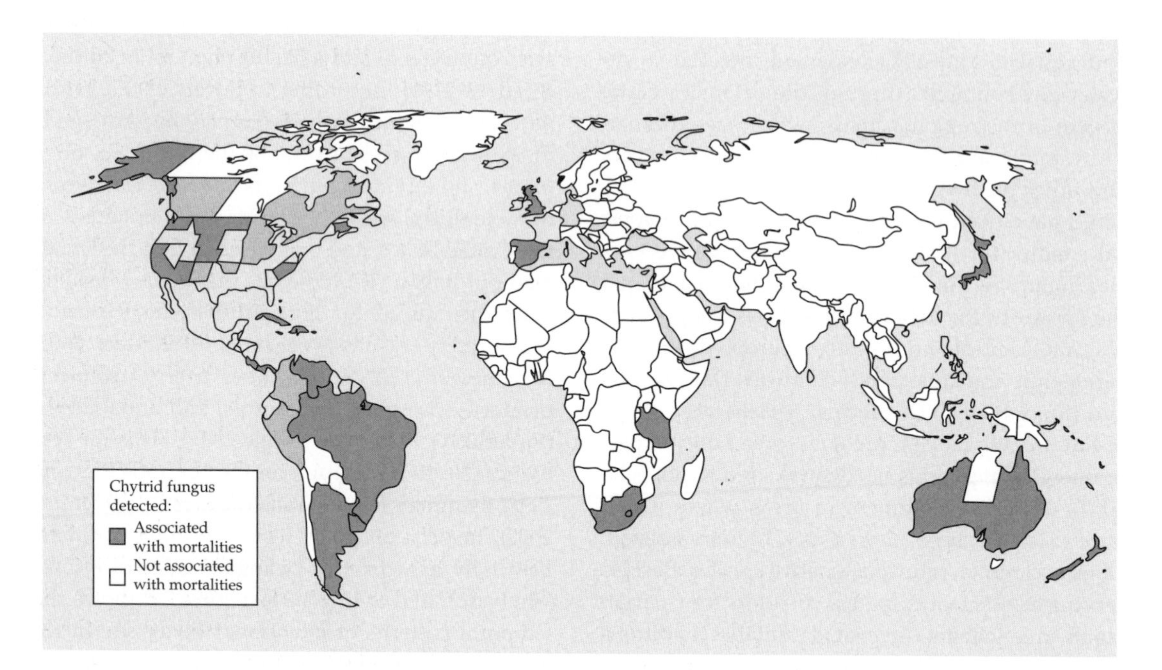

Fig. 2.8 Global spread of chytrid fungus in 2007. Redrawn and printed, with permission, from Dede Olson and Kathryn Ronnenberg, USDA (http://www.parcplace.org/images/BD_Map5.jpg).

of harmful non-native species. There are six elements to the framework proposed by Carlton and Ruiz: cause, route, vector type, vector tempo, vector biota, and vector strength.

Cause refers to why a species is transported, i.e. whether or not the transport was intentional (e.g. for food, medicine, biocontrol, horticulture, the pet trade). Intentional transport generally will be easier to monitor, although this is not necessarily the case. For example, intentional but illegal transport of species, as has often occurred in the pet trade, is usually done covertly.

Route refers to the geographic path followed during the transport of the species, from its origin to its destination. In this framework, routes are distinguished from corridors, which Carlton and Ruiz define as the actual physical conduit through which the species is transported, e.g. roads, railroads, walking paths, canals, shipping lanes.

Vector type refers to the physical vehicle that transports the propagules, such as trains, planes, automobiles, ships, airplanes, and individual people, e.g. hikers. The route describes potential areas of dispersal during the transportation process, but the likelihood of dispersal along the route is influenced by the nature of the corridor and the vector type. The vector type also can influence the likelihood that species are able survive the transport period.

Vector tempo refers to the temporal dynamics of the vector, i.e. the frequency, duration, and timing of the transported species. An increase in the number of times that a vector moves between an origination and destination site will be expected to increase the likelihood of a successful introduction (Lockwood *et al.* 2005). Duration of the transport is important since transported organisms may be more likely to survive shorter transport episodes. Finally, timing of the transport event is often critical since propagules are more likely to establish when introduced to a region during a time of the year when conditions are hospitable to the species.

Vector biota refers to the species being transported, and can be described in terms of diversity, density, and condition. Density, number of organisms per some unit space (e.g. m^2 of hull surface) or volume (e.g. m^3 of ballast water), is an important characterization since invasion success is widely understood as being greatly influenced by the number of arriving propagules, part of propagule pressure (Lockwood *et al.* 2005). Carlton and Ruiz emphasize that condition refers not only to the physiological status of the organisms but also, in some instances, to the life stage (e.g. larval, juvenile, adult), and point out that all three features of vector biota—diversity, density, and condition—can change during the transport period.

Vector strength refers to the number of established invasions at a specific site resulting from a particular vector during a specified time period, which Carlton and Ruiz (2005) characterize as the 'ultimate measure of invasions.' Carlton and Ruiz recommended that if vector strength can be adequately assessed, it should be used to prioritize management targets.

The proposed value of this conceptual framework is that it can facilitate the development of integrated vector management (IVM) systems. IVMs are defined by Carlton and Ruiz as programs that apply management strategies and technologies at multiple stages during the transport process, with the goal being to reduce or prevent the transport and release of living organisms. For more detailed discussion of this framework and IVMs see Carlton and Ruiz (2005).

Motivated by the same objective as Carlton and Ruiz, that being to facilitate the implementation of effective policies to prevent introductions from occurring in the first place, Hulme *et al.* (2008) presented an alternative conceptual framework of the introduction process (Fig. 2.9). Acknowledging the enormous variety and complexity of invasions, Hulme *et al.* argued that a simplified framework was needed for utility sake. As long as the framework was well-conceived, they believed that comprehensiveness would not be unduly comromised. In making their argument for simplification, they presented the 31 different types of introductions described by the Global Invasive Species Information Network (http://www.gisinetwork.org/):

acclimatization societies, agriculture, aircraft, aquaculture, aquarium/pet trade, biological control, contaminated bait, floating vegetation/debris, ornamental purposes, forestry, horticulture, ignorant possession, internet sales/postal service, landscape/fauna 'improvement', live-food trade, military, mud on birds, nursery trade, people sharing resources, road vehicles, seafreight, self-propelled, ship, ship ballast water, ship hull fouling, smuggling, stocking, botanical garden/zoo, translocation of machinery, transportation of domesticated animals, and transportation of habitat materials.

It is not difficult to imagine that such a detailed, and almost mind-numbing, characterization of invasions might overwhelm the fortitude of most policy makers. Thus, Hulme *et al.* proposed a simplified classification scheme. They proposed that non-native species enter a new region through one or more of three general ways, which they referred to as mechanisms: importation of a commodity, arrival of a transport vector, and natural spread from a neighboring region where the species had already established itself as a non-native species (Hulme *et al.* 2008). As shown in Fig 2.9, Hulme *et al.* described these three mechanisms as resulting in six primary pathways: release, escape, contaminant, stowaway, corridor, and unaided.

In Hulme *et al.*'s framework, release describes the intentional introductions of organisms into a new region. For example, this would include the release of game animals and biocontrol agents, as well as the intentional planting of non-native species in the landscape, e.g. for erosion control. Escape refers to the unintentional release of organisms originally introduced with the intent of keeping them in captivity or under control. Contaminant describes organisms that accompany the introductions of others, e.g. weed seeds in grain shipments, seaweed in oyster shipments, and parasites and pathogens that accompany their hosts, such as desired plant and animal species, during introduction. Other organisms, while not directly associated with intentional shipped organisms, are transported unintentionally on the transport vehicles—ships, planes, automobiles, and so on. These are what Hulme *et al.* (2008) describe as stowaways, and include organisms found in ballast water, in soil attached to the vehicles, and in shipping containers. The corridor pathway refers to organisms that disperse on their own but utilize corridors created by humans to do so, e.g. canals, roads, bridges, and tunnels. Finally, unaided describes species that disperse without any human facilitation. Since it is common to apply the term non-native to organisms that have been introduced by humans (1992 Convention on Biological Diversity, Richardson *et al.* 2000a), the unaided category might seem a questionable one. However, following an initial human-aided introduction, many species subsequently disperse into new adjacent regions independently. The term unaided would apply to these species.

It is important to recognize some of the terminology differences between the frameworks presented by Hulme *et al.* and Carlton and Ruiz (2005). Hulme *et al.* used the term pathways to describe the diverse set of six ways by which organisms can enter the natural environments of a new region. Carlton and Ruiz explicitly avoided using the term pathways, which they argued had been used too generally in the field and with too many different meanings, and which, consequently contributed to ambiguity and confusion in discussions and analyses. For example, they said the term has been used to describe four of the six phenomena in their framework: cause, route, vector, and corridor. However, Carlton and Ruiz did refer to paths, in which case the term was used specifically in reference to the geographic routes taken during the introduction process.

While both Hulme *et al.* (2008) and Carlton and Ruiz (2005) have proposed frameworks to aid in the development of effective invasion policy and control measures, the two frameworks differ in their focus and in their likely contributions. The framework proposed by Carlton and Ruiz should prove to be very useful for those trying to prevent or manage the introductions of particular organisms. By focusing attention on distinct elements of the introduction phenomenon, including the cause, the route, and the transport mechanism, as well as the temporal dynamics involving the propagule pressure, the framework can help managers and policy makers identify times and places during the dispersal and introduction process where intervention may be most effective.

The framework proposed by Hulme *et al.* should be most helpful to those charged with developing a comprehensive policy and management approach to non-native invasive species in

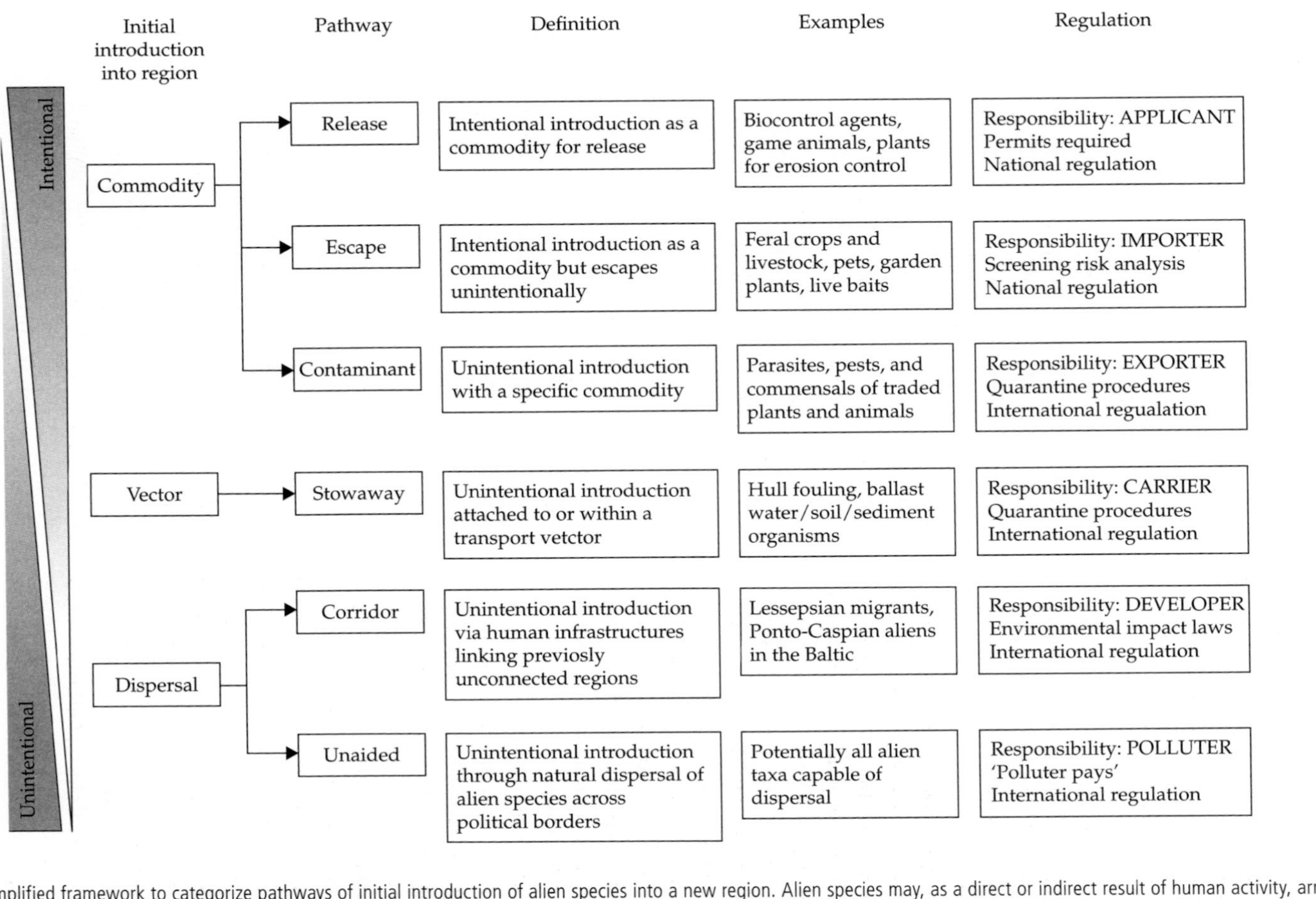

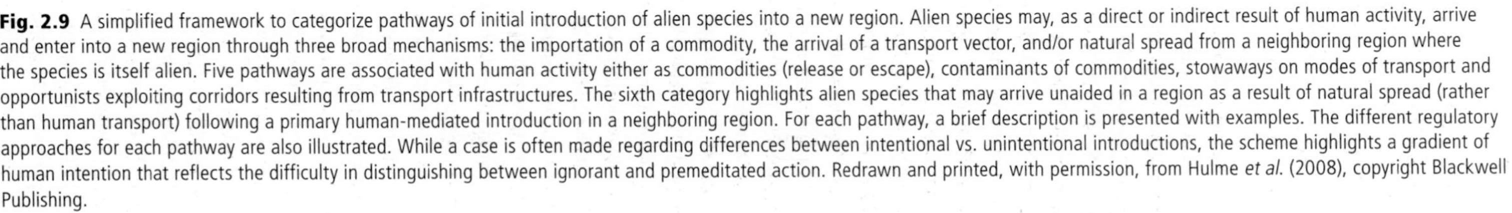

Fig. 2.9 A simplified framework to categorize pathways of initial introduction of alien species into a new region. Alien species may, as a direct or indirect result of human activity, arrive and enter into a new region through three broad mechanisms: the importation of a commodity, the arrival of a transport vector, and/or natural spread from a neighboring region where the species is itself alien. Five pathways are associated with human activity either as commodities (release or escape), contaminants of commodities, stowaways on modes of transport and opportunists exploiting corridors resulting from transport infrastructures. The sixth category highlights alien species that may arrive unaided in a region as a result of natural spread (rather than human transport) following a primary human-mediated introduction in a neighboring region. For each pathway, a brief description is presented with examples. The different regulatory approaches for each pathway are also illustrated. While a case is often made regarding differences between intentional vs. unintentional introductions, the scheme highlights a gradient of human intention that reflects the difficulty in distinguishing between ignorant and premeditated action. Redrawn and printed, with permission, from Hulme *et al.* (2008), copyright Blackwell Publishing.

general. Although most intervention strategies will need to be fine-tuned for individual species, Hulme *et al.* (2008) showed that certain categories of introductions will tend to call for particular policy/management approaches. For example, they argued that intentional introductions should take place through the issue of permits and licenses, following detailed risk assessment. With respect to the problem of escapees, the authors said that while individual purchasers need to be educated as to the problems associated with escapees, policies should be implemented that place a substantial amount of the responsibility on the importers. In this case, even though the importers are not normally responsible for the escape, Hulme *et al.* said that they could be required to import only species that had been determined to be low risk with respect to their impacts, if and when individuals escape. The application of this approach could be considered for the pet and horticultural industries. A similar approach could be taken with respect to contaminants, i.e. placing the responsibility on the importer to ensure that products are not contaminated. Comparable policies could be implemented requiring carriers to make sure that they are not transporting stowaways.

Hulme *et al.* (2008) also emphasized that their framework helped to identify similarities and differences in the introduction process among different taxa. For example, on the basis of an assessment of European non-native species, corridors were found to be much more strongly associated with the introduction and spread of aquatic than terrestrial organisms. This was believed to be due to the important role played by human-made canals in the dispersal of many aquatic species. Another difference was found between terrestrial and aquatic plants, with introduction of the former more likely to be due to intentional releases. Aquatic plants were more often found to be introduced via escape or as stowaways. In general, European-introduced vertebrates were more often the result of intentional releases compared to invertebrates, which were more likely introduced via contamination. Finally, pathogenic microorganisms, fungi, and parasites were usually introduced as contaminants (Hulme *et al.* 2008).

Are introduction rates increasing?

Except in cases where introductions are intentional, it is very difficult to know when introductions occur. Most dispersal events probably fail to establish any individuals, or at least not enough to produce a population that persists for a substantial length of time. We are not aware of a large number of unintentional introductions, simply because the individuals do not persist long enough and/or in abundance enough for us to notice them before they disappear. Only when the new populations are abundant, long-lasting, and/or producing a noticeable impact do we discover them.

Although not as well known as species–area curves, species–time relationships predict that the number of species recorded in an area should increase over time (Preston 1960, Magurran 2007). Preston argued that three factors could account for the species–time relationship. The first was simple sampling effect. As in the case of the species–area relationship, more extensive sampling should document more species. The other two factors would actually add new species to the pool over time: one through ecological processes, e.g. succession, and the other through evolutionary processes, i.e. speciation. In the case of species introductions, the number of non-native species documented should be expected to increase over time due to the same or comparable factors, including a sampling effect, an increase in population sizes of the establishing populations, and the introduction of additional species.

It is often suggested that the rate of introductions, and hence propagule pressure, has been increasing in certain environments and regions (Leppäkoski and Olenin 2000b, Wonham and Carlton 2005). However, a good case has been made that the exponential-like curves resulting when cumulative discoveries are plotted as a function of time can be explained as an artifact of the establishment and discovery process (Costello and Solow 2003, Solow and Costello 2004; Fig. 2.10). Recognition of new species requires successful dispersal, successful establishment, and then also detection (Wonham and Pachepky 2006). On the basis of their analysis, Wonham and Pachepky (2006) showed that constant introduction and establishment success will

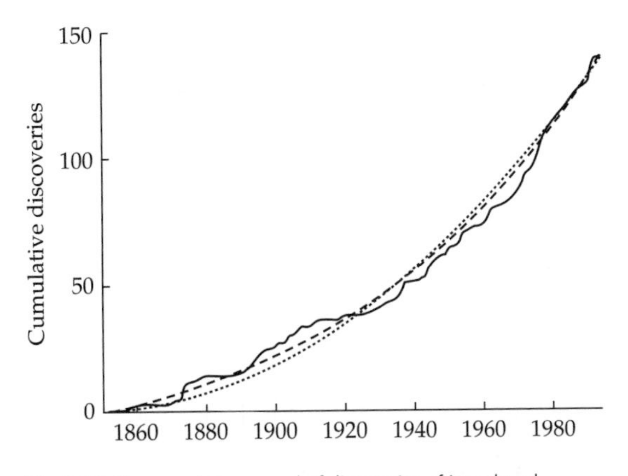

Fig.2.10 The cumulative record of discoveries of introduced species in the San Francisco estuary, California, USA, 1850–1995 (solid line). Also illustrated are fitted values allowing for an increased introduction rate (dashed line) and assuming a constant introduction rate (dotted line), showing little difference in the expected cumulative discoveries with introduction rates increasing or remaining constant. Redrawn and printed, with permission, from Solow and Costello (2004), copyright Ecological Society of America.

also yield the commonly observed exponential distributions of introduction trends. Even if there is just a single dispersal event, an exponential discovery curve would be expected to result since the number of populations that have grown large enough to be noticed will be very small shortly after the dispersal event, but over time more and more will reach the detection threshold (Costello and Solow 2003). Solow and Costello emphasize that their finding does not mean that rates of introductions have not been increasing. Certainly, given the increase in the travel of humans and their commerce in the past several hundred years, it is difficult to imagine that introduction rates for many organisms have not gone up.

Summary

The first challenge, or filter, faced by any traveler is surviving the trip. In many instances, the vicissitudes of the journey pose too great an obstacle and most, if not all, dispersers die *en route*. No matter how receptive might be the new environment, if the dispersing propagules never reach it, there can be no invasion. The intensity or magnitude of dispersal to a site is often described as propagule pressure, which encompasses both the number of propagules in a single dispersal event and the number of dispersal episodes. It should probably also encompass propagule quality, since many propagules of low quality pose less pressure than a few high-quality propagules. Williamson (1996) concluded that propagule pressure plays a central role in determining the success or failure of an invasion, and there has been considerable evidence collected during the past decade to support Williamson's claim. Both experimental studies and studies of the geographic distributions of non-native species have demonstrated the importance of dispersal limitation and propagule pressure in the invasion process.

Although the term 'dispersal' may suggest a rather simple, one-dimensional event, in fact dispersal is usually a complex, multi-factored process. The nature of the dispersal vector, the pathway followed during dispersal, the duration of the dispersal process, conditions encountered along the way, and the nature and quality of the propagules (e.g. stage of the life cycle and overall physiological vigor) are all factors that can determine whether or not the propagules survive the transport in good enough condition, and in sufficient numbers, to have a chance to successfully establish in the new environment. While there is no question that the propagule pressure involving some species is increasing in certain regions, one should be cautious in making such claims based solely or primarily on increased detection rates of individuals or populations, since the latter can be an artifact of the discovery process. In any case, even if dispersal to a new region is successful, an invasion is not ensured. The newly arrived individuals need to establish themselves in their new environment.

CHAPTER 3

Establishment

For an individual to successfully establish, defined as persisting long enough in the new environment to reproduce, it needs to accomplish four tasks. It needs to find an environment with abiotic conditions (e.g. temperature, salinity, moisture) it can tolerate. It needs to be able to access resources necessary for its maintenance, growth (if a juvenile or a species with indeterminate growth), and reproduction. For out-crossing species, it needs to find a mate, or at least its gametes need to find the gametes of a mate. And, it needs to avoid pre-reproductive mortality. The ability of an organism to achieve these tasks is going to be greatly influenced by the traits it possesses.

Establishment and traits

Kolar and Lodge (2002) compared the life-history traits, as well as other factors such as habitat requirements, of non-native fish that had been introduced into the Great Lakes of North America. They found that species that exhibited higher growth rates and greater tolerance of temperature and salinity were more likely to successfully establish. Marchetti *et al.* (2004) examined the success of fish introductions in California on the basis of particular traits, e.g. body size, type of parental care, and physiological tolerance to changes in water quality, as well as other factors such as trophic status, prior invasion success, and propagule pressure. They found that all variables contributed to the effectiveness of a predictive model; however, both physiological tolerance and body size were among the best predictors for establishment. In a study of non-native fish introductions in the Colorado River basin, Olden *et al.* (2006) found that, as a group, the non-native fish differed from the native species in a number of ways, including showing less dependence on fluvial conditions to complete the life-cycle, preference for slow currents and warmer water, faster maturation, and smaller and more rapidly developing eggs. They argued that these traits may enable them to take advantage of novel river conditions produced as a result of human actions on and along the river, particularly dam construction, which has substantially altered flow regimes. The tolerance of the predatory water flea, *Cercopagis pengoi*, to a broad range of salinity and water temperatures is believed to have facilitated its establishment in diverse regions of the Baltic Sea (Telesh and Ojaveer 2002). The ability of several species of non-native crayfish to establish and spread in large numbers in Europe is partly attributed to their larger body and chela size, faster growth, increased thermal tolerance and fecundity, and more aggressive behavior than the native species (Lindqvist and Huner 1999, Westman 2002).

Many studies that have examined the relationship between traits and invasion success have involved plants. Although a number of studies have concluded that non-native and native plants do not exhibit pronounced differences in traits or life-histories (Thompson *et al.* 1995, Williamson and Fitter 1996, Meiners 2007), other studies have documented correlations between certain traits and establishment success. For example, in a study of South African Iridaceae, van Kleunen and Johnson (2007) compared traits of species that had become naturalized elsewhere with those of species that had been introduced elsewhere but which failed to establish. They looked at seed mass, seedling emergence time, and early growth rates of 30 naturalized species and 30 congeneric species that had not naturalized and found that, although seed size did not differ between the two groups, the species that had successfully naturalized elsewhere exhibited more rapid emergence and a higher emergence rate. Since naturalization success in the Iridaceae has been found to be positively associated with plant size, it was expected that rapid emergence time would likewise contribute to establishment success (van Kluenen and Johnson 2007).

Losos *et al.* (2000) and Kolbe and Losos (2005) showed that *Anolis* lizards exhibit phenotypic plasticity with respect to the length of hind limbs and that limb length was influenced by the nature of their climbing substrate. This plasticity may have facilitated the lizards' ability to establish themselves in new Caribbean environments (Losos *et al.* 1997). While phenotypic plasticity in *Anolis* leg length was documented, it is also known that leg length in *Anolis* is highly heritable (Losos *et al.* 2004, Kolbe and Losos 2005). Thus, both genetic adaptation and rapid phenotypic adaptation may contribute to the ability of these lizards to establish in new sites.

Phenotypic plasticity involving behavior has been found to influence establishment success in other animals as well. In birds, behavioral flexibility has been found to increase establishment success of non-native species (Sol and Lefebvre 2000, Sol *et al.* 2002). Sol *et al.* (2005) reviewed more than 600 avian introduction events and found that species with larger brains, relative to their bodies, tended to be more successful in establishment. With other evidence showing that the ability of birds to respond effectively to novel conditions was also associated with proportionally larger brains, the authors concluded that the increased establishment success of species with proportionately larger brains was due to the increased ability of these species to adapt behaviorally to novel environments.

Darwin (1859) believed that newly arrived species would have a more difficult time establishing and persisting if they are closely related to the resident species, due to the likelihood of increased competition (Darwin's naturalization hypothesis). This hypothesis rests on the assumption that closely related species will be ecologically similar. While this may be true in many cases, the hypothesis inevitably will be confounded by evolutionary divergence and convergence with respect to the traits of the species under investigation, i.e. closely related species may differ substantially in key ecological traits, while more distantly related species may be quite similar ecologically. The naturalization hypothesis also assumes that inter-specific competition is a primary determinant of community assembly, although it is well-known that other events and processes can overwhelm competition effects (Davis *et al.* 2000, Sax and Gaines 2003, Ricklefs 2005, Stohlgren *et al.* 2008a). These factors may partly explain why empirical support for the naturalization hypothesis has been mixed.

Rejmánek (1996, 1998) and Strauss *et al.* (2006a) provided evidence in support of the notion that ecological novelty facilitates the establishment and spread of a new species. Studying Californian grasses, they found that introduced invasive species tended to be less related to the native grasses in the community than were introduced non-invasive grasses, concluding that the benefits of being phylogenetically dissimilar to the long-term residents may involve new ways to utilize resources, enemy escape, and/or novel 'weapons.' However, Mitchell *et al.* (2006) reviewed the role phylogenetic similarity might play in invasion success and concluded that a high degree of relatedness between the new and resident species could either facilitate or deter establishment of the new species, depending on the circumstances. Specifically, they noted that obstacles associated with a high degree of relatedness may include increased competition and shared predators, herbivores, and pathogens (Blaney and Kotanen 2001, Parker and Gilbert 2004). On the other hand, Mitchell *et al.* (2006) noted that other studies have shown that a high degree of relatedness may confer benefits as well, including shared mutualists and increased likelihood of tolerance to the physical conditions of the environment (Daehler 2001b, Duncan and Williams 2002).

Ricciardi and Mottiar (2006) analyzed fish introductions in several independent regions and found the data supported neither side of the argument, concluding that taxonomic affiliation is not a useful predictor of fish invasion success. Ultimately, Mitchell *et al.* (2006) argued that invasion success is not likely to be consistently related to phylogenetic relatedness. Rather, in a more nuanced approach, they concluded that Darwin's naturalization hypothesis would be expected to be supported in cases where the negative effects of shared enemies and increased competition outweighed the positive impacts of shared mutualisms and favorable abiotic conditions (Fig. 3.1).

Dietz and Edwards (2006) have proposed that invasion biologists consider the establishment process as consisting of two stages. They suggest that the first stage consists of the new species establishing in an area on the basis of their existing traits, which enable them to take advantage of available resources. Dietz and Edwards argued that subsequent persistence is due more to subsequent adaptation to the new environment, adaptation that can be either genetic or non-genetic. The difference between these proposed two stages is likely not as distinct as Dietz and Edwards suggest. While it is true that evolutionary adaptation cannot occur immediately because it is an intergenerational process, phenotypic adjustments can be made directly upon arrival. In some instances, species may be able to respond to novel conditions right away through phenotypic plasticity, e.g. changes in energy allocation patterns, growth forms, and, in the case of animals, behavior. Flexibility in diet and nest selection, and in habitat selection in general, are behaviors that have been found to be particularly associated with successful introductions of birds (McLain *et al.* 1999, Cassey 2002, Sol *et al.* 2002), suggesting that broad ecological tolerance can enhance establishment (Duncan *et al.* 2003, Labra *et al.* 2005). Indeed, findings such as these indicate that the addition of new species to an environment, and the integration of these species into existing food webs, do not necessarily have to involve any evolutionary fine-tuning of population interactions. Rather, complexity can be constructed in ecological time (Taylor 2005).

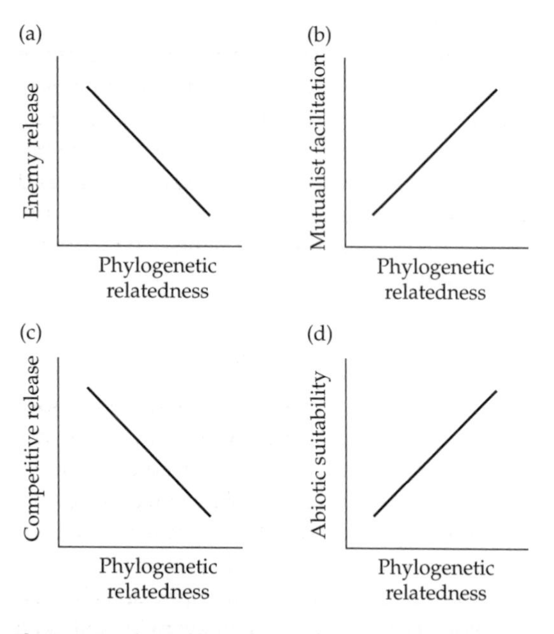

Fig. 3.1 Hypothesized dependence of four proposed mechanisms for biological invasions on the phylogenetic relatedness of an introduced species to resident species, integrated across all those species. (a) The contribution of release from natural enemies to invader demographic success is predicted to be greater in communities of resident species less related to the introduced species, assuming that enemies are phylogenetically specialized. (b) The contribution of resident mutualists to invader success is predicted to be greater when the introduced species is more closely related to resident species, assuming that mutualists or their benefits are phylogenetically specialized. (c) The contribution of competitive release to invader success is predicted to be lower in communities of resident species more related to the introduced species, assuming that more related species have greater niche overlap. (d) The contribution of a suitable abiotic environment to invader success is predicted to be greater in communities of resident species more related to the introduced species, assuming that more related species are adapted to similar abiotic conditions. (a–d) To the degree that each of these assumptions is violated, the slopes of the hypothesized relationship would approach zero. Redrawn and printed, with permission, from Mitchell *et al.* (2006), copyright Blackwell Publishing.

Establishment and invasibility

The ability, or inability, of an environment to permit establishment has been termed invasibility. Thus, invasibility describes the susceptibility of an environment to the colonization and establishment of new species (Levine and D'Antonio 1999, Lonsdale 1999, Davis *et al.* 2000). The term invasion resistance is often used in place of invasibility. Essentially the inverse of one another, both terms describe the same phenomenon: the susceptibility of an environment to colonization and establishment of new species. That invasibility is an important factor in accounting for documented patterns of invasion, is hardly a new insight. For at least 150 years, ecologists, naturalists, and biogeographers have observed that some environments are more easily colonized and populated by non-native species than others. In Chapter 13 of the *Origin*,

Darwin concluded, 'an intruder from the waters of a foreign country would have a better chance of seizing on a new place than in the case of terrestrial colonists.'

The term invasibility has been widely used and accepted by researchers and practitioners, thankfully mostly without the confusion, and often controversy, that has plagued other terms in the field. Nevertheless, there are a few very important caveats that must be remembered when using the term. First, invasibility is not a static condition of an environment; rather it fluctuates (Davis *et al.* 2000). Since invasibility is influenced by the environment's biotic and abiotic events and processes, as these change, so will the environment's invasibility. Second, since environments do not behave as uniform entities, resource fluctuations occur at different times and to different extents in different areas within a single environment. Even at a given moment in time, a particular environment will not exhibit a single level of invasibility, but will instead manifest different invasibilities in different places throughout the environment (Davis *et al.* 2000, Stachowicz and Byrnes 2006, Melbourne *et al.* 2007). Third, the invasibility of an environment, or portion of an environment, varies from species to species, from genotype to genotype, and even from phenotype to phenotype of identical genotypes (Davis *et al.* 2005a). The same environment at the same point in time may be quite invasible to one organism type but quite resistant to colonization and establishment by another type. Fourth, invasibility is a fundamental condition of all environments (Davis *et al.* 2005a). Whether rich with species or devoid of any life form whatsoever, virtually all environments exhibit some susceptibility to colonization and establishment of a new life form. Williamson (1996) came to the same conclusion, describing all communities as invasible, though some more than others.

Williamson (1996) placed more importance on propagule pressure than invasibility in accounting for the success or failure of different invasion episodes, and others have also emphasized propagule pressure when accounting for invasion success or failure (Lockwood *et al.* 2005, Rejmánek *et al.* 2005a). However, it is clear that both propagule pressure and invasibility of the new environment play important roles in determining the outcomes of species' introductions (Norden *et al.* 2007). In a comprehensive analysis of the marine literature, Lester *et al.* (2007) found that range size was only loosely associated with dispersal ability, and not at all in some cases. Lester *et al.* concluded that range size is clearly also being influenced by other factors besides dispersal, such as those involving resources and other habitat characteristics, the same factors believed to influence invasibility (Davis *et al.* 2000, Shea and Chesson 2002).

Theories to account for invasibility have variably focused on the diversity of the resident community (species diversity and/or functional group diversity), resource availability (particularly its temporal and spatial heterogeneity), physical stress, and enemies and mutualists of the arriving species, with some of the theories invoking more than one of these factors.

Invasibility and diversity

The diversity–invasibility hypothesis holds that increased species-richness should confer a higher degree of invasion resistance to an environment, and thus that invasibility should be inversely correlated with diversity. Although Elton (1958) is often accorded the authorship of this theory, its roots, like those of so many other theories, actually can be found in the *Origin of Species* (Darwin 1859). The reasoning behind Darwin's assertion that freshwater environments should be more invasible than terrestrial environments is because 'the number [of kinds of inhabitants] even in a well-stocked pond is small in comparison with the number of species inhabiting an equal area of land, the competition between them will probably be less severe than between terrestrial species.'

The diversity–invasibility hypothesis, as presented by ecologists in recent years (Knops *et al.* 1999, Naeem *et al.* 2000, Shurin 2000, Stachowicz *et al.* 2002, Fargione and Tilman 2005, Maron and Marler 2007), is grounded in traditional niche theory and ultimately is resource-based, the reasoning being that fewer empty niches would be available in species-rich environments (complementarity), meaning that fewer resources would be available to new arrivals. Hence, new organisms would be prevented from colonizing due to biotic resistance. According to this line of reasoning, which is the same used by Darwin in his naturalization hypothesis, colonists must be sufficiently dissimilar to the residents, i.e. able to occupy some uninhabited niche space, in order to successfully establish (Stachowicz and Tilman 2005). In some studies, the diversity–invasibility hypothesis has been reformulated so that diversity refers to functional diversity, as opposed to species diversity (Symstad 2000, Fargione *et al.* 2003, Xu *et al.* 2004, Britton-Simmons 2006, Perelman *et al.* 2007; Fig. 3.2). The notion of community saturation is implied in this niche-based argument, with diverse and invasion-resistant communities assumed to be closer to saturation than species-poor and highly invasible communities.

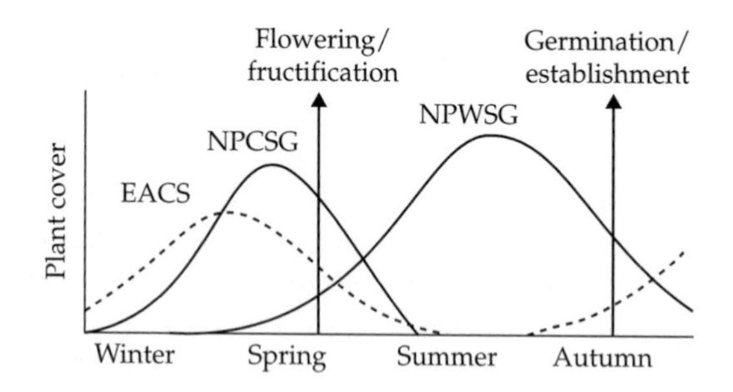

Fig. 3.2 Schematic representation of phenological patterns for major native and exotic plant functional groups. EACS, exotic annual cool-season species; NPCSG, native perennial coolseason grasses; NPWSG, native perennial warm-season grasses. The vertical lines highlight two critical periods for the regeneration of common exotic species, and how they overlap with native species growth patterns. Perelman *et al.* (2007) suggest that the negative response of exotic richness to native warm-season grasses found in humid prairies, chiefly reflects interference of summer grasses with seedling recruitment of exotic annuals during autumn. Redrawn and printed, with permission, from Perelman *et al.* (2007), copyright Blackwell Publishing.

Findings from small-scale experiments

The applicability and generality of Elton's theory has been the subject of intense controversy in the field in recent years, with a large number of field experiments and censuses of natural communities conducted to test the diversity–invasibility hypothesis. Supporting results have come mostly from studies using small-scale constructed communities involving plants (Knops *et al.* 1999, Levine 2000, Naeem *et al.* 2000, Fargione and Tilman 2005, Maron and Marler 2007), zooplankton communities (Shurin 2000), and sessile marine invertebrates (Stachowicz *et al.* 2002; Fig. 3.3). Few experimental tests of the diversity–invasibility hypothesis have been conducted with mobile animals. One such study was by France and Duffy (2006), who tested the effects of mobile crustacean grazers on invasibility in flow-through seagrass mesocosms. They found that, on average, increased diversity of resident grazers reduced the abundance and biomass of the grazers they introduced.

In addition, some small-scale plot studies done in natural settings have also supported the diversity–invasibility hypothesis. MacDougall (2005) seeded plants in small savanna plots that had experienced different burn frequencies and found that the establishment success was lowest in the high-diversity plots and that resource availability (light and bare ground) was higher in the low-diversity and more invasible plots. In a small-scale garden plot study of tall goldenrod, *Solidago altissima*, Crutsinger *et al.* (2008) showed that intra-specific diversity can also influence invasibility, finding that invasibility was reduced by high stem density and that plots with greater intra-specific genotypic diversity exhibited higher stem densities. In a review of the literature, Olyarnik *et al.* (2008) concluded that experimental studies in marine and terrestrial systems have generally found a negative effect of increasing diversity on invasion success.

Olyarnik *et al.*'s (2008) conclusion notwithstanding, the results from a number of small-scale experiments have not supported the diversity–invasibility hypothesis. Robinson *et al.* (1995) found that species-rich plots in a winter annual California grassland were more invasible to hand-seeding than species-poor plots. In a study in a New Zealand mountain beech forest, Wiser *et al.* (1998) found that *Hieracium lepidulum*, a non-native perennial herb, was more likely to colonize species-rich small plots than species-poor ones. Using experimental microcosms of microbes, Jiang and Morin (2004) found a positive relationship between diversity and invasibility. In an observation study of sessile marine invertebrates at a small spatial scale (0.1 m^2), Dunstan and Johnson

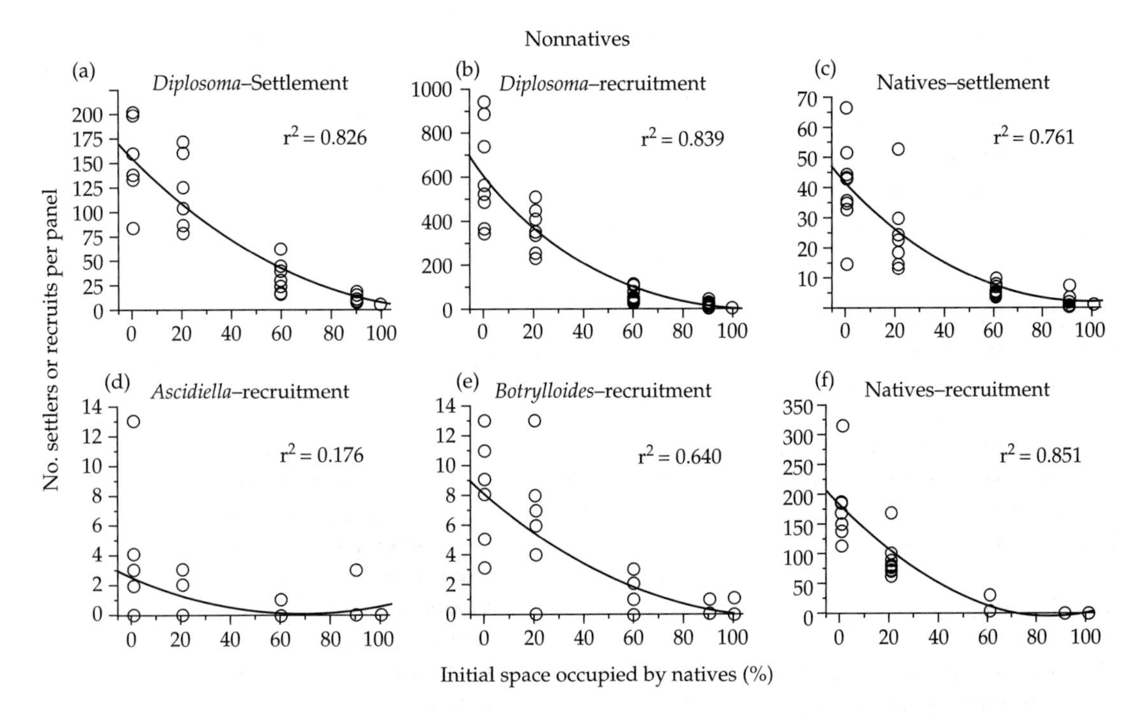

Fig. 3.3 Effects of the direct manipulation of space on the settlement and recruitment of native and non-native sessile invertebrates. Analysis by quadratic regression demonstrates that space availability reduces the short-term settlement and recruitment of the nonnative species: (a, b) *Diplosoma*, (d) *Ascidiella*, and (e) *Botrylloides*, and of (c, f) native species ($r^2 > 0.64$ for all except *Ascidiella*, for which recruitment was <4 individuals/100 cm^2 for all but one settlement panel). Redrawn and printed, with permission, from Stachowicz *et al.* (2002), copyright, Ecological Society of America.

(2004) found that invasibility was lowest in the species-poor experimental environments. Interpreting their results, which were supported in a simulation model, Dunstan and Johnson (2006) concluded that in this system, invasibility was highly correlated with community variability (change in the relative abundance of species in the community), with a decline in the abundance of a species being associated with an increase in the limiting resource of this system—space. They argued that in the low-diversity environments, one or two species were able to spread over the surface with little subsequent fluctuation in abundance, thereby monopolizing the limiting resource in this environment, and leaving few colonization opportunities. However, in the species-rich environments, one or two species were not able to monopolize the entire space, and fluctuations in the abundance of the many species made space periodically available to colonizers. A similar finding was made by Bezemer and van der Putten (2007), who found that naturally assembled vegetation plots exhibited higher extinction and colonization rates than less diverse plots. If local extinctions, or even declines in abundances of one or more species, create windows of opportunity for a new species, e.g. through making available key resources (Davis *et al.* 2000), and if the population stability of individual species declines with increasing diversity (Tilman 1996), then one should not be surprised that increased species-richness is often associated with increased invasibility.

Findings from larger-scale and non-experimental studies

Most non-experimental studies conducted in natural settings have not found an inverse correlation between diversity and invasibility, but instead find either no association between the two variables or a positive correlation. Studies to date of aquatic environments have turned up little evidence to support the diversity–invasibility hypothesis. Havel *et al.* (2005a) reviewed current knowledge regarding invasibility of freshwater reservoirs and concluded there was no evidence yet to support the diversity–invasibility hypothesis. In a comprehensive study of the distribution in 171 US lakes of the non-native and invasive cladoceran, *Daphnia lumholtzi*, Havel *et al.* (2005b) found that the distribution of *D. lumholtzi* was not associated with native zooplankton species-richness, but instead was associated with lake size and fertility, with the species more likely to be found in larger and phosphorus-rich lakes. Zaiko *et al.* (2007) sampled the benthic communities in 16 sites in the Baltic Sea and found a positive correlation between native and non-native richness.

Most large-scale species inventories of upland plant communities have found positive relationships between the number of native and non-native species (Stohlgren *et al.* 1999, Brown and Peet 2003, Wiser and Allen 2006). Based on an analysis of more than 15,000 relevés in Catalonia, Vilá *et al.* (2007) concluded that native plant species-richness was not a good predictor of non-native species-richness, and that instead the latter was likely due more to environmental and invasion event factors (e.g. propagule pressure) than on biotic interactions. Studies of riparian vegetation have also not supported the diversity–invasibility hypothesis (Richardson *et al.* 2007). In a study of invasibility and diversity of plants across a disturbance gradient, Belote *et al.* (2008) found little or no evidence that invasibility to non-native plants was inhibited by high species-richness of native species. In fact, native and non-native species responded similarly following disturbances, with both groups more likely to establish following more intense disturbances. Belote *et al.* concluded that their findings were much more consistent with hypotheses based on resource availability and disturbance than biotic resistance.

Problems with the diversity–invasibility hypothesis

The diversity–invasibility hypothesis is appealing in its simplicity, its logic within a niche-based paradigm, and its implicit affirmation of the value of diversity. However, during the past decade, the diversity–invasibility hypothesis has been intensively studied and discussed, and many of these activities have revealed a number of shortcomings with it. I believe that a central problem with the hypothesis is that the conditions in which it is most likely to be supported (very stable and homogenous conditions, which would permit a diverse set of species to sequester the majority of resources) are unlikely to occur in most natural settings. Most of the evidence supporting the diversity–invasibility hypothesis has come from experiments conducted with constructed environments, in which diversity and, frequently, other environmental conditions are manipulated and artificially controlled, and often lacking in much spatial heterogeneity or temporal fluctuations. Under these conditions, with 'all else equal' (Fridley *et al.* 2007), it does not seem particularly surprising that resource availability, and hence invasibility, would often be inversely associated with species-richness.

For some, the negative relationship found between invasibility and diversity at small scales was never very convincing, since most support for the diversity–invasibility hypothesis came from constructed plant systems, and sampling and weeding effects were argued as substantially affecting the results in these experiments (Huston 1997, Wardle 2001, Rejmánek *et al.* 2005b). Some of these concerns were addressed in subsequent experiments designed specifically to control for these effects (Fargione and Tilman 2005, Maron and Marler 2007). Nevertheless, there was still the problem that the results from the experimental systems, i.e. the negative correlation between invasibility and diversity, did not reliably correspond to the patterns found in natural systems, even at comparable small scales (Robinson *et al.* 1995, Wiser *et al.* 1998, Sax 2002, Cleland *et al.* 2004, Dunstan and Johnson 2004, Stachowicz and Byrnes 2006, Stohlgren *et al.* 2006b, Fridley *et al.* 2007, Belote *et al.* 2008).

Some have concluded that, although certainly competition can reduce the local abundance of species, competition is often not the dominant force structuring natural communities and, as a result, it rarely limits immigration or causes extinctions

(Davis 2003, Sax and Gaines 2003, Ricklefs 2005, Stohlgren *et al.* 2008a). Also, the fact that introductions have increased species-richness in so many communities and regions throughout the world (Gido and Brown 1999, Rosenzweig 2001, Davis 2003, Sax and Gaines 2003, Bruno *et al.* 2004), seems clear evidence that very few natural environments are species-saturated (Sax *et al.* 2005c, Smith and Shurin 2006) and that virtually all environments are invasible to some degree (Williamson 1996). Vermeij (2005) argued similarly, claiming most species are able to adapt, or at least cope and persist, as other species come and go. Thus, some have questioned whether the term 'community saturation' has any justifiable ecological meaning at all (Sax *et al.* 2005c). With respect to plants, Stohlgren *et al.* (2008a) referred to species saturation as a myth and presented considerable evidence that plant communities are not saturated with species at scales as small as 100 m^2. Harrison (2008) argued that saturation in plant communities might still occur at very small spatial scales, e.g. 1 m^2. However, it should always be possible to demonstrate saturation if one makes the spatial and temporal scale small enough, e.g. large enough for only one or a few individuals. Moreover, it is not clear what saturation at very small scales would reveal about community assembly processes at larger scales (Stohlgren *et al.* 2008b).

Many studies have shown that, with respect to invasibility and biotic resistance, species composition matters more than species-richness. Using experimental phytoplankton communities that were created through the serial introductions of different species, Robinson and Edgemon (1988) showed that the invasibility is greatly dependent on the species composition of the communities. In a study of benthic macro-algal communities, Arenas *et al.* (2006) concluded likewise. Lennon *et al.* (2003) found that the invasibility of a freshwater system of a non-native zooplankter, *Daphnia lumholtzi*, was more likely in high zooplankton diverse systems, and that the invasive ability of *D. lumholtzi* was negatively correlated with the abundance of another cladoceran species, *Chydorus sphaericus*. Emery and Gross (2007) found that the identity of the dominant species in grassland mesocosms significantly influenced invasibility, and Dunstan and Johnson (2004) argued that a community's resistance to invasion will be determined more by the properties of the species present than by any aggregate community property, such as species-richness.

It is not difficult to imagine that a community of 10 ecologically very similar species might utilize fewer types of resources than a community of 5 ecologically very different species; in which case, the more diverse community would be expected to be more invasible than the less diverse one. Thus, the nature of the species pool in experimental studies of the diversity–invasibility hypothesis would be expected to influence the results. If constructed communities are populated by randomly assigning species from a pool consisting of ecologically diverse species, then more diverse communities will tend to exhibit more extensive complementary resource use than less diverse ones, and then, all else equal, reduced invasibility. However, if the species pool consisted of mostly ecologically very similar species, then diversity and invasibility probably would not be as predictably related, even with all else being equal. With respect to plant studies of the diversity–invasibility hypothesis, there is another concern regarding the use of constructed species pools. In nature, plants seem to be distributed in highly non-random patterns (Gotellie and McCabe 2002). Thus, creating communities by randomly assigning species from a designated species pool would not be a good surrogate for natural and historical assembly processes (Rejmánek *et al.* 2005b).

Another shortcoming of the diversity–invasibility hypothesis is that, in nature, other processes may commonly overwhelm any diversity effect. Levine (2000) emphasized that processes operating at larger spatial scales, e.g. propagule pressure, may overwhelm neighborhood processes such as competition. Abiotic factors may also supersede any neighborhood biotic effects. For example, in a small-plot experimental study of herbs in a savanna environment, MacDougall and Turkington (2006) found that functional similarity did not reduce recruitment, which instead was influenced much more by the environmental filters created by fire suppression. In fact, due to the stringent environmental filter of fire suppression, MacDougall and

Turkington found that recruitment and coexistence of plant species introduced into a fire-suppressed oak savanna depended more on species being functionally similar than different. Paavola *et al.* (2005) found that non-native species that have established in the Baltic Sea were most abundant in mesohaline regions, where species-richness of the native species was lowest, suggesting support for the diversity–invasibility hypothesis. However, the non-native species were also distributed primarily in mesohaline waters in their native region, and thus the inverse relationship between native and non-native species-richness may be completely coincidental, the result of opposite adaptations to the saline environments by the two sets of species, i.e. the native species adapted to either oligohaline or polyhaline waters, while the non-native species are most adapted to mesohaline waters (Paavola *et al.* 2005).

During the latter half of the twentieth century, hypotheses developed to describe community assembly were commonly based on the assumption that competition was the primary structuring mechanism. The diversity–invasibility hypothesis is among this group. However, recent emphasis on the importance of facilitation in community assembly (Bruno *et al.* 2003, Valiente-Banuet and Verdú 2007) has changed perspectives considerably. Rather than new species imposing challenges and constraints on the resident species, they may bring with them opportunities. Since facilitation often occurs between species that are phylogenetically distant (Valiente-Banuet and Verdú 2007), because introduced and long-time resident species often have not shared close evolutionary histories, the possibility of native and non-native species coexisting might be enhanced due to facilitation. This also suggests that an increase in the number of native species would not necessarily be expected to increase biotic resistance, since one or more of the additional native species may function more as facilitators for the introduced species than as competitors (Stachowicz and Byrnes 2006).

In a recent meta-analysis, Levine *et al.* (2004) concluded there was little evidence to support the idea that biotic resistance can prevent the colonization and establishment of new species, although the spread and impacts of the species can be moderated by biotic resistance (Fig. 3.4). Unless newly colonizing species are operating under a different set of ecological processes than the native species, it is inevitable that naturally diverse environments would be highly invasible. Providing speciation events are not the primary origins of new species at a site; the resident species had to have colonized the site at some point in the past. Importantly, for high diversity at a site to persist over time, the invasibility of the site must remain high, unless the majority of the species are clonal and thereby not dependent on subsequent establishment by propagules. From a propagule's perspective, it matters little whether one was transported to a site from a long distance or one was propagated on site. In either case, the propagule faces the same fundamental challenges of establishment. In both instances, the propagule is matched against the invasibility of the environment. Thus, whether the maintenance of high diversity at a site is due mainly to constant shuffling of different species, with new species coming in as resident species go extinct, or to the ongoing successful recruitment of resident species, invasibility of the environment must be high. Based on experimental and observational data, many have concluded that native and non-native species respond similarly to environmental drivers (Sax 2001, Labra *et al.* 2005, McKinney and Lockwood 2005, Meiners 2007). Thus, it is to be expected that environments susceptible to the colonization and establishment of native species will also be susceptible to the colonization and establishment of non-native species.

An effort at resolution

In a 2007 paper in the journal *Ecology*, advocates of both sides of the diversity–invasibility argument tried to reach some consensus. Specifically, the two sides tried to resolve the ostensible 'invasion paradox,' the finding that native and non-native species-richness are often negatively correlated at small scales in experimental and theoretical studies, while positively correlated in large-scale observational studies (Fridley *et al.* 2007). Readers of the Fridley *et al.* paper will likely differ in their assessment of how successful the authors were in reaching consensus. While the article clearly

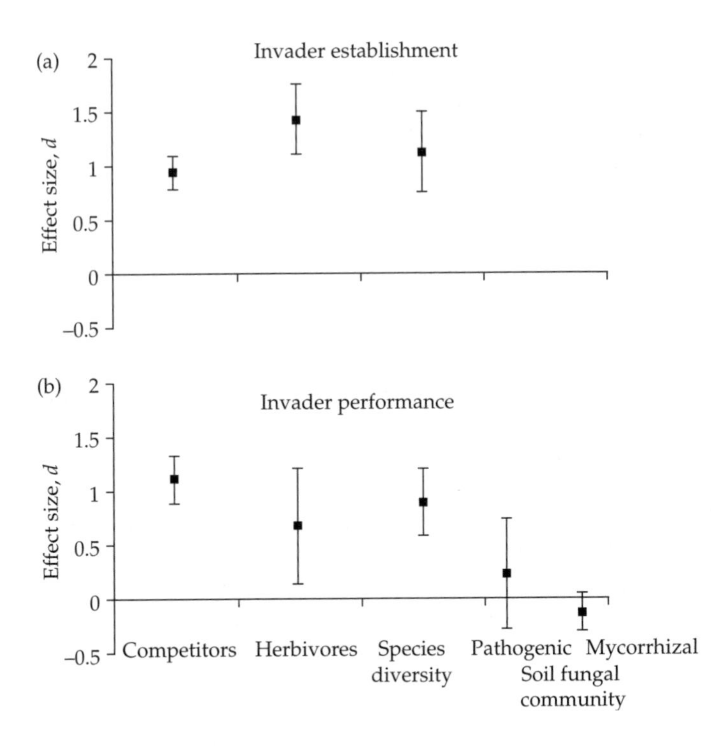

Fig. 3.4 Effects of resident competitors, herbivores, species diversity, and soil fungal communities on invader (a) establishment and (b) individual performance variables. Points show means bracketed by 95% confidence intervals. Effects of the soil fungal community are presented for individual performance variables only because too few studies examined establishment. Redrawn and printed, with permission, from Levine *et al.* (2004), copyright Blackwell Publishing.

tried to provide some support and legitimacy to the diversity–invasibility theory, it also seemed to highlight the weaknesses and limitations of the theory. For example, the acknowledged fact that negative richness relationships are not regularly found outside of small experimental plots would seem to suggest that control of species composition by niche-partitioning seldom dominates at most spatial scales under natural conditions. Moreover, arguing that, 'all else equal', species-rich environments should be more resistant to species-poor environment seems only to emphasize the limitations of the theory, since if there is one thing we know about the natural world, it is that all else is almost never equal. In another effort to affirm the value of the diversity–invasibility theory, the authors stated:

> If control of [species] composition by niche partitioning remains high compared to these extraneous factors [disturbances, resource pulses, immigration rates], NERR's [native-exotic richness relationships] should be negative.

Today, most ecologists think of disturbances and non-equilibrial dynamics as inherent processes of natural communities (Holt 2005). Thus, the use of the term 'extraneous' to describe natural events and processes such as disturbances and resource pulses seems oddly Clementsian, and a forced effort to position niche-partitioning processes at the center of community assembly.

It is an interesting question as to whether an invasion paradox really exists. As the authors in the 2007 article point out, some environments seem to exhibit a positive NERR at both small and large scales. Moreover, negative NERRs at small scales may simply be a statistical artifact (Fridley *et al.* 2004, Herben *et al.* 2004). A primary purpose of the 2007 paper seemed to be to develop a theory of invasibility, or at least an approach to studying invasibility, that accommodates a competition, niche-based approach, which has been most clearly manifested in the diversity–invasibility theory. However, rather than successfully developing an integrated approach that might give the diversity–invasibility theory more life, the paper seemed more to highlight the inadequacies of the diversity–invasibility hypothesis. No doubt, readers will disagree on this point.

Actually, it may not be difficult to resolve the controversy surrounding the diversity–invasibility theory. The key is in expressing the relationship in terms of resource availability. Consider the following proposed statement:

If the availability of limiting resources (defined in terms of the needs of the arriving individuals) is negatively associated with species-richness at a particular scale, then one would expect invasibility also to be negatively associated with species-richness at this scale, providing invasibility was being constrained primarily by resource availability, and not to some other factor such as the presence or absence of enemies or mutualists.

The limited value of this statement lies in the fact that the conditions necessary for this statement to be true, i.e. a consistent inverse relationship between resource availability of limiting resources and species-richness, is unlikely to reliably and persistently occur under most natural conditions, at any scale. This would help explain why the diversity–invasibility hypothesis is so often not supported under natural (non-experimental) conditions (Sax 2002, Cleland *et al.* 2004, Dunstan and Johnson 2004, Stachowicz and Byrnes 2006, Stohlgren *et al.* 2006a, Belote *et al.* 2008, Stohlgren *et al.* 2008a).

When considering coexistence of native and non-native species, the spatio-temporal framework under consideration will greatly influence whether one determines that coexistence has or has not occurred. If one is working at a small local spatial scale with a very long time horizon, then coexistence will be much less common, and when it does occur, will normally require strong equalizing and/or stabilizing forces. However, if one considers the new mixed communities over a more modest time-scale and a larger spatial scale, then coexistence will be much more common, often even in the absence of strong equalizing or stabilizing forces. In a practical sense, all that is necessary for coexistence to occur is that circumstances and processes do not result in the rapid loss of a species from the community. Coexistence should not, and cannot, in realistic terms, mean forever. Recently introduced species and long-term residents may often be able to live sympatrically for centuries, or even longer in some cases. While events and processes may not permit indefinite or permanent sympatry, conditions often required in niche-based models, not referring to such species as coexisting seems to me to belie common sense.

There is no question that it would be easier to assess invasibility if there were some community aggregate variable that would be reliably associated with invasibility. More than a decade of intensive research has shown that species-richness is not this variable. This is not a new observation. Skepticism and dissatisfaction with the diversity–invasibility hypothesis have not just emerged in the past few years. Huston (1994), Rejmanek (1996), and Williamson (1996) all concluded that there was little evidence to support the notion that increased species-richness reduces invasibility of an environment.

Invasibility, resources, and environmental heterogeneity

Fluctuating resources

Constrained by the second law of thermodynamics, all living organisms must be able to access and sequester resources in order to maintain their physiological processes. The importance of resource availability in colonizations and invasions has been recognized for some time (Huston and DeAngelis 1994). Huston and DeAngelis argued that if resources were not limiting due to spatial and temporal variation in their abundance, then new species might be able to colonize and persist along with the resident species. This idea was further developed in 2000 and presented as the fluctuating resource availability theory of invasibility, which stated that pulses of resources should be expected to increase the invasibility of an environment (Davis *et al.* 2000). The important role that resource pulses may play in affecting community and ecosystem processes in general is a topic currently receiving considerable attention (Yang *et al.* 2008).

While it is widely recognized that the theory of fluctuating resource availability stresses the role played by the temporal heterogeneity of resources, it has often been mistakenly characterized as addressing only temporal heterogeneity, and neglecting the importance of spatial heterogeneity (Stachowicz and Byrnes 2006, Melbourne *et al.*

2007). In fact, the 2000 paper explicitly emphasized that fluctuations in resource availability frequently occur patchily in space, e.g. due to small-scale disturbances, such as those produced by burrowing animals, grazing, or drought. Thus, the resource fluctuations to which the hypothesis refers were intended to be considered part of a combined temporal and spatial framework, i.e. 'fluctuating availability of resources in space and/or time will lead to a fluctuation in the intensity of competition which may prevent competitive exclusion from occurring' (Davis *et al.* 2000). This same point was reiterated in a subsequent related paper (Davis 2003), which described the fluctuating resource availability theory as 'emphasiz[ing] spatiotemporal variability in habitat characteristics (e.g. availability of resources).'

Originally proposed in the context of terrestrial plants, the fluctuating resource availability theory has been tested and evaluated in hundreds of studies, in a wide range of environments with many other types of organisms (including terrestrial and marine plants, marine benthic organisms, freshwater vertebrates and invertebrates, and microbes) and it has proven to be strikingly robust at multiple spatial scales (e.g. Thompson *et al.* 2001, Bertness *et al.* 2002, van der Velde *et al.* 2002, Jiang and Morin 2004, Havel *et al.* 2005a, James *et al.* 2006, Stachowicz and Byrnes 2006, Williams and Smith 2007). Resource changes on a global scale may also influence establishment of certain species in particular regions. For example, it is hypothesized that increasing concentrations of atmospheric CO_2 may reduce water stress in some plants, thereby possibly permitting some non-native species to establish in drier habitats (Dukes 2000). Similarly, ongoing nitrogen inputs from atmospheric deposition may facilitate the invasions of some species by increasing resource availability (Davis *et al.* 2000, Hobbs and Mooney 2005). In some cases, fluctuations in resource availability may occur due to seasonal variation in resource uptake, with periods of low resource uptake possibly creating windows of opportunity for new species to enter the environment (Stachowicz *et al.* 2002, Stachowicz and Byrnes 2006, Olyarnik *et al.* 2008).

Although the fluctuating resource availability theory of invasibility has been quite successful in its generality, there are certainly exceptions to it. For example, Lennon *et al.* (2003) found that invasibility of a zooplankton community declined with nutrient enrichment because of the corresponding substantial increase in abundance of another cladoceran, which was believed to reduce the establishment success of other zooplankton, although the mechanism was not identified. Some studies have shown that, while increases in invasibility are associated with increases of certain resources, invasibility is not increased, and in some cases even decreased, with the increase of other resources (Kolb and Alpert 2003, Gross *et al.* 2005). It is important to remember that it is the increase in the availability of the *limiting* resources that is expected to increase invasibility. Thus, one would not necessarily expect an increase in invasibility by increasing a particular resource, since it is possible that increasing one resource, e.g. nutrients, might decrease the availability of another resource, e.g. light, the latter which may be the limiting factor in the system. It is not difficult to imagine other situations in which one would not find a correlation between resource availability and invasibility. In saline environments, adaptation to high salinity may trump any differences in responses to other resources, e.g. nitrogen (Kolb and Alpert 2003). In situations where resident predators, herbivores, or pathogens kill virtually every arriving immigrant, resource availability will be irrelevant. MacDougall and Wilson (2007) found that widespread seedling herbivory by rodents and lagomorphs presented a major obstacle to colonization by new species in the northern Great Plains. It does not matter how many resources are available if you can't live to enjoy them.

Environmental heterogeneity

As mentioned above, efforts to resolve the apparent scale-dependence of the diversity–invasibility relationship have focused primarily on the role played by environmental heterogeneity in resources. There has been consistent agreement that environmental heterogeneity, both in space and time, should increase invasibility (Huston and DeAngelis 1994, Davis *et al.* 2000, Shea and Chesson 2002, Huston 2004, Davies *et al.* 2005, Arenas *et al.* 2006, Dunstan and Johnson 2006, Renne *et al.* 2006, Melbourne *et al.* 2007). Empirical support for the hypothesis that temporal and spatial heterogeneity are primary drivers of diversity in environments has come from marine, freshwater, and terrestrial systems (Huston 1994, Huston and DeAngelis 1994, Davis and Pelsor 2001, Turnbull *et al.* 2005, Dornelas *et al.* 2006, Lepori and Hjerdt. 2006, Stohlgren *et al.* 2008a).

By definition, hypotheses based on environmental heterogeneity contrast with the neutral model (Hubbell 2001), since the latter assumes that the environment is homogeneous. However, invasibility theory built around environmental heterogeneity can easily accommodate the portion of the neutral model that deals with organism traits. For example, Davis (2003) pointed out that the fluctuating resource availability model was similar in certain aspects to the neutral model in that it does not require new species to be ecologically different from resident species in order to successfully colonize a new environment. Integrating resource heterogeneity in space and time with features of what is now called the neutral model is not a completely new idea. It was portended by lottery-based models of community assembly proposed more than thirty years ago, which integrated disturbance with random colonization (Sale 1977).

Disturbances often contribute to a temporally and spatially heterogeneous environment (Connell 1978, White and Pickett 1985, Lepori and Hjerdt 2006). One way this can happen is through an increase in species aggregation that has often been found to occur following a disturbance (Couteron and Kokou 1997, Potts 2003, Davis *et al.* 2000). In these instances, although most individuals of disturbance-sensitive species are killed, patches of individuals survive, due to the fact that the full impact of the disturbance often does not reach all areas, either purely fortuitously or because the characteristics of certain areas provide protection from the disturbance (Davis *et al.* 2005b). If disturbances are frequent and/or very intense, then virtually all the individuals of disturbance-sensitive species are killed and the extent of species aggregation in the community declines (Rebertus *et al.* 1989, Davis *et al.* 2005b). Figure 3.5 illustrates a hypothetical relationship between disturbance frequency/intensity and community-wide aggregation patterns suggested by these studies. This hypothesis could be considered a corollary to the intermediate-disturbance hypothesis (Connell 1978), the corollary stating that intermediate levels of disturbance are expected to maximize community-wide patterns of aggregation, or pattern diversity (Rebertus *et al.* 1989, Davis *et al.* 2005b). In areas where individuals are killed by the disturbance, more resources would be made available, thereby increasing the invasibility of the environment (Davis *et al.* 2000). However, if disturbances are too frequent and/or too intense, then, although there may be abundant available resources, invasibility would be expected to decline since many species could not survive and persist under this disturbance regime. This actually suggests a possible contributing mechanism to the original intermediate-disturbance hypothesis (Connell 1978). By increasing spatial heterogeneity through the killing of some residents, disturbances occurring at intermediate levels (frequency and/or intensity) create patches of available resources while not imposing excessive obstacles to establishment, thereby leading to increases in both invasibility and species diversity.

In some cases, disturbances may create a land/seascape mosaic consisting of different aged patches, i.e. in different stages of succession (Clark 1991, Lertzman *et al.* 1996, Fuhlendorf and Engle 2004). In their study of grassland herbs, Renne

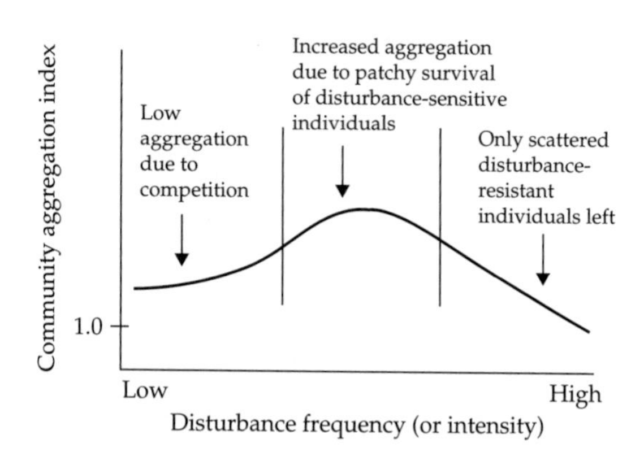

Fig. 3.5 Graphical representation of a model predicting how community-wide patterns of aggregation in a forest are hypothesized to vary as a function of the frequency or intensity of disturbances. Invasibility would also be expected to peak at intermediate frequencies/intensities of disturbance, as would species diversity (intermediate disturbance hypothesis). See text for a more detailed explanation. Redrawn and printed, with permission, from Davis *et al.* (2005b), copyright Opulus Press.

et al. (2006) appropriately characterized the fluctuating patchiness as a 'shifting invasibility mosaic.' In other instances, disturbance may create patches differing fundamentally in kind. In the latter instance, increase in invasibility (and diversity) of the larger environment is believed to occur through the increase in the variety of smaller scale habitat types distributed patchily across the landscape due to the disturbances (Havel *et al.* 2005a). Both types of patchiness are often used as examples of patch dynamics (Pickett and White 1985). Disturbances and the patch-dynamics effects of disturbances have been invoked as primary drivers of invasibility and diversity in marine, freshwater, and terrestrial ecosystems (Shea and Chesson 2002, Leibold *et al.* 2004, Davies *et al.* 2005, Seabloom *et al* 2005, Arenas *et al.* 2006, Renne *et al.* 2006, Williams and Smith 2007; Fig. 3.6). If invasibility is normally enhanced by periodic disturbances that create an environmental mosaic, then one might expect that very frequent, or ongoing, disturbances that tend to homogenize the environment would reduce invasibility. Lohrer *et al.* (2008) documented exactly this phenomenon in a soft-sediment marine system. In this case, large burrowing echinoids (spatangoid urchins) act as bioturbators, constantly moving and mixing the sediments, which inhibits the colonization of other species (Lohrer *et al.* 2008).

In their review of the invasibility of freshwater reservoirs, Havel *et al.* (2005a) noted the high degree of spatial heterogeneity associated with reservoir systems, including the upstream riverine environments, downstream lacustrine zones, and the reservoir itself, much of this heterogeneity due to differences in disturbance regimes, particularly fluctuating water levels due to drawdowns. They concluded that, together, the spatial and temporal heterogeneity in the physiochemical environment and resources provide a myriad of diverse colonizing opportunities.

As described in Chapter 2, economic development often increases propagule pressure for a country through the increase of trade. Economic development can also enhance the establishment of arriving propagules through physical disturbances of the landscape, which can free up resources. Also, eutrophication can supply both terrestrial and aquatic environments with additional resources that can be captured by the new species. Increased disturbance rates and eutrophication are believed to be contributing to the substantial increase in establishment of non-native species being experienced by China, as the country is undergoing unprecedented economic growth (Ding *et al.* 2008).

It should be remembered that while spatial and temporal heterogeneity often occur together, e.g.

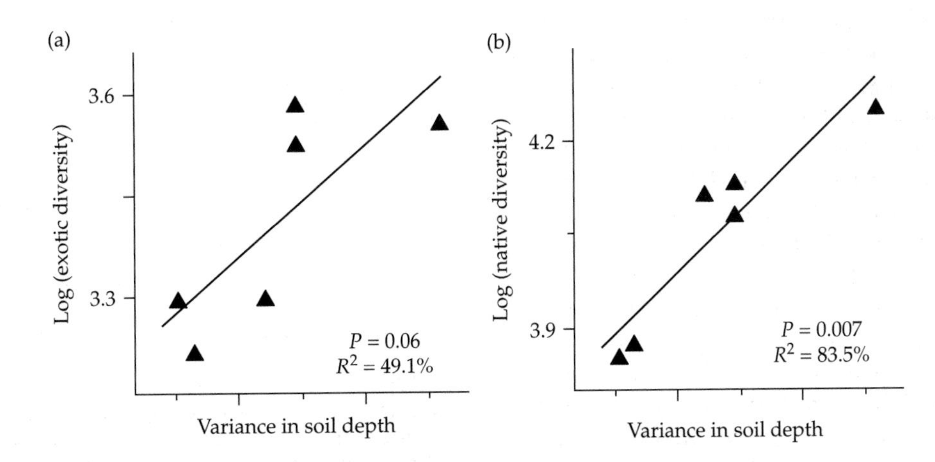

Fig. 3.6 Relationship between the variance of soil depth and (a) non-native diversity, and (b) native diversity. Redrawn and printed, with permission, from Davies *et al.* (2005), copyright Ecological Society of America. (Note: Although the relationship between soil depth and non-native plant diversity was clear in this case, it is not clear how robust this relationship is. In a separate study, MacDougall and Turkington (2006) did not find a relationship between soil depth and diversity of non-native plants.)

periodic small disturbances in an environment will create a spatial patchiness, this is not always the case. That is, spatial heterogeneity can occur without temporal fluctuations, and temporal fluctuations can occur without spatial heterogeneity. For example, in some cases, spatial patchiness may be more the result of some fundamental and more permanent factors (e.g. geological features in terrestrial systems), in which case the patchiness would not necessarily be associated with temporal fluctuations. The environment containing these permanent patches might nevertheless still be more invasible than a permanent homogenous environment, since the invasibility of some of the patches may be greater than that of the homogeneous environment. Conversely, an entire environment may periodically be subject to large-scale events, such as floods, fires, or hurricanes, or various human transformations of the landscape or seascape, the effect of which would be a new single large patch that largely lacked spatial heterogeneity, but which provided available resources and/or other different environmental conditions. For example, many coastal marine areas have experienced high invasion rates, and it has been hypothesized that this may be due to the prevalence of human disturbances in these environments, which may free up resources by killing/removing native species/biomass and/or by altering physical conditions (Olyarnik *et al.* 2008). Depending on the extent of these anthropogenic disturbances, they could either create a spatially heterogeneous marine environment or a new and largely homogeneous one, either of which may be more susceptible to invasion than was the environment prior to human disturbance. Castilla *et al.* (2005) suggested that one reason that Chile's coastal waters had not yet experienced a high level of invasion may be because the coast contains comparatively few sheltered bays, the areas in which one would expect human development and disturbance to most likely occur.

It is also important to remember that disturbances are events and not mechanisms. If one asserts that disturbances facilitate invasions, one is making a statement of observation not of mechanism. The possible mechanism(s) that might facilitate an invasion include changes in community composition, ecosystem processes, and/or propagule supply that occur as a result of the disturbance. Different types of disturbances will impact different processes, and even the same disturbance may facilitate invasions of multiple species for different reasons, i.e. the mechanisms involved in the respective introductions may differ in each instance.

Of course, no matter how much a disturbance increases the invasibility of an environment, if the increase in resource availability produced by a disturbance does not coincide with an episode of incoming propagules, no invasion will occur (Davis *et al.* 2000, Olyarnik *et al.* 2008). In a study of the establishment and spread of *Berberis thunbergii*, in woodlands of central Massachusetts, USA, DeGasperis and Motzkin (2007) concluded that the species established in the area in the region in the early twentieth century by colonizing recently abandoned agricultural fields. Although some spread has occurred into adjacent woodlands, the authors argued that the abandonment of the fields provided a 'window of opportunity' (Johnstone 1986), of which the species, already present in the area, could take advantage. DeGasperis and Motzkin used their findings to emphasize the importance of knowing the particular disturbance history of a region in order to understand current distribution patterns of non-native species.

Invasibility and the study of diversity

An important result of the efforts to explain invasibility in terms of environmental heterogeneity has been the realization that the study of invasibility is much the same as the study of diversity (Shea and Chesson 2002, Levine *et al.* 2004, Davis *et al.* 2005a, Rejmánek *et al.* 2005b, Melbourne *et al.* 2007). The importance of this recognition cannot be overemphasized. In many ways, invasion ecology had become dissociated from other specialty research areas, and even from core ecological theory (Davis *et al.* 2001, 2005c). The appreciation that the study of invasibility, along with dispersal, is really the same enterprise as the study of diversity, means that one should not search for unique explanations to account for the processes and patterns of invasions. Instead, one should apply the same conceptual toolbox ecologists have developed to explain patterns of diversity.

The reason invasibility theories based on environmental heterogeneity have tended to be more robust and reliable predictors of invasibility than the diversity–invasibility hypothesis is probably because they are based on a more realistic paradigm of the natural world. One might say that the theories emphasizing the temporal and spatial heterogeneity of resources are more Aristotelian in nature, while the diversity–invasibility hypothesis is more fundamentally Platonic. The former acknowledge the 'imperfections' of nature, its constant and idiosyncratic bumpiness and jitteriness, where change and uniqueness, not stasis and uniformity, are the norm. The diversity–invasibility hypothesis rests on a much smoother view of nature, one where consistency, equilibrium, and generality rule. Ecologists have always disagreed over which is the better model, and probably always will. My own view is that the latter is not a particularly good model of the natural world, which often has seemed not particularly obliging in following the rules ecologists have prescribed for it.

Invasibility and physical stress

Extreme physical stress of an environment would be expected to reduce invasibility to all species except those possessing adaptations to the physical stressors, and for which the environment's invasibility would then be determined more by other factors. Extreme temperatures and precipitation, either high or low, obviously would preclude the colonization and establishment of all species except those able to tolerate these conditions. Rejmánek (1989) observed that plant communities in mesic environments tend to be more invasible than those in xeric conditions, presumably because drought stress reduces germination and seedling survival in most terrestrial plants. Abiotic stressors found to reduce invasibility in plants include waterlogged soils (Woolfrey and Ladd 2001, Rood *et al.* 2003) and sediment type (Dethier and Hacker 2005). High levels of salinity can reduce the invasibility of some aquatic systems (Moyle and Marchetti 2006; Fig. 3.7). For example, when the salinity of coastal waters is reduced by increases in rainfall or other freshwater input, the establishment success of some non-native wetland plants has been found to increase (Minchinton 2002, Deithier and Hacker 2005). However, if physical stress harms the native species more so than the new arrivals, an increase in physical stress could increase invasibility. In freshwater and some marine environments, low levels of dissolved oxygen make the aquatic environment intolerable to many species, sometimes favoring non-native species (Jewett *et al.* 2005, Paavola *et al.* 2005).

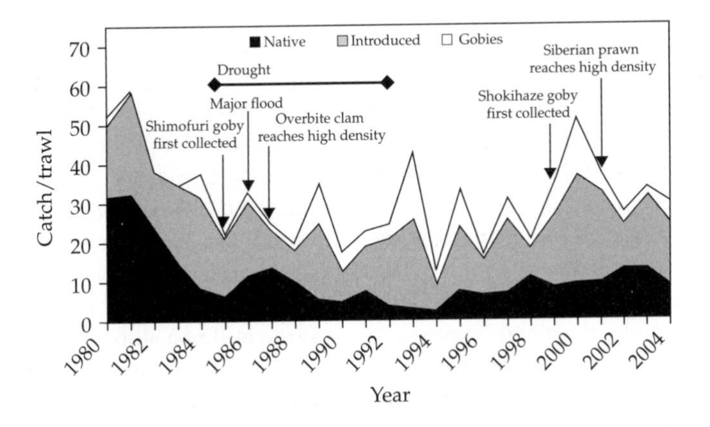

Fig. 3.7 Trends in the populations of native and non-native fishes in brackish Suisun Marsh, California, 1980–2004. The trends are based on the average number of fish caught per trawl, assuming 22 trawls per month for all months of every year. Gobies are graphed separately because the three goby species are non-native and the numbers are dominated by the shimofuri goby, which invaded in the 1980s. Noted on the figure are a major drought, which increased salinities in the marsh, and the first record of the shimofuri goby (arrow on the left). Redrawn and printed, with permission, from Moyle and Marchetti (2006), copyright American Institute of Biological Sciences.

Very frequent and intense disturbances can also impose physical stressors that would prevent most species from colonizing and persisting. Although such disturbances may free up resources, comparatively few species would be sufficiently disturbance-tolerant to be able to exploit those resources (Lepori and Hjerdt 2006). In some cases, the introduction of particular species may have such a large impact on the physical processes of an environment that the modified physical processes end up overwhelming any biotic and resource filters that may have been important prior to the introduction of the species. In these instances, the environment can be transformed from a principally biotically controlled environment to one primarily controlled by physical processes. The ability of some introduced species to change disturbance regimes, e.g. increasing fire frequency, is a good example of this phenomenon (D'Antonio and Vitousek 1992). Another is the invasion of non-native earthworms into the Great Lakes forests of North America (Frelich *et al.* 2006). Earthworm species in the genus *Lumbricus* consume the litter and duff layer in the forests, dramatically altering physical soil properties, including temperature and density, the latter influencing permeability and soil water levels. These physical changes are making these environments less invasible to some of the native forbs (Frelich *et al.* 2006; Fig. 3.8) and, at least in the short term, seem to be reducing herb diversity. However, what may cause duress in one species may be welcomed by another. While the worms appear to be substantially reducing the invasibility of the forest floor for many native species, the bare soil patches represent unutilized resources that non-native species, such European buckthorn *Rhamnus cathartica*, and garlic mustard, *Alliaria petiolata*, may be able to exploit (Frelich *et al.* 2006).

The flip-side of physical stressors is the absence of them, i.e. when the physiology of a new colonizer is quite compatible with the new physical environment. For example, the range expansion of a number of tropical and subtropical marine species into temperate regions is thought to be due partly to increased water temperatures in the temperate regions, which have made the waters more invasible (Perry *et al.* 2005). In addition to the high propagule pressure and possible low degree of biotic resistance, Paavola *et al.* (2005) suggested that the high invasion incidence of aquatic organisms in brackish water seas was likely partly due to the wide salinity gradients in these waters which allow for a greater range of environments for establishing species. Evaluating the research to date, Paavola *et al.* found that most of the non-native aquatic species that had established in the Black Sea, Baltic Sea, and Caspian Sea, had done so in the mesohaline zone, which, for most of the species, was their primary zone of occupation in their native environment. Although most abundant in this zone, Paavola *et al.* also concluded that newly established species were rather tolerant of a wide range of salinity levels. This wide range in salinity tolerance is likely a particularly important trait during the transport process, since salinity levels in ballast can be quite variable.

A similar argument has been made for the recent establishment in Antarctica and surrounding islands of many non-native organisms, including microbes, fungi, plants, and animals (Frenot 2005). While the tremendous increase in propagule pressure during the past two centuries, due to human activity in the Antarctic region, is a major part of the explanation for this phenomenon, it is also believed that warming temperatures in some areas has also played a large role by reducing physiological barriers (Frenot 2005). A similar influx of new species is expected in Antarctic shallow-water benthic environments, where communities dominated by slow-moving invertebrates and epifaunal suspension feeders have persisted for millions of years (Aronson *et al.* 2007). Conspicuously, mostly absent from this indigenous community have been fast-moving and durophaguous (skeleton-crushing) bony fish, sharks, and crabs (Aronson *et al.* 2007). However, in 1986 adult brachyuran crabs, *Hyas araneus*, a spider crab from the northern hemisphere, was recorded off King George Island, possibly transported via ballast water (larvae) or on the hull of a ship (adult) (Tavares and De Melo 2004). Aronson *et al.* argued that the cold Antarctic waters historically have prevented non-native species introduced in a similar way from establishing, but that warming waters are removing physiological barriers, thereby reducing the invasibility of this environment. They point out that this

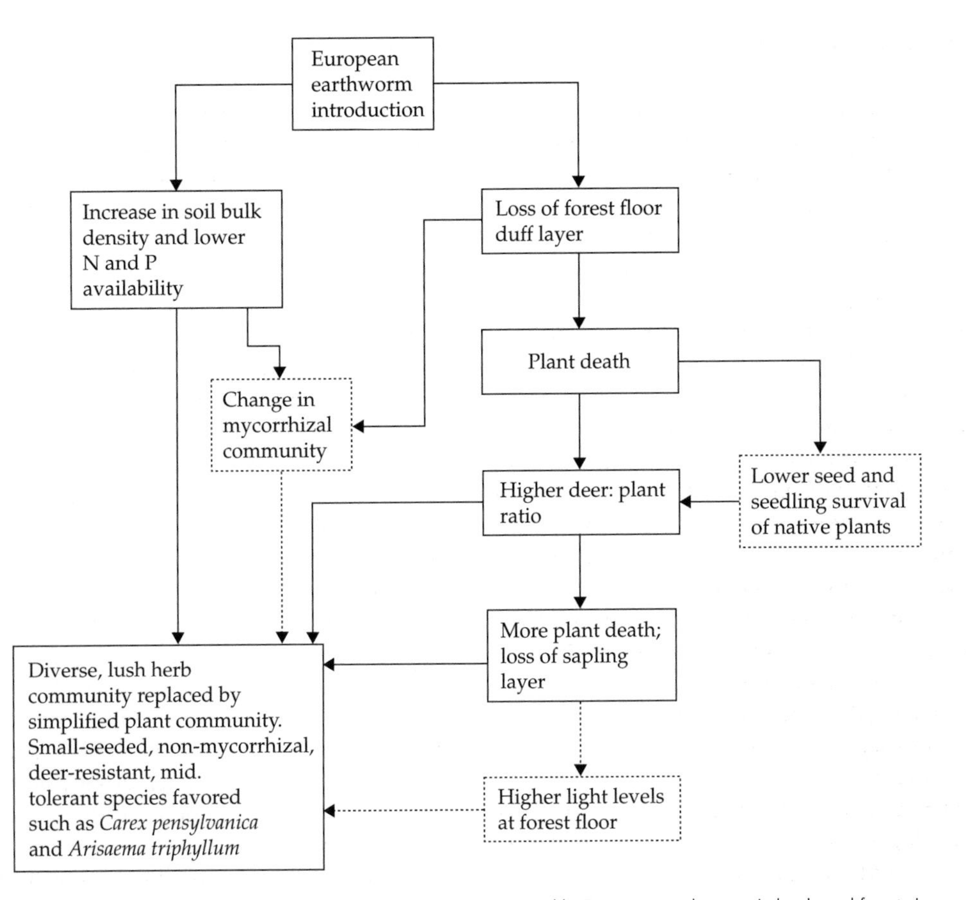

Fig. 3.8 Conceptual diagram for changes in plant community composition caused by European earthworms in hardwood forests in Minnesota, USA. Dashed boxes and arrows indicate hypothesized processes and connections with little data at this time. Redrawn and printed, with permission, from Frehlich *et al.* (2006), copyright Springer.

warming will also enable other species to colonize the shallow-water Antarctic benthos on their own, e.g. fish and shark species currently inhabiting adjacent waters.

Not surprisingly, the physical suitability of the environment for a species can be a good predictor of invasibility with respect to that species (Williamson 1996). As a result, there has been much interest in the use of climate models and species distribution models to predict the susceptibility of certain regions to the colonization and establishment of species from other regions in the world (Guisan and Zimmermann 2000, Guisan and Thuiller 2005). These models have also been used to predict future changes in invasibility of particular regions with expected shifts in climate zones due to climate change (Peterson *et al.* 2002, Midgley *et al.* 2003). However, there are well-known weaknesses of these models, including the fact that factors other than climate may be limiting a species in its native range (Guisan and Thuiller 2005). Thus, the existing physical/climatic conditions of the native environment cannot be assumed to represent the actual physical/climate envelope for a species (Davis *et al.* 1986, Sax *et al.* 2007).

Invasibility and enemies

That enemies such as predators, parasites, herbivores, and pathogens can exert top-down control populations has been widely demonstrated (Hairston *et al.* 1960, Paine 1966, Terborgh 1992), and thus it is not surprising that the ability of new species to colonize and establish in new environments has often been attributed to the absence of natural enemies in the new environment. The Enemy Release Hypothesis (ERH) has been a leading hypothesis to account for the high invasibility of some environments (Crawley 1987, Williamson 1996, Keane and Crawley 2002). The ERH holds that by leaving its specialist enemies behind, newly introduced species will enjoy superiority over the native species which must withstand attack by both their specialist and generalist enemies. The assumption is that the native generalist enemies will either attack the new colonists less or at least not more than the native species. Tella and Carrete (2008) pointed out that while parasites are typically viewed in the role of enemy in the ERH, many are actually introduced species themselves, encountering their own 'enemies', sometimes other parasite or pathogen species, but often the chemical (plant) and immune (animal) defenses of their hosts. Thus, consistent with the ERH, introduced parasites may gain a foothold by infecting naïve hosts.

Despite the fundamentally straightforward reasoning of the ERH, results of experiments and other field studies have been mixed (Keane and Crawley 2002). A number of studies of both plants and animals have documented that some introduced species do experience less parasitism (including herbivory) and reduced pathogenic infection than native species (Klironomos 2002, Mitchell and Power 2003, Torchin *et al.* 2003, Torchin and Mitchell 2004, Williams and Smith 2007, Rodgers *et al.* 2008). Hierro *et al.* (2006) found that while disturbance facilitated the abundance and performance of *Centaurea solstitialis* in California and Argentina, as well as in its native Eurasian range, it responded better outside its native range. The authors also found that native soil microbes suppressed growth more than did the soil microbes from California and Argentina, leading the them to hypothesize that escape from soil pathogens may contribute to the strong positive response of non-native species to disturbances outside of their native range. Van der Putten *et al.* (2007a) found that the soil biotic community exerted neutral to positive feedback on the non-native grass, *Cenchrus biflorus*, while it produced neutral to negative feedback on two native grasses.

These and other studies notwithstanding, other findings have not supported the ERH hypothesis (Schierenbeck *et al.* 1994, and Parker and Gilbert 2007). Liu *et al.* (2007a) found that, although an invasive species of *Eugenia* (Myrtaceae) experienced less herbivory than a native congener in Florida, the level of herbivory of the invasive species did not differ from that of a non-native congener that was not invasive, leading the authors to conclude that enemy release alone cannot account for invasiveness in *Eugenia*. This study points to the potential problem of only comparing non-native invasive species with native species, and emphasizes the value of comparing non-native invasive species with non-native species that are not invasive. In a study of 12 phylogenetically related vines, including native, non-native invasive, and non-native non-invasive, Ashton and Lerdau (2008) found that all species were susceptible to herbivory by mammals and insects, and concluded that differential enemy attack could not explain the success of the invasive species. However, a greenhouse experiment of simulated herbivory showed that the invasive species were better able to compensate for herbivory than either of the other two groups, suggesting that enemy tolerance, rather than enemy release, may contribute to the success of these species (Ashton and Lerdau 2008).

It is interesting that the emphasis has mostly been on the absence of natural enemies. Just as likely, it would seem, would be the possibility that resident enemies in the new environment would attack the new colonizers, which, lacking in the appropriate defenses for these enemies, would fail in their colonization effort. Gilbert and Parker (2006) emphasized that native pathogens might still play an important role in the biotic resistance of an environment. Parker and Gilbert (2007) described several reasons why enemy release may not be an important factor in accounting for differences in invasibility:

(1) many enemies may have broad host ranges and thus may easily be able to accommodate new species in their 'diet', particularly if the new species have native relatives;

(2) many introduced species may be accompanied by their native enemies;
(3) the introduced species may encounter cosmopolitan enemies.

More and more instances of colonization and establishment being thwarted by resident enemies have been reported in recent years (Agrawal and Kotanen 2003, Colautti *et al.* 2004). In a meta-analysis of 63 manipulative field studies, Parker *et al.* (2006) found that native herbivores not only do not tend to avoid introduced plant species, they tend to suppress them. Native predators may also inhibit the establishment of non-native prey. In the coastal waters of New England, de Rivera *et al.* (2005) found that a native predatory crab, *Callinectes sapidus*, is inhibiting the spread of the European green crab, *Carcinus maenas*, thereby providing biotic resistance to the spread and establishment of the invasive *C. maenas*. In a meta-analysis review, Levine *et al.* (2004) found little evidence to support the contention that the success of non-native plant species in establishing in an environment was primarily due to the absence of enemies.

Of course, it is also possible that the ability of a species to colonize and establish in a new region might be inhibited by the presence of non-native enemies that preceded their arrival. In Hawaii, an introduced cricket, *Teleogryllus oceanicus*, is being parasitized by a non-native fly, *Ormia ochracea* (Zuk *et al.* 1998). The flies are phonotactic, or acoustically oriented, in their search of hosts. They primarily parasitize male crickets, which they locate by eavesdropping on the crickets' mating calls. Due to the intense selection pressure (males are eventually killed by the parasitism), evolution has rapidly produced a new silent-type male. As reported by Zuk *et al.* (2006), singing males, although much less common, had not disappeared from the population yet. In turn, sexual selection has favored a new mating strategy for the silent males. The silent males are still able to mate by clustering around a singing male and intercepting females attracted by the calling male. It remains to be seen if selection will continue to cause a decline in the number of calling males or whether a polymorphic mating strategy might be maintained through density-dependent selection. If the number of calling males continues to decline and no new mating strategy evolves, it would seem the extinction of the Hawaiian cricket population is inevitable. In either case, it is obvious that the invasibility of its new environment is declining due to its interaction with a new enemy, which is also non-native. A comprehensive evaluation of the spread of the genus *Pinus* throughout the world (Richardson 2006) similarly showed that the establishment and persistent of *Pinus* species was inhibited by already present non-native

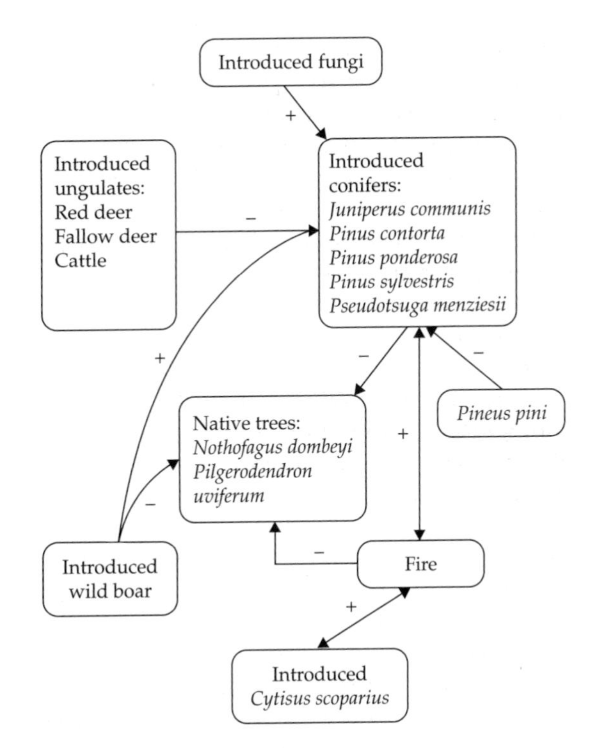

Fig. 3.9 An example illustrating the complex interaction between biotic factors in mediating the fate of introduced pines. The diagram summarizes interactions between introduced pines and key components of the community at Isla Victoria (Nahuel Huapi National Park, Argentina), a large island dominated by native *Nothofagus* and *Austrocedrus* forest, with old plantations of many introduced tree species. Alien pines benefit from introduced fungi, wild boar, and fire (whose occurrence is favored by another introduced plant). Pine regeneration is limited by introduced ungulates and the introduced insect pest *Pineus pini*. The introduced conifers have a negative impact on native tree species. Redrawn and printed, with permission, from Richardson (2006), copyright Czech Botanical Society (figure modified from Simberloff *et al.* 2003), copyright Springer.

herbivores (Fig. 3.9). The fact that introduced species can contribute to the biotic resistance of the environment, thereby influencing the success of subsequent new arrivals, is supported by findings that the sequence of introductions can influence the eventual community composition (Duncan and Forsyth 2006).

In many cases, introduced species may bring their enemies with them, e.g. pathogens that become introduced via infected arriving hosts. However, the fact that the host species does not escape its enemies does not necessarily mean the invasive potential of the species is reduced. In fact, a recently introduced species may benefit by introducing a new enemy into its new environment, if the enemy negatively impacts native species more than the non-native one. For example, the success of the introduced gray squirrel, *Sciuris carolinensis*, in Europe is believed to be partly due to the decline of the native red squirrel, *S. vulgaris*, a decline partially due to the latter's increased susceptibility to a virus introduced by *S. carolinensis* (Tompkins *et al.* 2003). Ricciardi (2005) described several other similar examples, including freshwater fish introduced into Australia that were accompanied by parasites, which resulted in declines of some native species (Dove 1998), and the introduction of the American crayfish, *Pacifastacus leniusculus*, along with a fungal parasite, *Aphanomyces astaci*, the latter of which decimated many native European crayfish species (Reynolds 1988).

Blumenthal (2005) proposed a hypothesis intended to integrate the fluctuating resource availability theory with the ERH. Blumenthal argued that species adapted to high resource conditions, i.e. those particularly positioned to take advantage of pulses or patches of under-utilized resources, would particularly benefit from leaving their enemies behind. Blumenthal's theory was challenged by Reinhart (2006), who argued that these species would likely attract resident generalist herbivores in the newly colonized community, which would thereby be expected to impose considerable environmental resistance to colonization and establishment (Agrawal and Kotanen 2003, Parker and Hay 2005; Fig. 3.10).

Without question, it is likely that the successful colonization and establishment by some species is due to the fact that they left their native enemies behind, with the new environment not providing new enemies to take their place. However, it is just as likely, perhaps even more so, that many introduction episodes are thwarted by enemies in their new home. Moreover, top-down controls are often not the primary drivers affecting population dynamics. While all species must always

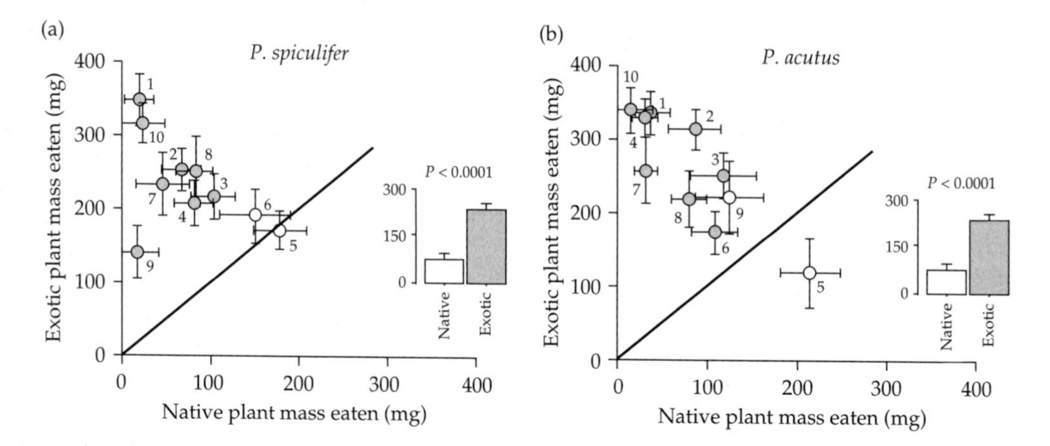

Fig. 3.10 Mean (+SE) plant biomass consumed by the native crayfishes (a) *Procambarus spiculifer* and (b) *P. acutus* when offered a choice between phylogenetically paired (either congeneric or confamilial) native and non-native freshwater plants. Gray circles were statistically significant individual feeding assays ($P < 0.05$, paired *t*-tests. Insets are overall means between native and non-native plants, with standard errors corrected for the nesting factor. Numbers refer to the taxonomic pairs. Redrawn and printed, with permission, from Parker and Hay (2005), copyright Blackwell Publishing.

gain access to resources in order to persist, it is not always necessary for a species to escape its enemies to do so. Thus, the correlation of invasibility with enemy abundance is likely to be a weak one, with enemy presence or abundance important in some instances but not others.

Invasibility, facilitation, and mutualisms

It is well-known that the persistence and abundance of many species depends on mutualisms or commensalisms with other species (Bertness and Callaway 1994, Bruno *et al.* 2003, Crain and Bertness 2006). The importance of facilitation in the establishment phase of non-native species has been appropriately emphasized in recent years (Bruno *et al.* 2005, Badano *et al.* 2007, Milton *et al.* 2007, Olyarnik *et al.* 2008). In their studies of alpine plants in the Chilean Andes, Badano *et al.* (2007) and Cavieres *et al.* (2007) emphasized the nurse effects of the native cushion plant, *Azorella monantha*, on two non-native species, *Cerastium arvense* and *Taraxacum officinale*. They found that the performance of both non-native species was enhanced when the plants were growing within patches of *A. monantha*, and that the facilitative effect increased with altitude. In fact, at higher elevations, *C. arvense* was only found associated with the cushion plants, indicating that the facilitative effects of *A. monantha* actually extended the altitudenal range of *C. arvense*. Providing a very different set of abiotic conditions than those associated with adjacent areas of bare soil and rock, *A. monantha* is believed to moderate a variety of abiotic conditions, including temperature, moisture, and nutrient availability (Arroyo *et al.* 2003, Badano *et al.* 2007). At high elevations in Colorado, whitebark pines have been infected by the non-native white pine blister rust, *Cronartium ribicola*. Due to warmer winters in recent years, the mountain pine beetle, *Dendroctonus ponderosae*, a native species that normally attacks mid-elevation lodgepole and ponderosa pines, has extended its range upward and it is now attacking whitebark pines. One consequence of its arrival has been that it is serving as an effective dispersal vector for the rust, thereby enhancing its establishment (Petit 2007).

The importance of facilitation in the establishment of non-native plants was also documented in a study of shrub establishment in arid South African savanna, which found that the establishment of fleshy-fruited and bird-dispersed shrubs was facilitated by the presence of trees growing in the savanna (Milton *et al.* 2007). Both native and non-native trees facilitated shrub establishment, and both native and non-native shrubs benefited from the facilitation by the trees. However, the non-native shrubs were more dependent on the tree facilitation, since non-native fleshy-fruited shrubs were only found growing beneath trees, while some native fleshy-fruited shrubs were found growing in open areas in the savanna (Milton *et al.* 2007). Non-native earthworms have been found to facilitate the establishment of giant ragweed (*Ambrosia trifida*) by burying their seeds, thereby protecting them from mice predation (Regnier *et al.* 2006). Although introduced herbivores were found to inhibit the establishment of *Pinus* species (Richardson 2006), other non-native species facilitated its establishment, including introduced fungi and the wild boar (Fig. 3.9). In many marine environments, resident species, both native and non-native, often facilitate the introductions of new sessile species by providing establishment surfaces for them (Schwindt and Iribame 2000, Stachowicz and Whitlatch 2005, Wonham *et al.* 2005). In other instances, marine residents can facilitate new introductions by reducing physical stress, e.g. from heat or dessication (Olyarnik *et al.* 2008)

In a review of the literature, Bruno *et al.* (2005) found that direct facilitative interactions occurring during introductions were just as common and important as competition and predation. These facilitative effects included dispersal of seeds and fruits of non-native species by native animals (Bossard 1991, Vilá and D'Antonio 1998), mycorrhizal associations (Richardson *et al.* 2000b), and environmental modification that facilitated the successive establishment of other species. For example, Castilla *et al.* (2004) found that the introduction of an ascidian, *Pyura praeputialis*, into rocky-intertidal habitats in Chile increased local invertebrate richness four-fold by providing a more heterogeneous physical structure. Lugo (2004) found that non-native tree species, which were first to colonize

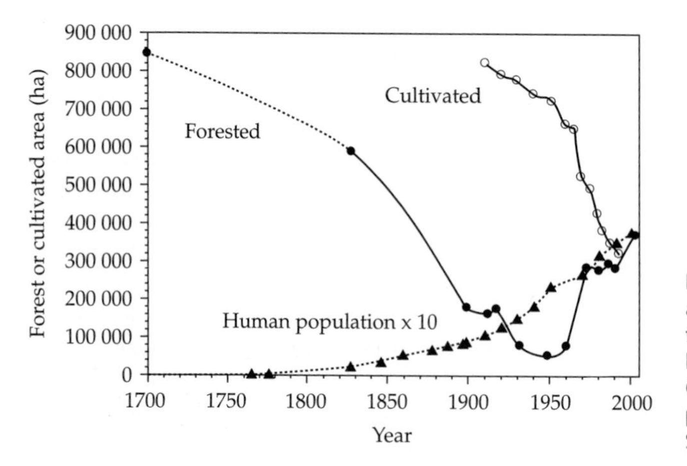

Fig. 3.11 Change in the area of forest cover and active agriculture in Puerto Rico. Forest area data are from USDA Forest Service inventories and agricultural land-use data from the USDA Natural Resources Conservation Service. Redrawn and printed, with permission, from Lugo (2004), copyright Ecological Society of America.

deforested areas in Puerto Rico (Fig. 3.11), facilitated the eventual re-establishment of native trees. Non-native species may also facilitate the introductions of additional non-natives (Simberloff and von Holle 1999). However, there is little reason to think that non-native species would be more likely to benefit other non-natives more than the native species. In their literature review, Bruno *et al.* (2005) did not find any evidence that non-native species are more likely to facilitate non-natives than native species. They also noted that many non-natives will negatively impact other non-natives.

While an already introduced species may sometimes facilitate the subsequent establishment of other non-native species, this does not mean that the facilitation was necessary for the latter species to successfully establish. For example, Ricciardi (2005) concluded that there is little evidence that facilitation plays an important role in introductions of non-native aquatic organisms, which appear to be primarily the result of propagule pressure and physical habitat conditions.

In many instances, facilitation takes place as part of a mutualistic interaction between the incoming non-native species and resident species or other incoming non-native species. Richardson *et al.* (2000b) reviewed this phenomenon with respect to plants and concluded that most plant-pollinator and plant-disperser relationships are not tightly coevolved, meaning that introduced plant species are often able to establish and spread even if unaccompanied by their native mutualists by taking advantage of native pollinators and seed dispersers. The fact that many animals and plants involved in pollinator and/or dispersal mutualisms have evolved to be quite opportunistic in their relationships undoubtedly partly accounts for the frequent invasion success of many of the participants, including both the plants and the animal pollinators and dispersers.

One would assume that species involved in exclusive, or nearly exclusive, mutualistic relationships in their native environment would have difficulty establishing in a novel environment without their mutualistic partner. The naturalization of Euglossine bees in Florida, USA, questions this assumption (Pemberton and Wheeler 2006). In their native tropical environments, Euglossine bees engage in a well-known mutualism involving orchids, in which male bees collect floral fragrances (believed to be used during courtship) and in doing so also pollinate the plants. This is an obligatory mutualism from the plant's perspective, since the bees are its only pollinators; however, apparently it is not obligatory for the bees, since they have been able to establish and persist in Florida by gathering fragrances from other plant species (Pemberton and Wheeler 2006).

The geography of establishment

Geographic patterns of invasions can provide clues regarding possible factors driving and facilitating establishment. While any conclusions regarding

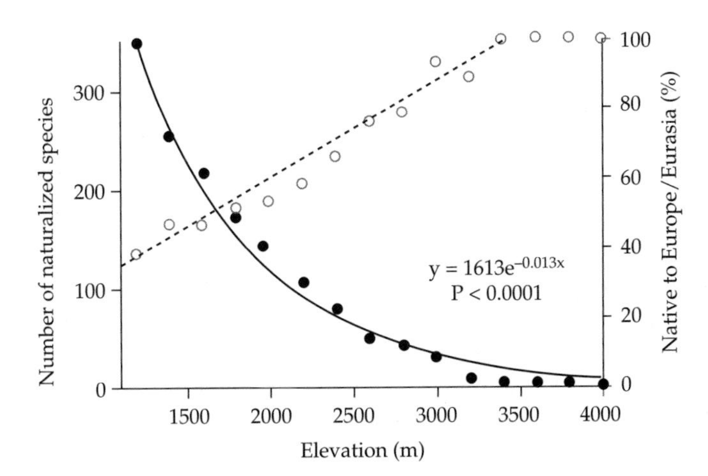

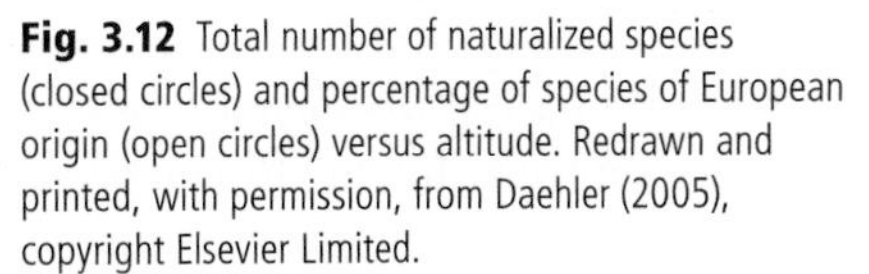
Fig. 3.12 Total number of naturalized species (closed circles) and percentage of species of European origin (open circles) versus altitude. Redrawn and printed, with permission, from Daehler (2005), copyright Elsevier Limited.

possible mechanisms will be based on correlative analyses, this can be an excellent first step in trying to understand the underlying mechanisms. Specifically, correlations arising from such studies can form the basis of hypotheses for subsequent experimental studies, as illustrated in the following examples.

Chytrý *et al.* (2008) compared the floras of non-native species in three areas of Europe: the Czech Republic (subcontinental), Catalonia (Mediterranean-submediterranean), and Great Britain (oceanic). They classified species into neophytes (post-1500 introductions), archaeophytes (pre-1500 introductions), and natives, and calculated the proportion of each group in 35 habitats. Given the different climate regimes, it is not surprising that the non-native floras of the three regions were quite dissimilar. In a correspondence analysis, the habitats were clustered by countries, indicating that the between-habitat similarity within regions was greater than the between-region similarity of the same habitat. Despite these large regional differences, distribution patterns were consistent among regions. In all three regions, nutrient poor habitats that seldom experienced disturbance, e.g. mines, heathlands, and high mountain grasslands, experienced very low rates of invasion. High proportions of non-native species were found in frequently disturbed environments that experienced fluctuating levels of resource availability. And, neophytes tended to be common in the same habitats as the archaeophytes. Taken together, these findings point to the importance in plant introductions of habitat characteristics, particularly resource availability (Chytrý *et al.* 2008).

A particularly fruitful approach to studying patterns of plant invasions has been to document patterns along elevation gradients in montane regions. In a study of Hawaiian montane flora, Daehler (2005) found that the number of non-native species declined exponentially with altitude (Fig. 3.12). At the same time, the percentage of non-native species that were of temperate origin increased linearly, until at high elevations virtually all species were from temperate regions (Europe and Eurasia). Hypotheses raised to account for the inverse relationship between altitude and number of non-native species invoked species–area relationships (i.e. high altitude sites represented smaller areas) and environmental stress (higher sites tended to be colder, drier, and exposed to higher levels of solar radiation) (Daehler 2005). Significantly, the same exponential decline in species-richness with altitude exists for native species in the same study sites, making it unlikely that the reduced richness of non-native species at high altitudes was substantially due to dispersal limitation (Daehler 2005). The primary hypothesis generated from the finding that the proportion of non-native species of temperate origins increased with altitude was that declining tolerance of low temperatures among non-native plants of tropical origin restricted their ability to establish at the higher sites. However, Daehler pointed out that, although several hypotheses have been proposed to account for the transition from tropical to temperate species with increasing altitude, little experimental evidence exists to test the hypotheses. Part of the difficulty in interpreting results from comparative altitude studies is that while some variables are intrinsically tied to altitude (e.g. atmospheric pressure and temperature), others are not (e.g. moisture, wind, seasonality, and land use), often making it difficult to discern the extent to which observed changes are truly due to altitude or to one of the latter variables, which may have happened to correspond to altitude in a particular study (Körner 2007).

Studies in the Swiss and Australian Alps have also found that the number of non-native species declines with altitude (Becker *et al.* 2005, McDougall *et al.* 2005). In the case of the Swiss plants, native species did not exhibit the same strong altitude–richness relationship (Becker *et al.* 2005), contrasting with Daehler's findings in Hawaii. This raises the possibility that low propagule pressure might account for the reduced non-native plant species-richness at high Swiss elevations. The fact that the maximum altitude reached by a non-native species was positively associated with time since introduction (Becker *et al.* 2005), is consistent with a dispersal-limitation hypothesis. At the same time, the longer the period since introduction, the greater are the adaptive opportunities for a species, which could also account for the observed establishment pattern. At this point, adequate data are not available to assess the relative importance of these hypotheses (Becker *et al.* 2005).

Guo *et al.* (2006) conducted a comprehensive comparison of the ranges of non-native plants reciprocally introduced between eastern Asia and North America. The two regions were selected because they share similar climates and habitats. The researchers found that proportionally more plants from eastern Asia had become established in North America than vice versa, and that the eastern Asian species inhabited larger areas in their introduced range than did North American plants in eastern Asia. Since the flora of eastern Asia is more species-rich than that of North America, Guo *et al.* concluded that the results are consistent with the notion that increased diversity confers biotic resistance (Elton 1958). However, Guo *et al.* also found that in both eastern Asia and North America, areas with more non-native species also contained more non-native species, certainly weakening the biotic resistance argument. The authors also pointed out that widespread human travel and migration within eastern Asia has become common just in the past few decades, whereas it has been commonplace in North America for a much longer period of time. This raises the likely possibility that differences in propagule pressure may account for much of the observed geographic patterns documented. In any case, the study by Guo *et al.* (2006) illustrates both the value and limitations of undertaking comprehensive geographic surveys of non-native species.

Spatial heterogeneity has been shown to be one of the principal shapers of biodiversity patterns (Huston 1994, Turner 2005). In a study of vegetation in the Rocky Mountain National Park, Colorado, USA, Kumar *et al.* (2006) investigated the relationship between the distributions of both native and non-native species in the context of 13 measures of spatial heterogeneity (e.g. mean patch size, edge density, and mean nearest neighbor distance). They found that the distributions of both native and non-native were associated with particular types of environmental heterogeneity, but that the association was stronger for non-native species. They also found that the non-native species-richness was auto-correlated at the landscape scale, while species-richness was not found to be auto-correlated for native species. Kumar *et al.* proposed several possible explanations, or hypotheses to account for these findings. They suggested that the auto-correlation in non-native species-richness may be a result of the seed dispersal patterns or other spatially structured ecological processes. While the answer likely lies somewhere in this explanation, it is not clear why similar processes do not also produce auto-correlation patterns of native species-richness. As for the stronger association of non-native plants with spatial heterogeneity, the authors suggested that this may be due to the fact that the non-native species have not yet fully dispersed throughout the landscape. For example, the stronger relationship between non-native species and patch size and edge density (e.g. m of edge per ha) may be because dispersal of non-native species are particularly strongly influenced by edges and disturbances (Kumar *et al.* 2006).

In a review of the geographic distributions of several taxonomic groups, Sax (2001) documented that considerably fewer non-native species had become established in tropical environments than in temperate ones. This contrasts with the well-known latitudinal pattern for non-native species, in which species-richness generally increases with decreasing latitude. However, outside the tropics, the distributions of non-native species did tend to follow the latitudinal gradient, with progressively fewer non-native species at higher altitudes (Sax 2001). In some instances, whether or not geographical patterns are found, depends on the spatial scale of analysis. Stohlgren *et al.* (2005, 2006a), examined the distributions of non-native plant species in the 48 conterminous states in the US and found that, although there was a statistically significant latitudinal correlation, it was biologically meaningless given the low amount of variation explained by latitude, 1%. Much better predictors of non-native plant richness within this latitudinal extend were factors such as human population density and specific biological and environmental variables, including native plant species-richness, potential evapotranspiration, elevation, and bird species-richness, the latter of which has been found to be a reliable surrogate for habitat diversity, integrating elements of productivity, habitat heterogeneity, and levels of disturbance (Jarnevich *et al.* 2006).

Qian and Ricklefs (2006) also examined the geographic patterns of the floras of the United States and Canada, and came to a different conclusion with respect to the predictability of non-native species. They concluded that while the distributions of native plant species were strongly associated with environmental variables, such as elevation and climate, non-native species were only weakly linked to these variables. Instead, they found that the best predictor of non-native species-richness was human population density. In some respects, the results of the two studies are not as different as the conclusions suggest. Human population density was the variable second most correlated with non-native species-richness in the Stohlgren *et al.* (2005) study and the magnitude of the variation explained by environmental factors was similar in both studies.

Summary

In 1983, SCOPE listed three questions that were to guide research and thinking on this topic. The first addressed traits of 'invaders', the third addressed management, and the second challenged researchers to identify site properties that determine whether an environment would be susceptible to invasion or not. In his 1996 book, in which he reviewed progress since the SCOPE project began, Williamson downplayed the roles of invasibility and species traits, emphasizing much more the importance of propagule pressure in determining whether an invasion is successful. While findings have led most researchers to agree with Williamson that with sufficient propagule pressure, virtually all environments are invasible to some extent, research since 1996 has clearly documented substantial variation in the invasibility of environments, both in space and time.

As illustrated by the above discussion, the study of invasibility is really the study of diversity (Shea and Chesson 2002, Davis *et al.* 2005a, Rejmánek *et al.* 2005b, Ejrnæs *et al.* 2006, Renne *et al.* 2006), and both are ultimately about coexistence (Huston and DeAngelis 1994, Levine *et al.* 2004, Davies *et al.* 2005, Melbourne *et al.* 2007). Because invasibility is really about community assembly, efforts to account for invasibility are inevitably influenced by ecologists' basic assumptions about the natural world. Those who perceive the world in a more deterministic way are likely to emphasize the role of biotic interactions, often in the context of niche theory and equilibrial systems, and to emphasize the fundamental similarities among different systems and environments. On the other hand, where this group sees smoothness and predictability, another group sees bumpiness and uncertainty, emphasizing differences in processes among systems and environments, and thereby according local uniqueness greater importance than universality. In some ways, the difference between these two groups is similar to the difference between 'lumpers' and 'splitters' in systematics. It is an interesting question as to whether an ecologist's fundamental view of the world is derived more from ecological data and experience, or one's basic personality and predisposition, which may incline some individuals to put their faith in universal principles, e.g. predictable systems of biotic interactions, while others are prone to see the overriding importance of history, stochasticity, and local idiosyncrasy. Of course, since ecology is a science, then empirical evidence should, over time, be able to resolve the debate over which paradigm is a better representation of the natural world. My own assessment of the various hypotheses and theories proposed to explain variation in invasibility strongly supports the merits of the bumpy paradigm. Hypotheses emphasizing the temporal and spatial heterogeneity of resources have, to date, proven to be much more robust and reliable than ones emphasizing biotic interactions, whether those be competitive (the diversity–invasibility hypothesis) or of a top-down nature (e.g. the enemy release hypothesis).

Based on the considerable empirical and theoretical work on invasibility that has been conducted in recent years, I believe it is possible to make the following conclusions with a high degree confidence:

- virtually all natural environments are invasible to some degree;
- very few natural communities appear to be saturated with species (this is really just a different way of stating the first bullet);
- the most reliable predictor of invasibility to date has been resource availability, with both temporal and spatial variation in resources shown to be the primary mechanisms by which pools of resources are made available to new colonists;
- enemy- and facilitator-related processes can be important in accounting for invasibility in some instances, but neither has proven to be as reliable a predictor of invasibility as resource availability;
- while there may be instances in which diversity, including both species and functional diversity, influences resource availability, physical stress, or other factors affecting invasibility, diversity has not been shown to be a reliable predictor of invasibility under natural conditions at any spatial scale;
- the same process affecting invasibility are driving diversity.

CHAPTER 4

Persistence and spread

A founding population emerges from the establishment of the first arriving propagules. However, the establishment of a founding population may not necessarily be deemed an invasion, which typically involves persistence of subsequent generations and spread well beyond the original point of entry. The term spread could refer to two different types of expansion. It could refer to the incremental spatial spread of a single population as it grows in size, gradually covering a larger area. Or it could refer to saltatory spread, in which the original population gives rise to one or more new populations somewhere else in the land/seascape via dispersal of individuals from the former to the latter. Depending on the species, the spatial scale considered, and the question at hand, either or both types of spread may be relevant to an invasion biologist. However, in this chapter, the term spread is generally used to refer to saltatory spread and the production of additional populations.

Iterative dispersal and establishment episodes (persistence and spread)

As described in Chapter 2, the persistence of a species in a newly colonized area, and the spread of the species to other areas, is the result of repeated successful dispersal and establishment episodes of individual organisms that are, in most cases, primarily part of within-region dispersal episodes. These regionally produced propagules encounter the same basic challenges as the propagules that initially arrived in the region. Reise *et al.* (2006) emphasized this point in their assessment of the impacts of non-native species on European coastal waters:

> ...it makes no difference whether immigrants stem from adjacent waters, have crossed oceans with ships or continents through canals. In the wake of this overall immigration and emigration process, immigrants that stem from adjacent waters and those that have been introduced from distant waters may be assimilated likewise by the recipient biota without any general difference.

As in the case described by Reise *et al.*, in many instances, it is likely that the dispersing propagules in a region are of mixed origin, some arising from within the region and some originating from other regions.

It is well-documented that extent of spread is positively correlated with time since introduction (Wilson *et al.* 2007; Fig. 4.1). This finding may involve ecological and evolutionary events and processes, e.g. genetic or phenotypic adaptations by the new species to the new environment (Cox 2004, Dietz and Edwards 2006). However, a null model also predicts a positive correlation between spread extent and time since introduction. This argument is essentially the same one made by Stephen Jay Gould (1996) with respect to the trend toward increasing biological complexity over evolutionary time. Gould likened the process of the evolution of complexity from single-celled organisms to a random walk that begins at a wall. The fact that you end up some distance from the wall after a period of time does not mean there was any directionality, since only one direction (away from the wall, or increasing complexity in the case of evolution) was an option. For those introductions that occur at a single location, there are only two options available to the new immigrants, assuming they succeed in establishing: either they spread, or they remain confined to their point of colonization. Even if some species do not exhibit any spread over time, most do, and any regression analysis of a large number of species will yield a positive correlation between time and spread.

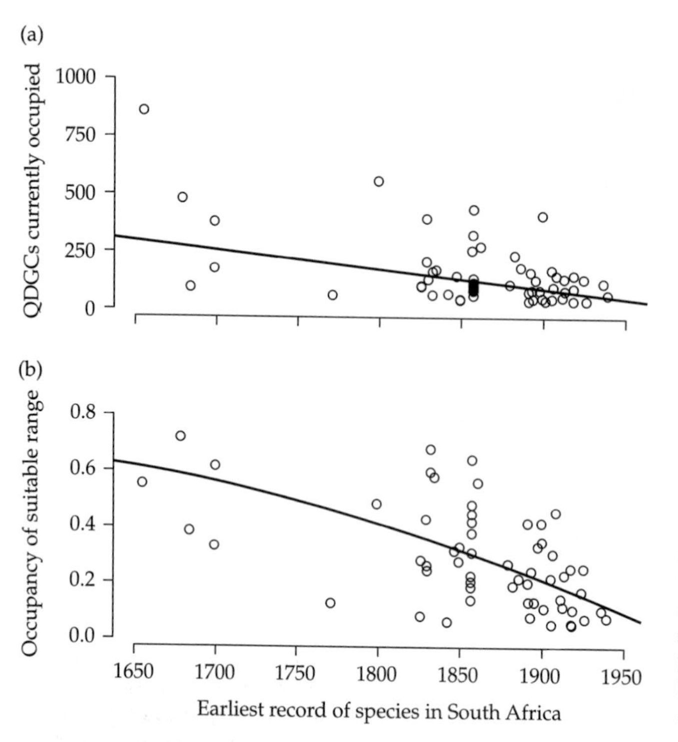

Fig. 4.1 How residence time affects the range size of invasive plants. (a) The relationship between range size [number of quarter-degree grid cells (QDGCs)] and time since introduction, $r^2 = 0.22$; (b) the proportion of the potential range that was occupied in 2000, $r^2 = 0.29$. Redrawn and printed, with permission, from Wilson *et al.* (2007), copyright Blackwell Publishing.

Following a long-distance dispersal event, usually mediated in some way by human activity, introduced species that successfully establish usually continue to disperse to some extent within their new region. The primary dispersal mechanisms of introduced species following establishment often varies with scale, e.g. global, regional, and local (Pauchard and Shea 2006). In some instances, humans continue to be the primary dispersal vector following introduction into a new region. For example, many freshwater organisms are dispersed from one freshwater system to another by recreational boaters (Johnson and Padilla 1996), and many non-native plant species are accidentally introduced into national parks by tourists (Macdonald *et al.* 1989, Lonsdale 1999). However, in many cases, the within-region dispersal of introduced species is accomplished mostly independent of direct human activity, and takes place using biologically traditional vectors, e.g. wind, water, and animals. For example, secondary dispersal of introduced zooplankton is believed to occur via connected waterways, resulting in systems located lower in the landscape being more likely to be 'invaded' than upstream systems (Havel and Medley 2006).

Kinlan and Hastings (2005) compared the rates of spread following initial introductions for both marine and terrestrial plants and for marine invertebrates and found that the spread rates of the marine species were substantially greater than those for the terrestrial plants (Fig. 4.2). The ability of water to disperse propagules farther, more easily than wind can disperse seeds, was believed to be a likely part of the explanation for this difference. An interesting finding from this study was that marine plants spread much faster than one would expect given the range in mean dispersal abilities, which was not the case with marine invertebrates. Since very long-range dispersers would typically be accompanied by few other individuals, these individuals would likely experience problems associated with low conspecific densities at their new site of establishment (Allee effect), which, Kinlan and Hastings suggested, might present less of an obstacle to some of the marine plants. Kinlan and Hastings concluded that the spread rates of species following their initial human-assisted introduction may be associated more with the occurrence of rare long-distance events than the average dispersal abilities of the species, particularly for species that possesses life-history traits

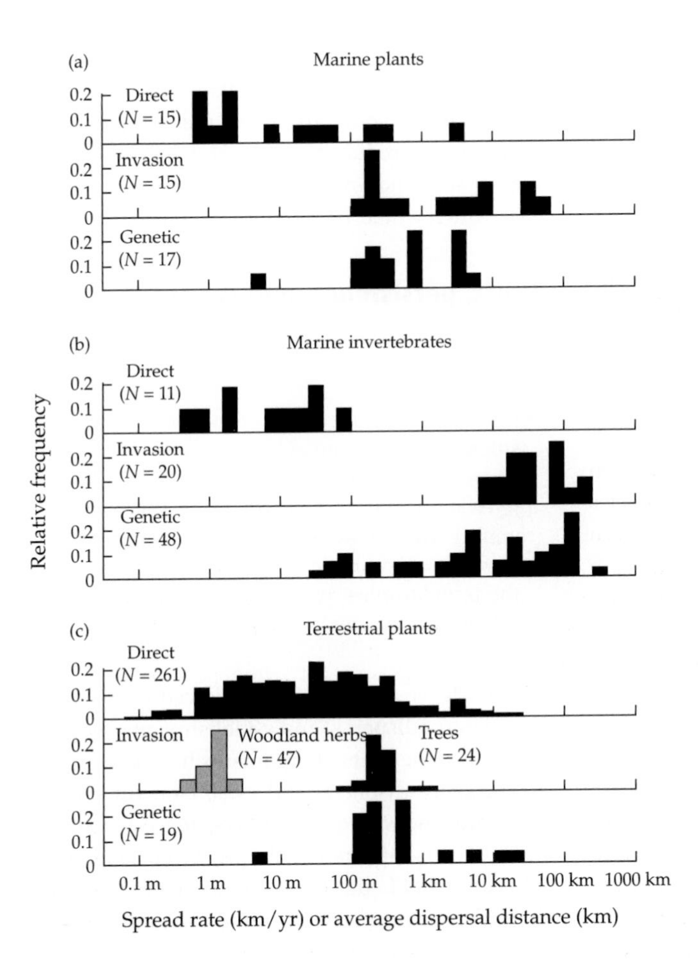

Fig. 4.2 A comparison of marine and terrestrial dispersal spread rates for observed invasions (middle histograms of each panel). Also included are measures of average dispersal distances obtained from direct observational studies (upper histograms) and dispersal rates inferred from population genetic structure (lower histograms of each panel). Redrawn and printed, with permission, from Kinlan and Gaines (2003) and Kinlan and Hastings (2005), copyright Sinauer Associates, Inc.

enabling individuals to establish and persist at very low initial densities. Based on their modeling efforts of invasive plants, Nehrbass *et al.* (2007) came to a similar conclusion, finding that the rate of spread in their model was determined primarily by the small number of long-dispersing individuals.

In some instances, secondary dispersal may occur through mutualistic relationships with other species (Richardson *et al.* 2000b). In Germany, the spread of *Prunus serotina* was found to be influenced by the abundance of trees in the landscape, which provided roosting sites for the avian dispersal agents (Deckers *et al.* 2005). Pond apple, *Annona glabra*, a non-native invasive tree in some Australian tropical forests, is typically water-dispersed. However, Wescott *et al.* (2008) showed that the southern cassowary, *Casuarius casuarius*, is a common disperser of the fruits in some areas. The dispersal by this large flightless bird substantially affects the invasive potential of *A. glabra*. Wescott *et al.* estimated that the far-ranging birds could disperse seeds more than 5 km. Moreover, since the birds can move fruits up-stream and across drainage boundaries, as well as to sites quite far from the aquatic habitats in which dispersal would normally occur, the birds significantly expand the range of habitats the tree can potentially colonize (Wescott *et al.* 2008).

Native leaf-cutter ants have been found to aid in the spread of non-native plant species along roads in Argentina (Farji-Brener and Ghermandi 2008). Specifically, the ants create refuse dumps along the roads consisting of organic debris they have removed from their nests. These refuse dumps constitute ideal growing sites for the non-native plants, which exhibit considerably higher densities and seed output on the refuse dumps than elsewhere in the landscape, an increase in performance believed to be the result of increased nutrient availability at the dump sites (Farji-Brener and Ghermandi 2008). Although secondary dispersal may not involve direct human involvement, i.e. humans are not moving the propagules, dispersal at the regional or local scale may still be greatly influenced by human activity. For example, as is the case in the leaf-cutter example, many plants are known to disperse along roadsides, which represent disturbance corridors (Pauchard *et al.* 2003).

If the abundance or characteristics of these traditional vectors change in the new environment, then the secondary dispersal of the introduced species would be expected to be affected. In the eastern United States, white-tailed deer, which forage both in suburban and adjacent natural areas, regularly disperse into the natural areas seeds of non-native species planted in the suburban environment (Vellend 2002). Deer populations have increased substantially in many areas during the past several decades, thereby significantly increasing the likelihood of dispersal of many non-native species into natural areas. The seep monkeyflower, *Mimulus guttatus*, a riparian species introduced into the UK in the early 1800s, dispersers via seeds and plant fragments, which are transported by water. Predictions of increased occurrence of high-flow events (flood events) in the UK, due to climate change, indicate that the invasive potential of *M. guttatus* is likely to increase in the future due to increased dispersal opportunities (Truscott *et al.* 2006).

While climate change may increase dispersal opportunities for some species, it may reduce them for others. In the case of many marine organisms, dispersal takes place in the juvenile stage when the planktonic larvae are dispersed by ocean currents. O'Connor *et al.* (2006) showed that the dispersal distance of these organisms was greatly influenced by development time, i.e. the amount of time the organism remains in the planktonic larval stage. Specifically, warmer ocean temperatures would be expected to reduce the dispersal capabilities of many marine organisms, since they would mature faster in the warmer waters, thereby shortening the amount of time spent in the plankontic stage. This finding has significance for existing native marine populations that may become increasingly disconnected from one another as dispersal capabilities decline. But it would also mean that warmer ocean currents may reduce the natural dispersal capabilities of some non-native and invasive marine species, which would be a beneficial consequence.

Traits, persistence, and spread

Note to readers

Many of the invasion researchers who have studied invasion spread have used the term 'invasive' in the more narrow sense, i.e. referring only to spread and not to any level of impact. Thus, in an effort to avoid confusion with this literature, in this section I have used the term 'invasive' to refer only to species exhibiting rapid spread, irrespective of impact.

If persistence and spread are viewed primarily as iterations of dispersal and establishment processes, then since the traits of organisms have been found to influence dispersal and establishment success, naturally one would expect persistence and spread to be likewise affected by the phenotype. For example, Kolar and Lodge (2001) found that invasiveness in birds was positively correlated with the number of broods per year, while invasiveness in some freshwater fish was associated with an increased ability to tolerate a wide water-temperature range. The extent to which particular traits are associated with spread has been a major focus of research in the field, particularly with respect to plants. Generally, these studies take one of two approaches. The first involves the comparison of traits of native and non-native species (Hamilton *et al.* 2005, Meiners 2007). The second confines itself to non-native species and compares traits of those that have become invasive with those that have not (Hamilton *et al.* 2005).

In a review of invasiveness in plants, Rejmánek *et al.* (2005b) concluded that traits that promoted reproduction and dispersal were particularly associated with invasiveness. For example, in a study of non-native invasive and non-native non-invasive pines, Rejmánek and Richardson (1996) found that increased invasiveness was strongly associated with three plant attributes, all associated with reproduction and dispersal: small seed mass, decreased time to reproductive maturity, and increased frequency of large seed crops. As Rejmánek and Richardson pointed out, all three characteristics are typically associated with r-selected species, not surprising given that the pine introductions often occurred in disturbed sites. The inverse relationship between seed size and invasiveness in pines is clearly not a general phenomenon. In fact, it may be more of an exception. Daws *et al.* (2007) compared the seed mass of 376 species in two plant families (Asteraceae and Poaceae) in California and found that the invasive Asteraceae species produced seeds 101% larger than the non-invasive species, while the invasive grasses produced seeds 68% larger than the non-invasive grasses. In a more phylogenetically diverse review of plant species encompassing 31 families, they found that the seed mass of plants in the invasive range tended to be greater than that in the native range, although there were numerous exceptions to this pattern (Dawes *et al.* 2007).

Daehler (2003) conducted an extensive literature review of studies that had compared performance of native and non-native invasive plants, performance being measured by variables such as survival, germination rate, growth rate, fecundity, and dispersal ability. He documented considerable variation, with native species outperforming the invasive species in some conditions and vice versa. However, Daehler concluded that, in general, the invasive species outperformed the native species when the natural disturbance regime had been altered and when resources were abundant. This conclusion is consistent with the findings of subsequent studies (Leishman and Thomson 2005, Leishman *et al.* 2007), which found that, compared to native plant species, non-native invasive species exhibited higher growth rates and survival in high nutrient conditions. These findings indicated that the invasive species would be at a disadvantage if resources were limiting. However, this may not always be the case. Funk and Vitousek (2007) measured short-term resource-use efficiency (light, water, and nitrogen) in 19 phylogenetically related pairs of Hawaiian plant species (one of the pair being native and the other an invasive non-native species) in resource-limited environments. They found that, as a group, the non-native invasive species utilized light and nitrogen resources more efficiently than the native species. (No difference was found for water-use efficiency.) This is an important finding since it demonstrates that not all invasive species must depend on an abundance of underutilized resources to establish, persist, and spread (Davis *et al.* 2000, Funk and Vitousek 2007).

In a 2007 review of the literature on plant traits and invasiveness that involved comparisons involving multiple species, Pyšek and Richardson concluded that vigorous vegetative growth and early and extended flowering were strongly and consistently related to invasion success. However, they also conceded that while there are some consistent findings, the importance of many traits remains ambiguous. For example, while pollen vector (wind or animal) was found to be important in some comparisons involving native and non-native plants (Williamson and Fitter 1996), Pyšek and Richardson did not find this to be a reliable predictor of invasion success. They similarly concluded that the breeding system, e.g. monoecious vs dioecious, was not a reliable predictor of invasiveness. Pyšek and Richardson (2007) also reviewed studies using congeneric pairs of species (one being invasive and the other not) and concluded that these studies were better able to identify more traits associated with invasiveness. Higher fecundity, increased growth rates, greater water- and nutrient-use efficiency, increased dispersal ability, and earlier and/or extended flowering were the attributes most consistently found to be exhibited by the invasive species in the congener studies. A comprehensive review of the ecology of garlic mustard, *Alliaria petiolata*, a highly invasive non-native species in North America, concluded that its success was due to the combined effects of many factors, including early phenology, high fecundity, and high dispersal ability (Rodgers *et al.* 2008).

Arthington and Mitchell (1986) asserted that invasiveness in aquatic plants tended to be associated with vegetative and rapid reproduction. In general, most of the traits found to be associated with invasiveness in plants parallel those made by Baker (1965) in his famous review of weed characteristics. Additional evidence supporting the relationship between growth rate and invasiveness comes from a comparative analysis of leaf traits involving native and non-native invasive plant species (Leishman *et al.* 2007). They found that, as a group, and in disturbed areas, the invasive species exhibited greater specific leaf area and greater foliar N and P on a mass basis, traits associated with increased grow rates. Of course, exceptions are not uncommon. Although naturalized and invasive plant species have often been found to exhibit faster growth rates (Grotkopp *et al.* 2002, Burns 2004, 2006), van Kluenen and Johnson did not find this to be the case in their study of South African Iridaceae. It is important also to remember that while a particular trait may prove beneficial to an organism at a particular point in the process, its value may be negligible, or even detrimental, at another time (Pyšek and Richardson 2007). For example, in their study of freshwater fish, Kolar and Lodge (2002) found that while fast growth was associated with successful establishment, slow growth was associated with rapid spread.

A number of studies have been conducted to determine the role that phenotypic plasticity may play in plant invasiveness and some have found that invasive plant species do tend to exhibit higher levels of phenotypic plasticity (Daehler 2003, Richards *et al.* 2006). In some cases, phenotypic plasticity seems to provide a general superiority, one that provides benefits in most conditions (Brock *et al.* 2005, Rodgers *et al.* 2008). However, in many cases, phenotypic plasticity seems primarily advantageous only in certain conditions, such as during periods of increased resource availability (Hastwell and Panetta 2005). A comparison of purple loosestrife, *Lythrum salicaria*, native populations (Germany) with non-native and invasive populations (North America), found that the invasive populations exhibited greater growth and reproductive plasticity in response to changing water and nutrient regimes, which may partly explain the invasive abilities of this species in North America (Chun *et al.* 2007). In a study of several native and non-native invasive aquatic plant species, Hastwell *et al.* (2008) found that the invasive species were more responsive to increases in nutrient supply, tending to accumulate more biomass and an increase in photosynthetic area. Richards *et al.* (2006) referred to plants exhibiting general superiority due to their increased plasticity as 'jacks of all trades', while they described plants superior in a subset of habitats as 'masters of some' (Fig. 4.3).

While numerous studies have shown that phenotypic plasticity can facilitate species' spread, this certainly does not mean that invasive species should be expected to exhibit greater phenotypic plasticity, since other factors may account for the invasiveness in many species. In a test for plasticity in two genera of plants, *Crepis* and *Centaurea*, which contained species of varying degrees of

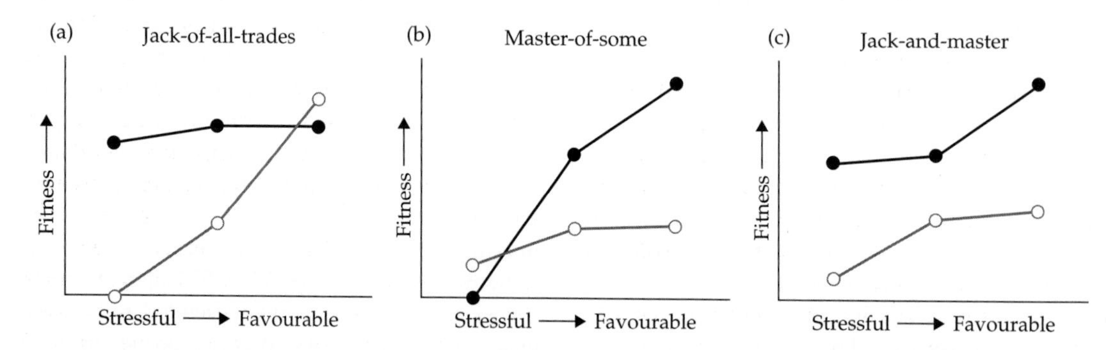

Fig. 4.3 Expectations for fitness plasticity of invasive (black line) vs non-invasive (grey line) genotypes/populations/species must qualitatively resemble one (or both) of two patterns: (a) invasives have more robust fitness in the face of stressful environmental conditions, possibly conferring greater ecological breadth (Jack-of-all-trades); or (b) inavsives are better able to respond with increased fitness in favorable conditions, possibly allowing for higher population densities under favorable conditions (Master-of some); (c) it is also conceivable to envision a fitness norm-of-reaction that has characteristics of both robustness and responsiveness (Jack-and-master). Redrawn and printed, with permission, from Richards *et al.* (2006), copyright Blackwell Publishing.

invasiveness, Muth and Pigliucci (2007) found that trait responses to various stresses (low phosphorus and water) and opportunities (abundant phosphorus and water) were highly variable and largely idiosyncratic with respect to the level of invasiveness exhibited by a species. They acknowledged that while phenotypic plasticity likely does play an important role in some species invasions, it will be very difficult to detect any robust general patterns, if they even exist.

Rehage *et al.* (2005) investigated whether behavioral plasticity may partially account for the invasive capability of two species of species of mosquitofish. In controlled experimental conditions, they compared the behavior (foraging efficiency) of two invasive species, *Gambusia holbrooki* and *G. affinis*, with that of two less-invasive congeners, *G. geiseri* and *G. hispaniolae*, in response to novel competitors and predators. In this case, no consistent differences in behavioral responses were found between the invasive and less-invasive species. However, in all instances, the two invasive species exhibited increased foraging efficiency compared to their less-invasive congeners, suggesting that the general superior foraging ability of *G. holbrooki* and *G. affinis* plays a more important role in their success than behavioral plasticity.

The emphasis on phenotypic plasticity in the invasion process has refocused attention on development, particularly the way in which development may be influenced by different ecological conditions. This suggests the relevance to invasion biology of an emerging area of research some have referred to as 'eco-devo' (Sultan 2007). Eco-devo is described as enhancing studies of phenotypic plasticity by adding an explicit focus on the molecular and cellular mechanisms of environmental perception and gene regulation underlying developmental processes (Ackerly and Sultan 2006). Sultan (2007) suggested that eco-devo should be able to contribute to our understanding of the mechanisms behind the establishment success of certain non-native invasive species.

It is important to remember that persistence and even domination at a site by non-native species does not necessarily mean that the non-native species are better adapted or competitively superior than the native species that historically occupied the site. In a field experiment in a grassland environment, Seabloom *et al.* (2003) found that the native perennial species were actually better competitors than the non-native annuals, capable of of reducing soil water, soil nitrogen, and light levels more than the non-native species, and that the native perennials were able to colonize even dense stands of the non-native annuals and eventually reduce the abundance of the non-natives. In this case, the persistent dominance of the non-native plants was determined to be due to dispersal limitations of the native species, rather to any superiority on the part of the non-natives. This finding has important management implications, since it suggests that restoration of some native species in some environments may be quite feasible, possibly only requiring seed reintroduction (Seabloom *et al.* 2003).

Genetic obstacles to persistence and spread

For some species, accomplishing secondary dispersal and establishment does not depend on the presence of other conspecifics. However, for many species, e.g. social species and obligatory out-crossing species, absence of conspecifics is problematic. Thus, while some individuals initially may be able to establish themselves following a dispersal episode, maintaining the iterations of dispersal and establishment within the new environment (Fig. 2.1), a process often referred to as persistence, often fails due to various factors associated with the low density of conspecifics, such as inbreeding and the difficulty in finding mates (Courchamp *et al.* 1999), i.e. the Allee effect (Fig. 4.4). The term 'invasion pinning' has been used to describe the process by which range expansion is stopped due to the Allee effect (Keitt *et al.* 2001). In these instances, although dispersal beyond the edge of the range continues, population levels never exceed the threshold for ongoing persistence. Longer lag times, slower spread, and decreased establishment likelihood of non-native species would all be expected to result from an Allee effect (Taylor and Hastings 2005). This phenomenon may explain the pulsed nature of many invasions, i.e. that invasions often do not proceed on a continuous basis but instead proceed in a saltatory fashion. This was argued to

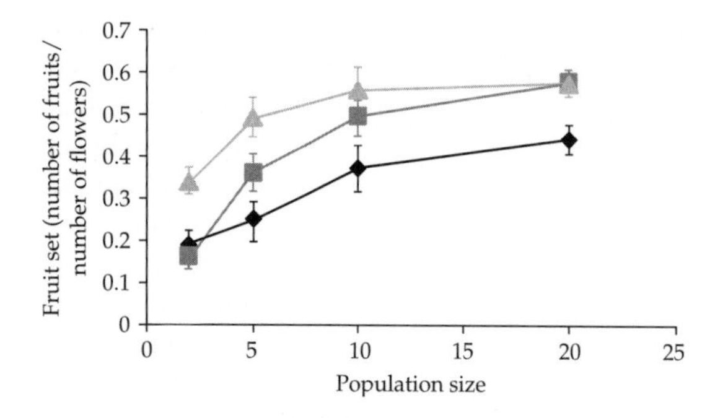

Fig. 4.4 The effects of population size and genetic relatedness on fruit set (number of fruits/number of flowers). Symbols (diamonds, full-siblings; squares, half-siblings; triangles, unrelated individuals) represent means across populations + SEM. Redrawn and printed, with permission, from Elam *et al.* (2007), copyright National Academy of Sciences, USA.

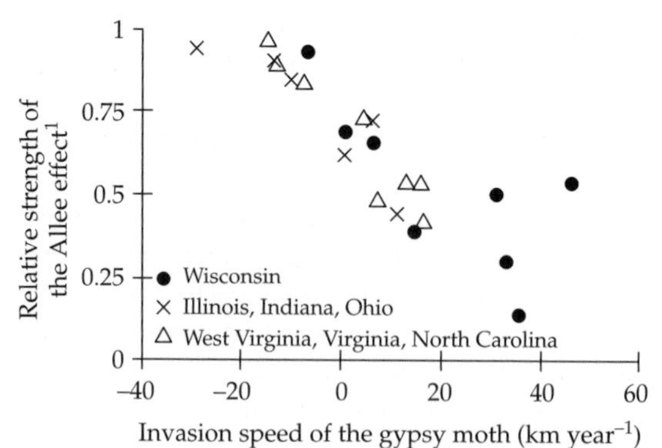

Fig. 4.5 Invasion speed of gypsy moths, *Lymantria dispar*, was found to be inversely associated with the strength of the Allee effect in three regions of the United States. As shown, at very high levels of the Allee effect, the front actually retreated. Redrawn and printed, with permission, from Tobin *et al.* (2007), copyright Blackwell Publishing.

be the case for the episodic spread of gypsy moths, *Lymantria dispar*, in North America (Johnson *et al.* 2006).

Using data obtained from more than 100,000 gypsy moth traps, set out and managed by the United States Department of Agriculture, Tobin *et al.* (2007) were able to empirically examine the extent to which spread of the moths was associated with density, and hence to determine the strength of any Allee effect. They documented an Allee effect throughout the current range of the moth in the United States, with the Allee effect being significantly and negatively correlated with invasion spread throughout its range (Fig. 4.5). More than this, Tobin *et al.* documented at least an order of magnitude variation in the Allee threshold among different regions. In Wisconsin, the threshold was calculated to be 2.2 moths per trap, while in West Virginia, Virginia, and North Carolina, the threshold was determined to be 20.7 moths per trap. Moreover, the investigators recorded temporal variability in the Allee threshold as well, with some regions exhibiting lower thresholds in some years than in others, and in some cases exhibiting no Allee threshold at all. In the case of gypsy moths, in which the females are flightless, it is believed that the difficulty in finding mates at low densities is a major cause of the Allee effect in this species (Sharov *et al.* 1995, Tobin *et al.* 2007). However the data collected by Tobin *et al.* could not account for the reasons behind the substantial temporal and spatial variation in the strength of the Allee effect documented in the study.

Despite its likely importance in some invasions (Davis *et al.* 2004), the Allee effect may not be a universal invasion phenomenon. A manipulative field experiment involving the introduction of a parasitoid wasp at different densities, found no evidence of an Allee effect but, instead, complete negative density-dependence in population growth (Fauvergue *et al.* 2007). Specifically, the researchers found that the probability of establishment was independent of initial population size and that the net reproductive rate was highest at low parasitoid densities. Fauvergue *et al.* suggested a variety of possible explanations for the lack of an Allee effect in this species, and possibly other parasitoids as well, including very efficient mate-finding abilities, even at very low densities, and the fact that parasitoids generally experience high levels of intra-specific competition for the host resources, which would impose high levels of negative density-dependence on population growth rates, overwhelming any demographic Allee effect.

At the edge of their ranges, species are believed to be commonly represented by small and scattered populations (Thomas and Kunin 1999), although actual population distributions are not well-described for most species, and data that do exist are mixed with respect to the size and geographic distributions of populations (Sagarin and Gaines 2002). Despite the lack of good empirical evidence, traditional theory holds that the peripheral populations are small partly because of either abundant or scarce gene flow into peripheral areas (Haldane 1956, Bridle and Vines 2007; Fig. 4.6). In the former case, high levels of gene flow from the centre of the range are believed to continually swamp any local adaptive processes, thereby preventing the increase of genotypes that might be more successful at persisting at the edge of the range, and even beyond (García-Ramos and Rodríguez 2002, Lenormand 2002). In the latter instance, it is thought that very limited gene flow into small populations maintains low levels of genetic diversity in these populations, which were likely already genetically depauperate due to the founder effect, thereby severely limiting the opportunities for local adaptation and hence spread into adjacent areas beyond the edge of the current range (Holt and Keitt 2000, 2005).

Given that a non-native species that has recently dispersed, or been transported, to a new region may also be considered to be at the range edge, it is not surprising that both explanations, genetic swamping from central populations and genetic impoverishment due to a founder effect combined with minimal subsequent gene flow, have been proposed as mechanisms that may impede the spread of a newly colonized species (Lenormand 2002, Bridle and Vines 2007). However, in some cases, selection at the edge of a species range may be sufficient to prevent genetic swamping from the central populations. Sanford *et al.* (2006) found that the cold water at the northern edge of the range of the mud fiddler crab, *Uca pugnax*, selected for faster development, thereby reducing mortality risk associated with the planktonic phase. They concluded that this selection increased the likelihood of ongoing northern spread. Similar selection on recently introduced species, i.e. favoring traits better adapted to the new environment, would also be expected to enhance the likelihood of subsequent spread of the species in its new environment.

Just as high levels of genetic diversity are generally viewed as beneficial to native populations,

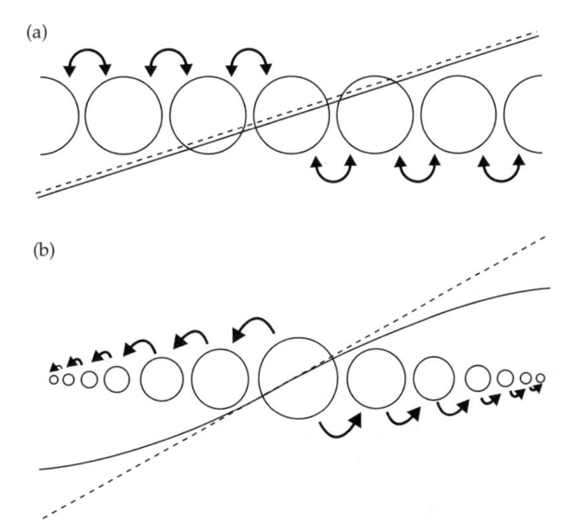

Fig. 4.6 Migration load and adaptation along selective gradients. (a) Range expansion without limit along a one-dimensional selective gradient. Here, the trait mean (solid line) at each point along the gradient matches the environmental optimum (dotted line) everywhere. Therefore, population fitness is high, population size is uniformly large (indicated by the size of the circles), and the species continually expands along the gradient. The arrows depict the direction and magnitude of migration between adjacent populations. (b) Range margins generated by migration load. In this case, the well-adapted central population is also the largest and sends out many migrants to adjacent populations (arrows). These immigrants prevent adjacent populations from reaching their trait optimum (the solid line is displaced from the dotted line), which reduces their fitness and, hence, their population size. These populations, in turn, send out migrants that are even less fit, further reducing the fitness and, therefore, the size of the more peripheral populations. Eventually, the trait mean of the peripheral populations is far from the optimum, and fitness is so low that population growth is negative, even with immigration. Redrawn and printed, with permission, from Bridle and Vines (2007), copyright Elsevier Limited.

high genetic diversity is thought to increase the likelihood of invasion success, including spread following initial establishment. Roman and Darling (2007) suggested four possible ways that increased genetic diversity can increase invasion success:

(1) an increase in the variety of genotypes increases the likelihood that at least some will be pre-adapted to the newly encountered environments;
(2) increased genetic diversity provides more raw material for natural selection, and hence increases the likelihood of adaptation to new local conditions;

(3) the mixing of genotypes from previously allopatric populations might increase fitness through over-dominance effects; and

(4) the mixing of previously allopatric genotypes may produce novel hybrid genotypes that are able to exploit the new environment.

Although it has been widely believed that a major obstacle to the persistence and spread of colonizing species was the low genetic diversity that was thought to characterize founding populations (Allendorf and Lundquist 2003), numerous recent studies have suggested that colonizing populations and species may not be as genetically depauperate as once believed (Wares *et al.* 2005, Roman and Darling 2007; Fig. 4.7). In fact, in some cases, recently introduced species have been found to exhibit greater genetic diversity than their native counterparts (Bossdorf *et al.* 2005). Empirical data and a better understanding of dispersal processes have made it clear that reduced genetic diversity should not be considered an expected attribute of newly established populations (Bossdorf *et al.* 2005, Wares *et al.* 2005, Roman and Darling 2007).

If propagule pressure refers to both the number of propagules in a single colonization event and the number of colonizing events (Lockwood *et al.* 2005), then an increase in propagule pressure involving either or both aspects should be expected to increase genetic diversity. The high genetic diversity of populations of the brown anole, *Anolis sagrei*, introduced into Florida from Cuba is considered to be the result of multiple introductions from different regions of Cuba (Kolbe *et al.* 2004). Multiple introductions are believed to account for similar findings of high genetic diversity that has been documented in many aquatic species, including both vertebrates and invertebrates (Roman and Darling 2007). Even in cases of a single colonizing event, loss of genetic diversity is a process that occurs between generations. As long as the founding population consisted of more than a few individuals, and providing the population rapidly increased in size, little genetic diversity would be lost (Austerlitz *et al.* 1997). The introduction of the marsh frog, *Rana ridibunda*, into England from Hungary involved only 12 individuals; yet due to rapid population growth, the established non-native population exhibited a level of genetic diversity similar to that of the source population (Zeisset and Beebee 2003). Based on a review of the relevant literature, Wares *et al.* (2005) concluded that introduced populations often retain as much as 80% of the genetic variation of the source population, or even more. Even when newly colonized populations do experience a genetic bottleneck, there are ways that species may still be able to persist and spread despite low genetic diversity. These include parthenogenetic and vegetative reproduction (Ren *et al.* 2005, Mergeay *et al.* 2006), and phenotypic plasticity (Geng *et al.* 2007).

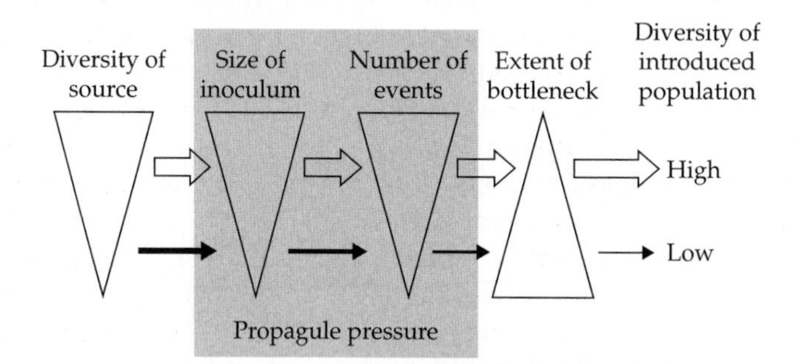

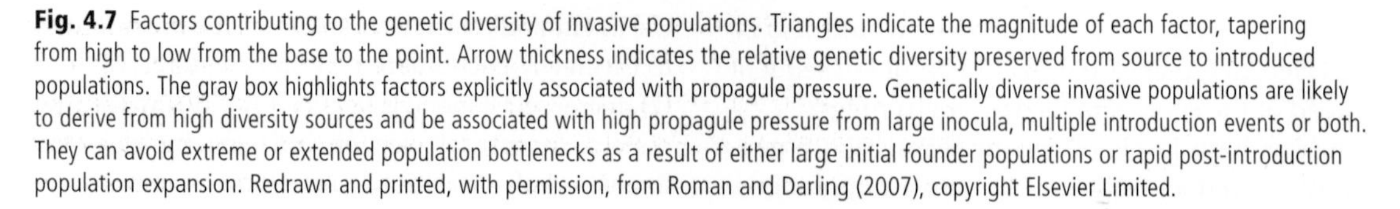
Fig. 4.7 Factors contributing to the genetic diversity of invasive populations. Triangles indicate the magnitude of each factor, tapering from high to low from the base to the point. Arrow thickness indicates the relative genetic diversity preserved from source to introduced populations. The gray box highlights factors explicitly associated with propagule pressure. Genetically diverse invasive populations are likely to derive from high diversity sources and be associated with high propagule pressure from large inocula, multiple introduction events or both. They can avoid extreme or extended population bottlenecks as a result of either large initial founder populations or rapid post-introduction population expansion. Redrawn and printed, with permission, from Roman and Darling (2007), copyright Elsevier Limited.

Opportunities for persistence and spread

Since community processes involving native species, such as succession, are influenced by landscape patterns (Prach and Řehounková 2006), we should expect that the spread of non-native species within their newly colonized region will be greatly affected by a variety of patterns and processes occurring at the land/seascape level. Range expansion is believed to be faster when dispersal is occurring from many smaller foci, than from a single large source site (Pyšek and Hulme 2005). If the non-native species tend to colonize disturbed sites, then the spatio-temporal patterns of disturbance, and any changes in these patterns, will influence spread accordingly (Hobbs and Huenneke 1992, D'Antonio *et al.* 2000, Vermeij 2005). Patch attributes and connectivity among patches will influence landscape spread of non-native species, just as they influence the spread and distribution of native species. For example, edge areas may experience altered levels of resource availability and non-native propagule pressure, particularly for patches with edges abutting human developed areas (Trombulak and Frissell 2000, McDonald and Urban 2006, Searcy *et al.* 2006). Both factors likely account for the fact that areas close to human development are normally more heavily occupied by non-native invasive species than more distant sites (Ohlemuller *et al.* 2006, Theoharides and Dukes 2007).

Since land-use practices affect landscape patterns, changes in land-use practices may affect spread of some invasive species. In California, the pathogen causing sudden oak death, *Phytophthora ramorum*, is believed to have been recently introduced into North America, although the origins are still somewhat uncertain (Martin and Tooley 2003, Ivors *et al.* 2006). The spread of this pathogen is thought to have been facilitated by fire-suppression policies, which have resulted in substantial increases in oak densities in many areas (Condeso and Meentemeyer 2007). In addition, a common alternative host for *P. ramorum*, bay laurel, *Umbellularia california*, is more likely to be infected in areas with continuous oak canopy, and this is also contributing to the spread of the disease among oaks (Condeso and Meentemeyer 2007).

High levels of genetic diversity are likely to favor invasion success in most cases; however, in certain, probably rare, situations, spread may be facilitated by low genetic diversity. For example, in ants, genetic differences between colonies are known to contribute to aggression between colonies, which tends to keep colonies separate and prevent fusion of them into super-colonies (Tsutsui *et al.* 2000). Thus, the low genetic diversity that has characterized the North American colonies of the Argentine ant, *Linepithema humile*, is believed to have permitted the establishment of huge colonies of this species (Tsutsui and Suarez 2003), which increases the inter-specific competitive abilities of the colonies (Holway and Suarez 1999).

Although a low level of genetic diversity is believed to have contributed to the invasiveness of the Argentine ant in North America by reducing aggressiveness between colonies, the invasiveness of the red fire ant, *Solenopsis invicta*, is thought to be due to an increase in genetic diversity following establishment (Tsutsui and Suarez 2003). In both instances, invasion success is believed to be at least partly due to reduced inter-colony aggression, which permits fusion of colonies and the creation of very large colonies that are competitively superior to the smaller colonies of other species. In the case of *S. invicta*, North American super-colonies are believed to be the result of a new allele associated with the general protein-9 (Gp-9) allozyme locus (Tsutsui and Suarez 2003). Allelic differences in this locus have been found to be associated with different colony structures in this species. Specifically, some colonies have a single queen (monogyne colonies), while others have multiple queens (polygyne) resulting in much larger and competitively superior colonies (Krieger and Ross 2002). Monogyne colonies produce queens that are homozygous at the Gp-9 locus (BB), and these queens leave their natal colony and try to found another colony elsewhere. Queens in the polygyne colonies are heterozygous (Bb) and hence can produce queen offspring that are BB, Bb, and bb. As described by Tsutsui and Suarez (2003), the new Bb queens either remain in their natal colony or leave and join another polygyne colony, while the BB queens are killed by the Bb workers if they try to join a polygyne colony (bb queens are reported to be rare). Krieger and Ross (2002) sequenced Gp-9 and found that it codes for a pheromone-binding protein, suggesting that the social structure of the respective colonies may be due to the ability, and inability, of colony members to recognize queens and hence to regulate their numbers.

For introduced species engaged in inter-specific competition, even a harmful event or process can be beneficial providing it harms the competitor more than itself. This phenomenon may contribute to the success of non-native annual plants in the California grasslands, historically dominated by native perennials. Borer *et al.* (2007) developed a model using field-based parameters that showed that the barley and cereal yellow dwarf viruses, which are common and persistent in these grasslands, negatively impact the native perennials more than the non-native annual plants, and thereby negate the competitive superiority otherwise exhibited by the native perennials. The reason for this differential impact is that the viruses are not transmitted inter-generationally, or vertically, e.g. via seeds; rather, the disease is transmitted horizontally, from plant to plant by aphids. This means that while the non-native annuals must be reinfected each year, infected native perennials remain infected from one year to the next, their growth and reproduction compromised. Moreover, the native perennials also serve as long-term reservoirs for the disease. Although serving as a host for the disease themselves, the non-native annuals increase and help perpetuate the pathogen loads of their competitors (Malmstrom *et al.* 2005).

It has been argued that temporally fluctuating resources, while able to permit the initial colonization of individuals into an environment (Davis *et al.* 2000), are not, by themselves, able to account for the continued persistence of the species in the new environment (Shea and Chesson 2002, Melbourne *et al.* 2007). According to this argument, there need to be some additional mechanisms that favor the colonizer, at least in some locations and/or times (Shea and Chesson 2002, Melbourne *et al.* 2007). Proposed mechanisms have included competition–colonization tradeoffs among species (Hastings 1980, Tilman 1994, Chase and Leibold 2003), storage effects (in which benefits acquired by the colonizer during good times are stored to enable it to persist during the bad times, i.e. when conditions favor the residents over the colonist) (Chesson 1994, 2000) and species sorting (in which different species exhibit higher fitness in different patch types) (Leibold *et al.* 2004). All of these proposed mechanisms were, to varying degrees, presented in the context of niche theory. In many respects, the persuasiveness of these arguments is influenced by the extent to which one believes in the value of niche theory as an appropriate paradigm for community assembly.

In fact, persistence and spread of a colonizing species do not require that it be more fit than residents at some place or time; they only require that they not be *less* fit than the residents, or at least not much less fit. Recent theoretical work and empirical findings have suggested that fluctuating resources may be able to contribute to long-term persistence of populations, even sink populations, because intermittent episodes of favorable growth permit immigrant populations to increase substantially during these periods, thereby increasing the likelihood of long-term persistence, despite a negative mean population growth rate (Gonzalez and Holt 2002, Matthews and Gonzalez 2007). This argument is similar to Chesson's storage effect. Thus, fluctuating environmental conditions should often be able to facilitate long-term coexistence. If a new species is ecologically different than any of the native species, then long-term persistence might be facilitated by fluctuating resources, even if the mean fitness of the new species is less than that of native species, as described by Matthews and Gonzalez (2007). If the new species exhibits a high degree of similarity to one or more native species, then long-term persistence could result due to the low displacement pressure exerted by other species (Hubbell 2001). In either case, coexistence may not be permanent, in the way that niche models normally require indefinite coexistence, but coexistence in the real world is never permanent. Coexistence in real life is just sympatry that has gone on long enough for ecologists to term it so.

Species' decline and range contraction

Although less discussed than rapid spread, another common dynamic of the invasion process is a decline in the abundance of the introduced species following a period of rapid spread and dominance (Aladin *et al.* 2002, Simberloff and Gibbons 2004, Reise *et al.* 2006). Since most species that have ever existed have gone extinct, it is reasonable to view both species' spread and species' contraction as inherent parts of the life-history of a species, sometimes referred to as the taxon cycle (Ricklefs and Bermingham 2002). Since spread presumably occurs when typical constraining factors are absent or limited, it is reasonable to believe that spread can occur rather quickly, as is believed to have occurred in many prehistoric, as well as recent, range expansions (Parmesan *et al.* 1999, Leppäkoski and Olenin 2000b). While findings have shown that non-native species can reach higher maximum abundances than native species of the same taxon (Labra *et al.* 2005), the taxon cycle perspective suggests that continual spread and high abundance levels of non-native species is not likely to be a permanent phenomenon. There are numerous examples of novel species and populations experiencing a dramatic decline following an initial period of success (Simberloff and Gibbons 2004). One explanation is that the declines might be due to a tradeoff involving reproduction and stress tolerance (Alpert 2006). According to this hypothesis, if initial invasion success was due to rapid growth and high reproductive output, which in turn were associated with low stress tolerance, then the population would suffer during times of inclement conditions. Alpert (2006) termed this idea the 'reckless invader hypothesis.' Some evidence for this argument comes from invasion experiments with protists, which showed that while establishment success in protists often was a fair predictor for long-term persistence, mismatches, e.g. high establishment success but low persistence success, occurred in a number of experiments (Weatherby 2000, Warren *et al.* 2006).

It is well-documented that after an initial population explosion in the new environment, introduced zebra mussels, *Dreissena polymorpha*, almost always experience a substantial decline in population size, although this decline may not be permanent either (Karatayev *et al.* 2002). The colonization of Lake Erie by zebra mussels in the late 1980s resulted in an initial population explosion of the mussels. However, native duck species, particularly greater and lesser scaup (*Aythya marila, A. affinis*), and bufflehead (*Bucephala albeola*) rapidly adopted *D. polymorpha* as a primary part of their diet, and Lake Erie populations of these duck species increased dramatically following the mussel invasion, from 38,500 waterfowl days for the scaup prior to the mussel introduction to 3.5 million by 1997 (4700 to 67,000 for bufflehead), resulting in mussel consumption equaling between 39 and 46% of the annual mussel biomass (Petrie and Knapton 1999). During this time, mussel abundance dropped by more than 70%, and Petrie and Knapton suggested that predation by these waterfowl was a likely contributor to this decline. In this case, the waterfowl populations responded both functionally (switching to a new prey) and numerically (waterfowl in the region dispersed to mussel-invaded areas of Lake Erie). With respect to plants, Thorpe and Callaway (2006) suggested that while some invasive plant species may benefit from positive feedback interactions with the soil microbial community, this benefit may not be permanent, and that over time, the development of more negative-feedback interactions, e.g. involving pathogens, may reduce the species' success.

Kondoh (2006) offered an explanation to account for the boom–bust cycle observed in many invasion histories based on asymmetric selection pressures operating on the new species and the long-term residents. Kondoh argued that, since the initial population size of the new species is usually going to be small, native individuals will seldom interact with them, whether that be through an intra- or intertrophic level interaction, meaning that the new species will exert little selection pressure on the native species. In contrast, individuals of the new species will likely be frequently interacting with native species, resulting in strong selection pressure on the new species. Under these circumstances, according to Kondoh, the recently introduced species, would adapt faster than the native species, which would enable it to spread and exploit the new environment (the boom phase). However, this 'mismatch in adaptive speed' would not be a permanent state of affairs, since once the new species became super-abundant, the situation would reverse, and the native species would experience stronger selection pressure than the new species, and hence biotic resistance to the new species would increase, resulting in a subsequent decline in the abundance of the new species (Kondoh 2005).

Using Kondoh's notion that the aspects of the contact experience between native and non-native species may contribute to a boom–bust cycle of an introduced species, one could imagine that delayed switching behavior on the part of a native predator to a non-native prey might also produce this effect. Switching, as described by Murdoch (1969), involves a predator switching to a more abundant species of prey. If the switch by a native predator to a new prey species was delayed, then the prey population might be expected to rise substantially before declining due to the heavy, but belated, predation by the native predator.

In some cases, the rise and fall in dominance of a recently introduced species is part of an ongoing series of such dynamics, in which an initially dominating species is replaced by another, which in turn is subsequently replaced by another, and so on. For example, as described by van der Velde *et al.* (2002), in portions of the Rhine River, introduced zebra mussels, *Dreissena polymorpha*, dominated the rocky substrates during the 1970s and 1980s. In 1987, a Ponto-Caspian filter feeder, *Chelicorophium curvispinum*, was introduced causing a severe decline in *D. polymorpha* abundance, and *C. curvispinum* rapidly assumed dominance. During the 1990s, new invertebrates came in and utilized the rocky habitats, sometimes replacing prior introduced species (Haas *et al.* 2002). In a study of a series of multiple-aged old fields, Kulmatiski (2006) found that some older fields remained dominated by non-native species, although the composition of the non-native species changed over time. Although dominance is often a temporary phenomenon, in some instances, a non-native species may contribute to a positive-feedback loop within the ecosystem that enables it to perpetuate its dominance, and no decline is exhibited, at least to date. The century-long domination of *Bromus tectorum* in the Great Basin of North America appears to be a good example of such long-term dominance (Mack 1981).

Predicting current and future ranges of non-native species

The most fundamental question driving community ecology and biogeography is likely why particular species are found where they are, and why they are not found in other places. Motivated by this question, many different types of models have been developed to predict species distributions, both current and future ones. As a group, these models are often referred to as species distribution models (SDMs). As described by Guisan and Thuiller (2005), SDMs are empirical models that relate field observations of distribution to environmental predictor variables using statistical or theoretically-derived response functions. In these models, there are three ways by which the environmental factors are typically believed to influence the species: they can impose physiological or ecological constraints, e.g. involving temperature, humidity, soil type; they can be associated with disturbance regimes (either natural or anthropogenic); and they can influence resource availability. The output of SDMs typically involves habitat suitability maps.

Climate matching or climate envelope or bioclimatic models are one common type of SDMs. In these models, climate data are used as the predictor variables and are associated with the distribution data of the species of interest. There are several limitations to climate envelope models of this type. If the association between the climate and distribution variables is purely correlational, then the models may poorly predict the suitability of environments in other regions or in the future, since the current species distribution may not include the full climatic range it can inhabit (Guisan and Zimmerman 2000, Crozier and Dwyer 2006). This could be because other factors are restricting its native range, such as competition, predation, and dispersal limitation, factors that may not be present in a newly established region. Bradshaw and Lindbladh (2005) found that the spread of *Fagus sylvatica* in southern Scandinavia over the past 4000 years was linked more to anthropogenic activities and fire, rather than changes in climate, suggesting another reason (disturbance dynamics) why strictly climate-based models may not be very successful at predicting the eventual spread of an introduced species (Sax *et al.* 2007).

A species may also not be currently occupying its full climatic tolerance because it is not in equilibrium with respect to climate limits. For example, in an analysis of the ranges of 55 European tree species, Svenning and Skov (2004) concluded that 36 species were not in equilibrium with the climate and were still exhibiting patterns of dispersal limitation since the last glacier. A similar situation may exist for many non-native species, since they have not had sufficient time in their new ranges to sample all the environments (Richardson and Pyšek 2008). Thus, bioclimatic models will be most useful when the actual native ranges of the species of interest are very close to their potential ranges limited by climatic constraints (Crozier and Dwyer 2006). Of course, a problem is that it is often very difficult to know the extent to which this is the case, although ecophysiology models may represent one way to evaluate this issue (Crozier and Dwyer 2006).

In their 2005 review of SDMs, Guisan and Thuiller concluded that most SDMs did not adequately incorporate dispersal, either not including it at all, or including it as non-limited, neither of which remotely describes most natural situations. If SDMs are going to have the chance to be reliable predictors of future changes in species distributions, it is clear that both dispersal dynamics (regional processes) and environmental factors (local processes) need to be included. Storch *et al.* (2006) developed such a model to account for global patterns of avian diversity, incorporating a local factor strongly associated with avian diversity, actual evapotranspiration, and regional dynamics (dispersal). This model was found to predict the actual patterns of avian diversity better than models using either of the factors singly.

Building on the progress made in the application of SDMs to species in general, invasion biologists have developed similar models to describe current patterns of distribution of non-native invasive species, as well as to predict possible future range spread of these species. The latter type of prediction typically characterizes risk analysis efforts, in which the goal is usually either to predict the potential invasive behavior of a species prior to its introduction, or to predict the eventual spread of a species after it has already been introduced into a region (Křivánek and Pyšek 2006). In some instances, the size of the native range of a species has been found to be correlated with the extent of spread by the species in a new region (Daehler and Strong 1993, Croci *et al.* 2007). The commonly proposed explanation for this correlation, when it is found, is that a species would likely have encountered, and presumably adapted to, a wider range of environmental conditions in a larger native range, and that hence it would be similarly tolerant of disparate conditions in the new region, normally meaning that it could spread over a larger area (Sax and Brown 2000).

Studies of birds and marine fish have shown that certain range limits are imposed by the ability of the organisms to tolerate high temperatures (Jiguet *et al.* 2006, Pörtner and Knust 2007), suggesting that increased tolerance of abiotic factors, such as extreme temperature, could facilitate range expansion in a new region. In their study of thermal tolerance in birds, in which they examined the ability of different French bird species to cope with the 2003 summer heat wave, (Jiguet *et al.* 2006) found that thermal flexibility, the ability to tolerate a wide range of temperatures, as measured by the range of temperatures normally experienced throughout its range, was a better predictor of successful coping than the thermal maximum typically experienced by the species. In other words, increased tolerance of this new environmental condition, the heat wave, was associated with greater thermal range, the difference between the maximum and minimum temperatures within the distribution of the species. Based on other analyses, Jiguet *et al.* concluded that the greater thermal tolerance of these species was due to the increased tolerance of individual populations and not to different populations exhibiting diverse tolerances. Although no data were available to account for the greater ability of thermally flexible species to tolerate thermal extremes, the researchers postulated that the individuals of these species may typically encounter a wide range of thermal microenvironments and have evolved to be relatively insensitive to temperature variations.

The study by Jiguet *et al.* (2006) could be said to describe the dynamics of an invasion in time rather than one in space. With respect to the French birds, they did not disperse to a new environment; the new environment came to them. In this case, the patterns of success and failure in this new environment were also explained by tolerance associated with the extent of the native range. However, the findings of the study emphasize that the customary measure of range size, area, may not always be the most meaningful and strongly correlated variable with respect to invasion success, whether in time or space. Although meaningful environmental variables, such as temperature, water availability, soil type, and habitat variety, are likely to correlate with range area, the range of the environmental fluctuations may prove to be better predictors than their surrogate, range area.

Climate matching, or climate envelope, models used by invasion biologists to predict the range expansion of non-native species often include other non-climatic variables, such as elevation and soil or plant type, or salinity for aquatic species. In the literature, these are often called 'niche-based' models (Peterson and Vieglais 2001, Guisan and Thuiller 2005, Chen *et al.* 2007). Once the niche has been characterized, the rules or characterization can be overlain on maps or GIS coverages to identify expected suitable regions for colonization and establishment by the non-native species (Peterson and Vieglais 2001). A basic assumption of most of these models is that the relationship between the species and its environment does not change over time and space, i.e. a condition referred to as 'niche conservatism' in the parlance of the niche paradigm. However, given the adaptive capabilities of organisms (Chapter 5), there is good reason to think that this assumption is unrealistic, and hence that predictions based on the assumption of niche conservatism may yield invalid and misleading results.

Using a climate-matching approach, Broennimann *et al.* (2007) found that spotted knapweed, *Centaurea maculosa*, had colonized regions in North America that were quite outside the climate associated with its native range. Because none of the native populations could be found in the climatic core of the North American 'invaded' areas, the researchers concluded that the shift in climate tolerance was real and not just related to a certain subgroup of the native populations. This meant that the shift occurred after introduction, although the explanations for this shift could not be determined. The shift could be due to evolutionary changes within the species, i.e. selection having favored adaptations to new climatic regimes (Urban *et al.* 2007), or to the release of certain biotic constraints that, in its native range, prevented the species from spreading into regions with climate regimes that the species could otherwise tolerate (Broennimann *et al.* 2007). Another possibility is that Europe simply lacks suitable habitats in regions that climatologically correspond to some of the inhabited regions of North America. As Broennimann *et al.* emphasized, their findings, along with the observations by others, that many species have been able to spread into areas with climatic regimes different, even substantially different than those in their native range (Mack 1996, Dietz and Edwards 2006), suggests that climate-matching models may be limited in their ability to predict the ranges of non-native invasive species (Hulme 2003). However, the climate regimes in the areas where *C. maculosa* initially established in North America were within the range of climates that characterize the species' range in Europe, and were predicted by the climate-matching model used by Broennimann *et al.*, meaning that climate-matching models may still be useful in identifying areas that could serve as the initial introduction and establishment sites (Broennimann *et al.* 2007).

The above possibilities notwithstanding, it is also possible that the climate range of a non-native invasive species could be less than the one it currently occupies in its native land. This could be due to new biotic constraints in the 'invaded' range that prevent it from expanding into regions it could otherwise tolerate from a climatic point of view, or it could simply mean that the species is still in its expansion mode. Wilson *et al.* (2007) modeled plant invasions in South Africa and showed that the extent of range spread was not only a function of environmental variables and the traits of the species but was also greatly influenced by the amount of time since the initial introduction, by the spatial distribution of potential suitable habitat, and by the extent to which humans transport the species throughout the landscape.

Summary

From the perspective of individual organisms, the invasion process involves only two fundamental processes—dispersal and establishment. Individuals produced within in the new environment following the initial introduction episode must continue to disperse and establish if the new population is to persist and spread. Thus, persistence and spread can be characterized as emergent properties at the population and metapopulation level, both arising from the two individual-based processes of dispersal and establishment. Following establishment, a species may spread utilizing the same dispersal vector(s) responsible for its initial dispersal event, or subsequent spread may rely on different vectors. While human activity plays a role in the initial dispersal event, subsequent spread may or may not depend on, or be influenced by, humans. In either case, empirical evidence and theory suggests that rate of spread is often influenced less by mean dispersal distances than the rare long-distance dispersal events exhibited by a few individuals.

Although it is difficult to generalize, traits promoting dispersal and reproduction have often been found to be associated with invasiveness, which is hardly surprising. Similarly, one would likely expect that phenotypic plasticity in a species would often facilitate its persistence and spread, and this also has been commonly confirmed, although exceptions are not uncommon. There has been considerable effort to apply species distribution models (SDMs) to invasive species as a way to predict likely future spread. In particular, climate-based SDMs have been a common choice among invasion biologists. To date, most of these efforts have met with limited success. There are many reasons why a species may be confined to an area smaller than its climatic tolerances would predict, including habitat availability and various biotic constraints. In addition, until recently, few SDMs incorporated dispersal dynamics. It seems quite clear that if SDMs are to have any good chance of predicting future spread of non-native invasive species, dispersal dynamics need to be included in the models.

Traditionally, various genetic constraints, e.g. low genetic diversity and genetic swamping, processes that would that impede the production and selection of adaptive genotypes, have been emphasized as obstacle to persistence and spread. While these genetic factors and processes have been found to be an impediment in some cases, they do not appear to be as consistently a barrier as once believed. Many studies have shown that recently established populations are often genetically quite diverse owing to the dynamics of the invasion process. This is an important finding since the nature and extent of evolutionary opportunities that a species may experience during the invasion will be directly influenced by the diversity of the species' gene pool.

CHAPTER 5

Evolution

It is frequently argued that one of the distinguishing features of invasions is that newly introduced species have experienced an evolutionary history separate from that of the long-term residents (Cox 2004). While it is often true that the new species did not interact with the species in its new environment prior to its arrival, it may have shared an evolutionary history with species phylogenetically related to some of the native species, e.g. congeners. Even in instances where the phylogenetic interactions may be new, the newly introduce species may have interacted with ecologically similar species. Thus, while invasions certainly do involve some new evolutionary interactions, it is probably prudent not to automatically assume a high degree of evolutionary novelty in the new interactions. Moreover, even if the evolutionary interactions are novel, they are not novel for long. As soon as a new species arrives in an environment, it begins to impose new selection pressures on the long-term residents, and vice versa. While the two sides may not have had a shared evolutionary history before, they do as soon as the new species arrives (Dietz and Edwards 2006). The recent emphasis on the role of evolution in the invasion process parallels a larger effort to integrate community ecology with evolution (Holt 2005). As pointed out by Johnson and Stinchcombe (2007), not only does the community help set the suite of selection pressures for the evolution of individual species, but the evolution of individual species can shape community-wide properties and processes.

Evolution during dispersal

By selecting for particular traits during dispersal, evolutionary processes begin affecting incoming species even before they arrive in their new environment. Following initial establishment, if subsequent spread of the species, occurs using the same dispersal vector(s) that introduced the species in the first place, then selection during the initial dispersal period will increase the likelihood of successful post-establishment dispersal. For example, if the primary reason a new plant was introduced into a region of the world was because of horticultural preference, then this same preference is likely to promote its spread within the region. On the other hand, traits favored during the initial dispersal process may be detrimental to subsequent dispersal, making these species less likely to spread, at least on their own. A tree introduced for horticultural purposes because it did not produce fruits would obviously lack the ability to spread on its own.

Evolution by the new species following arrival

A growing number of studies have shown that non-native species are able to evolve rapidly following introduction into a new region (Lee 2002, Maron *et al.* 2004, Callaway *et al.* 2005). For example, Phillips *et al.* (2006) argued that the increased rate of spread of cane toads, *Bufo marinus*, in Australia is due to an evolved increase in dispersal speed by the toad since their introduction more than seventy years ago. This is an interesting case in which the increase in leg length, which is associated with increased dispersal ability, may not actually be an adaptive response but simply the result of spatially-structured selection. Specifically, mating and reproduction occurs during the dispersal process, with those at the invasion front mating with one another, and the ones at the rear likewise mating with one another. Since those at the invasion front are naturally the fastest dispersers, there is strong directional selection for longer legs and fast dispersal among these toads. The toads in older populations exhibit shorter legs than those in newly established populations, suggesting that selection within a population is actually against dispersal, possibly due to tradeoff associated with dispersal, or the simple fact that the longer-legged toads are more likely to disperse from the established population, leaving the shorter-legged individuals to mate with one another. A similar argument based on population age was made for mean differences in dispersal abilities among populations of a North American native beetle (*Tetraopes tetraophthalmus*), which inhabits patches of milkweed (*Asclepias syriaca*), i.e. older patches contained beetles with low dispersal abilities compared to beetles in younger patches (Davis 1986).

Dispersal traits in plants have also been found to evolve following introduction to a new region. In a study of *Crepis sancta*, an introduced weed common in urban areas in France, Cheptou *et al.* (2008) found that plants produced two different types of seeds: a light wind-dispersed seed, and a heavier type, for which dispersal is much more limited. Plants growing in the pavement-dense urban environments produced a significantly higher proportion of heavier seeds than did plants growing in rural environments. Since the heavier seeds were more likely to remain in the parent habitat patch, the researchers concluded that there was a greater fitness benefit in the urban environment to produce short-distance dispersing seeds than in the rural environments. Whereas suitable habitat was likely available at variable distances in the rural environment, this was not the case in the urban setting, where pavement was the dominant substrate. Thus, urban plants that produced more heavy seeds achieved greater fitness. Based on genetic analyses, the researchers estimated that the observed evolution of increased production of heavy seeds occurred within five to twelve generations (Cheptou *et al.* 2008). The same pattern, the evolution of reduced dispersal ability, is believed to characterize the populations of many plant species after they have been introduced to islands (Cody and Overton 1996; Fig. 5.1).

The fact that some species are able to respond genetically to climate change (Balanyá *et al.* 2006, Parmesan 2006, Maron *et al.* 2007) suggests that climate matching may not be as critical for successful establishment and spread as often believed. Williamson (1996) and Storch *et al.* (2006) argued that climate matching, or climate-envelope models, have met with only moderate success, and Sax *et al.* (2007) concluded that climate alone is simply inadequate to predict the distribution of introduced species in a new region. The ability of new species to adapt to a new climate is likely one explanation for the fact that many new species are found in regions with climates quite different from that of their sites of origin. Similarly, the ability of aquatic organisms to adapt to different water conditions likely facilitates their ability to spread into new environments. For example, the grayling, *Thymallus thymallus*, was introduced into several lakes in Norway and within a century the species evolved into different ecotypes, with the type in each lake most adapted to the water temperature of their own lake (Koskinen *et al.* 2002).

In some cases, evolution in introduced species is due to interactions with other non-native species. By analyzing the levels of furanocoumarin toxins of herbarium specimens of North American wild parsnip, *Pastinaca sativa*, Zangerl and Berenbaum (2005) determined that the parsnip, which had been introduced into North America in the early 1600s, had increased its furanocoumarin levels within 20 years of the introduction of the parsnip webworm, *Depressaria pastinacella*, 250 years later. The degree to which the toxin levels increased was associated with the intensity of the interaction with the webworms, with *Pastinca* populations experiencing low levels of herbivory exhibiting less of an increase than those experiencing high levels of interaction with the herbivore. This resulted in a spatially patchy evolutionary response, supporting the geographic mosaic hypothesis of coevolution (Thompson 1994, 1999, 2005), which is characterized by 'hotspots,' where selection intensity is high, and 'coldspots,' where it is low. In a comparative study involving European populations of both species, Berenbaum and Zangerl (2006) found that the selection intensity of the webworm was uniformly lower, such that distinct hot and coldspots

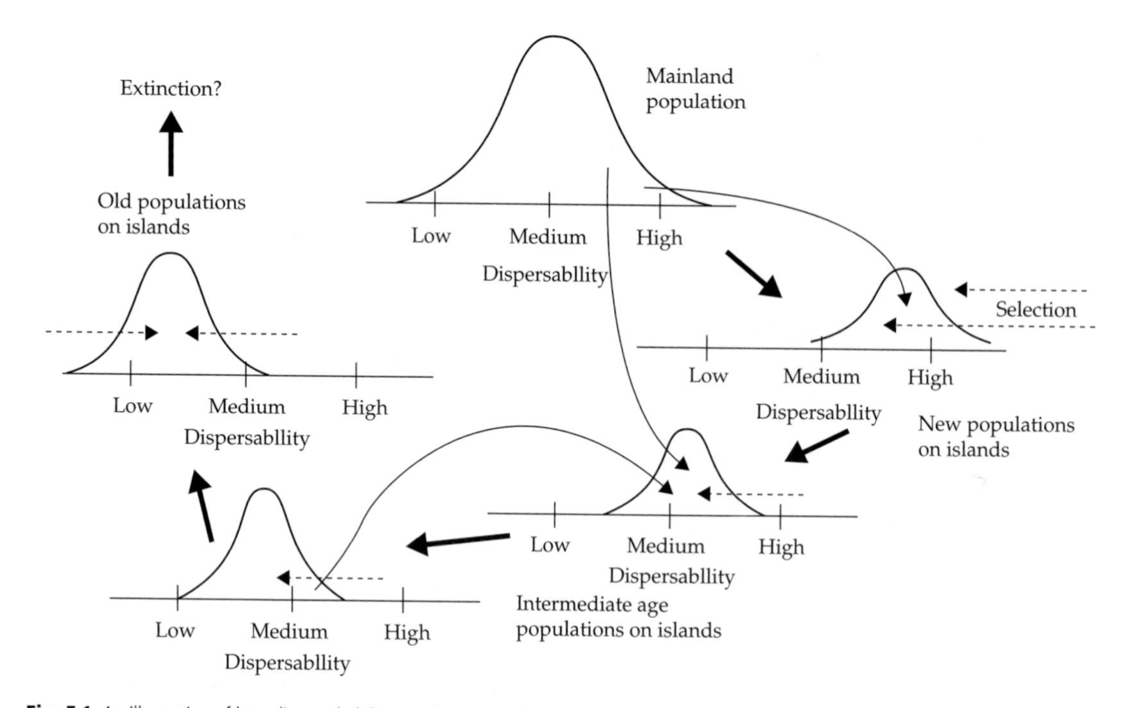

Fig. 5.1 An illustration of how dispersal ability may be expected to decline on island populations. Founders of new island populations are more likely to be quite vagile, which results in new populations exhibiting a high mean dispersal ability. Selection within the island population will be against long-distance dispersal. As long as dispersal to the island is a rare event, mean dispersal ability of individuals in the island population would be expected to decline over time. Redrawn and printed, with permission, from Cody and Overton (1996), copyright Blackwell Publishing.

were less common. This was believed to be due to the presence in Europe of an alternative, and preferred, host plant for the webworm, *Heracleum sphondylium*.

A prominent theory, which not only acknowledges the role of evolution during the invasion process, but actually invokes it as a primary driver, is the EICA (evolution of increased competitive ability) hypothesis (Blossey and Nötzold 1995). According to this hypothesis, if it is true that recently introduced species face fewer enemies (which may or may not be the case), then evolution should select for a reallocation of resources away from defense and toward increasing competitive ability. This is a sensible hypothesis, but results of studies that have tried to test the EICA hypothesis have been mixed (Bossdorf *et al.* 2005). In a review of plant studies that had tested for reduced resistance to specialist enemies, Maron and Vilà (2007) reported that approximately two-thirds of the studies had documented reduced resistance, also meaning that one-third did not. In their own study of St. John's wort, *Hypericum perforatum*, Maron and Vilà did not find evidence for reduced resistance to specialists. They also found considerable geographic variation in the level of defense (levels of defensive chemicals) exhibited by plants, presumably reflecting different selection pressures in different regions of their new land. Maron and Vilà (2007) concluded that the view of plants as being well-defended in their native lands and poorly defended in their new territory is an overly simplistic one.

In plants, increased competitive ability is believe to be often associated with plant size, and thus a number of studies have been conducted to determine if plants grow larger in their non-native range. In a review of EICA studies, Blumenthal and Hufbauer (2007) reported that nine species had been found to grow larger in their introduced range, seven showed no difference in size, and three grew larger in their native range. In their

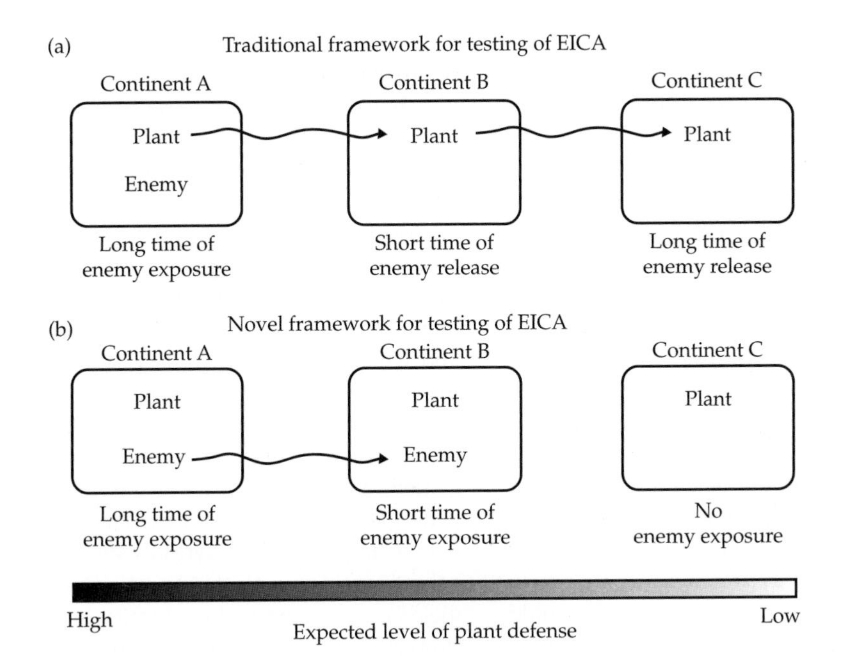

Fig. 5.2 The traditional testing of the evolution of increased competitive ability (EICA) hypothesis involves comparing individuals that have escaped their enemies for different periods of time (a). Another approach, proposed by Handley *et al.* (2008), involves a system in which the enemy is the traveler, following its host, either through accidental or intentional (e.g. biocontrol) introductions (b). Redrawn and printed, with permission, from Handley *et al.* (2008), copyright Ecological Society of America.

own study, Blumenthal and Hufbauer found that individuals from the non-native range tended to grow larger than those in the native range but that this only occurred in the absence of competition. If increased size is exhibited only under certain environmental conditions, then this may explain some of the reasons behind the mixed findings of other EICA studies. Dietz and Edwards (2006) argued that one reason for the mixed results is that researchers have not adequately recognized that the primary drivers of an invasion are likely to change with time, with evolution perhaps not being as important in the initial periods of the invasion, when fortuitous pre-adaptations may be more important. In any case, according to Dietz and Edwards (2006), only four plant species (*Lythrum salicaria, Sapium sebiferum, Silene latifolia,* and *Solidago gigantea*) have been found to exhibit the two conditions required to clearly support the EICA hypothesis, i.e. increased susceptibility to pests (reduced defenses) and increased competitive ability among individuals in the new region.

Taking a different approach to studying the EICA hypothesis, Handley *et al.* (2008) compared plant populations with a long history of enemy exposure with those that had only recently encountered the enemy and with those that had never encountered the enemy. Specifically, they studied populations of *Senecio vulgaris* in its native region (Australia), and in Europe and North America. In Australia, the plants have a long history of exposure to the rust fungus *Puccinia lagenophorae*. In Europe, the plants were believed to have been free of the fungus until approximately 50 years ago, and the fungus had not yet been introduced into North America at the time of the study. Thus, the authors predicted that the North American plants would exhibit the lowest level of plant defense, and in turn an increase in fitness, with the Australian plants exhibiting the highest level of defense and the lowest level in fitness (Fig. 5.2). However, contrary to their predictions, the researchers found no evidence for decreased levels of resistance to the fungus with reduced rust exposure time, nor an associated increase in fecundity, despite the fact that resistance appears to incur a significant fitness cost (Handley *et al.* 2008). The authors suggested that perhaps resistance to the rust is correlated with defense against other enemies, which are present where *P. lagenophorae* is absent.

In what might initially seem to be a counter-intuitive proposal, Bossdorf *et al.* (2004) hypothesized that the evolution of *reduced* competitive ability (ERCA) might benefit a non-native species under certain conditions. This hypothesis was proposed to help account for the very high densities exhibited by many North American garlic mustard, *Alliaria petiolata*, populations. Bossdorf *et al.* argued that for species that encountered greater intra-specific than inter-specific competition, selection might favor reduced competition, if high competitive ability incurred a significant cost. In this case, the less competitive individuals would be able to allocate more energy to other activities, such as reproduction.

Hybridization

In some instances, increased invasive potential comes not via genetic adaptation of the arriving genotype but from the creation of a new genotype through hybridization with a long-term resident species. In a review of the literature, Ellstrand and Schierenbeck (2000) documented 28 cases in which such hybridization appeared to have contributed to the invasiveness of a non-native population. In plants, such hybridization often results in a new species via allopolyploidy (Ellstrand Schierenbeck 2000, Mallet 2007). Significantly, not only is a new species produced through this process (since allopolyploids are typically reproductively incompatible with either parental genotype), but the species can possess a substantially novel genotype. This presents the possibility of a fortuitous match between the new genotype and its environment, potentially resulting in a species with substantial invasive potential. Two well-documented recent cases of allopolyploidy resulting in a new invasive species in Great Britain are *Senecio cambrensis* and *Spartina anglica*, two plant species that have successfully spread from their original sites in Wales and England (Ainouche *et al.* 2004, Abbott *et al.* 2005). *S. cambrensis* resulted from the hybridization of the native *S. vulgaris* with the introduced *S. squalidus*, while *S. anglica* is a tetraploid species derived from the hybrid of the European native cordgrass *Spartina maritima* and the introduced *Spartina alterniflora*. In San Francisco Bay, the introduced *S. alterniflora* has hybridized with a native cordgrass, *S. foliosa*, producing highly invasive hybrid plants that exhibit growth rates, fecundity, and environmental tolerances that exceed that of both parents, and which are believed to have contributed to its invasiveness (Davis *et al.* 2004).

Raphanus sativus (the invasive California wild radish) has been determined to be a diploid hybrid of the cultivated radish and the introduced congener, *Raphanus raphanistrum* (Hegde *et al.* 2006). It is believed that the hybrid's success in rapidly spreading is due to the novel genotype, which yields a new combination of traits, including early flowering and roots that, unlike the cultivated species, are not swollen and not as sensitive to disease and injury as are the swollen roots (Hegde *et al.* 2006). In cases where an introduced species hybridizes with a crop plant, it is possible that transgenes could become introduced into the gene pool of the non-native species (Jørgensen *et al.* 1998, Cox 2004). New genotypes can also be created through the hybridization of multiple non-native species (Gaskin and Schaal 2002, Cox 2004).

Hybridization of between an introduced and a native species threatens the genetic uniqueness of the native species through genetic introgression. Historically, ecologists have characterized this phenomenon as undesirable, emphasizing its homogenizing impact (Daehler and Carino 2001), often referring to the phenomenon as 'genetic swamping' (Olden *et al.* 2004), or, more pejoratively, as 'genetic pollution' (Daniels and Sheil 1999, Potts *et al.* 2003). In the case of the hybridization between *Spartina foliosa and S. alterniflora* (described above) *S. foliosa* is being threatened, not only by genetic dilution (back crosses between the hybrid and the native species are common) but by the higher reproductive success of the hybrids, which exhibit increased flower production and seed set and which have nearly extirpated the native parental genotype from San Francisco Bay's tidal marshes (Ayres *et al.* 2008).

The genotypes of native freshwater fish are believed to be particularly threatened by hybridization with introduced fish and the subsequent genetic introgression (Rosenfield *et al.* 2004). Hybridization between native and non-native species has also been documented in other freshwater animals, including crayfish (Perry *et al.* 2001) and cladocerans (Taylor and Hebert 1993). In a case some may view as justice being served, recent surveys of California wild radish populations have only found the invasive hybrid, *R. sativus*, leading to the conclusion that the hybrid has caused the extinction of the introduced parental type, *R. raphanistrum* through genetic dilution and subsumption (Hedge *et al.* 2006).

Epigenetic effects

If we did not already have enough on our evolutionary plate to consider, the possibility that invasiveness might be affected by epigenetic effects (Schierenbeck and Aïnouche 2006) gives us an additional phenomenon to ponder. Epigenetic effects are heritable phenotypic changes that do not involve changes in the DNA sequence, but may involve changes in gene expression, including the suppression or silencing of particular genes (Fig. 5.3). This could provide an alternative way for a species to adapt to a new environment, i.e. as opposed to selection for a different genotype. Epigenetic changes also have the potential advantage of being more easily reversible. Ecological research in this area is just beginning (Bossdorf *et al.* 2008), but as Schierenbek and Aïnouche (2006) emphasize, it will be important to examine changes in gene expression when looking for changes in invasiveness since epigenetic changes will not show up in gene sequence data.

Evolution by the long-term residents

Evolution is a two-sided affair when it comes to the new species and the long-term residents. While evolution of the new species may increase its ability to spread, it is also reasonable to expect that evolution by the long-term residents in response to the new selection pressures imposed by the new species may decrease the invasibility of the environment, making life more difficult for the new species (Carroll and Dingle 1996). Hubbell (2001) referred to non-native species that are able to break free from historical ecological restraints as 'rule breakers.' However, he also pointed out that, since the new species impose new fitness criteria, evolution will eventually even the playing field. Schlaepfer *et al.* (2005) made this same observation, emphasizing that native species are not necessarily permanent losers in a face-off with a recently introduced

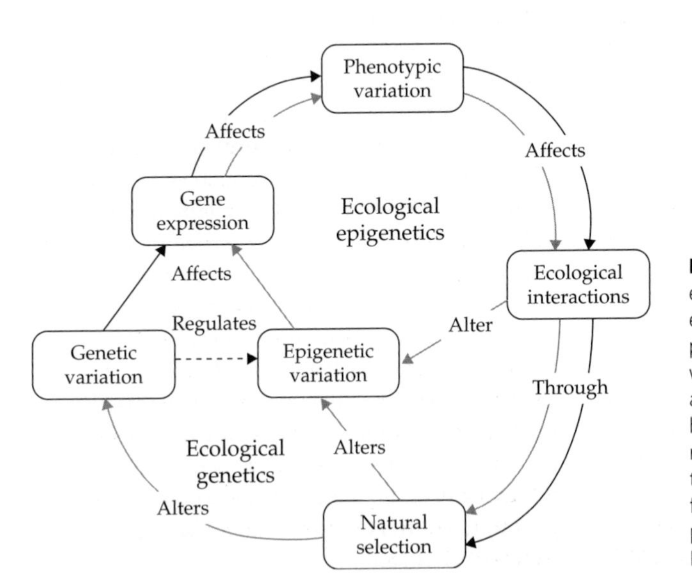

Fig. 5.3 Differences and similarities between ecological genetics (black arrows) and ecological epigenetics (grey arrows). On one hand, epigenetic processes may provide a second inheritance system, very similar to the genetic inheritance system, that allows evolution by natural selection. On the other hand, epigenetic variation, unlike genetic variation, may be altered directly by ecological interactions and therefore provide an additional, accelerated pathway for evolutionary change. Redrawn and printed, with permission, from Bossdorf *et al.* (2008), copyright Blackwell Publishing.

invasive species. Providing the native species have sufficient genetic variation, and the negative impact by the new species is not so extreme as to result in the extirpation of the natives before evolution can occur, one would expect an evolutionary and adaptive response by the native species. This could include changes such as an increase in competitive ability, increased resistance to a pathogen, or increased predator or herbivore defenses. For example, mussels in New England have been found to grow thicker shells in response to the establishment of the introduced predatory Asian shore crab, *Hemigrapsus sanguineus* (Freeman and Byers 2006). Experiments by Freeman and Byers showed that the mussels grew thicker shells simply in response to waterborne cues and that this evolutionary response occurred very rapidly, in less than 15 years following the introduction of the crab. Kiesecker and Blaustein (1997) found evidence that the introduction of bull frogs in California promoted adaptive changes in the predator detection and avoidance of abilities of red-legged frogs, *Rana aurora*, which were determined to be at least partly genetically-based.

Strayer and Malcom (2007) emphasized that long-term effects on native species by introduced species may be quite different than the short-term impacts following introduction. Strayor and Malcom based their argument on data showing that the pronounced decline in the 1990s of native bivalves in the Hudson River estuary, New York, which occurred following the introduction of zebra mussels, was not permanent. Between 2000 and 2005, populations of all native bivalve species increased and recruitment levels increased to pre-invasion levels. The reason for the population rebound of the native bivalves is not known, but Strayor and Malcom were able to rule out several hypotheses. No significant changes in filtration or fouling rates by the zebra mussels were documented; phytoplankton abundances and composition changed little after 2000; and there did not seem to be any spatial refuges available that could account for the reversal in the fortunes of the native bivalves (Strayor and Malcom 2007). Strayor and Malcolm raised the possibility that very strong selection pressures (annual loss rates were estimated to be between 19 and 57%) could have favored particular genotypes that were somehow more resistant to the effects of the zebra mussels. However, they lacked the data to evaluate this hypothesis.

The fact that evolution can begin to have an impact on the new host species immediately upon arrival emphasizes the dynamic nature of the relationship between the new arrivals and the long-term residents. Like most everything else in the natural world, this relationship is best characterized as one of flux. Current aspects of the relationship cannot be considered as static and enduring. Selection on vulnerable prey will favor the evolution of better defenses; selection should increase the competitive ability of poor competitors (and/or perhaps increase their resistance). Of course, evolution is an inter-generational process, and hence takes time. If a new species causes the extinction of a long-term resident species very quickly, or if long-term residents cause the extinction of a small founding population, there will not be time for evolutionary accommodations to occur. While natural selection may permit a long-term resident species to adapt to a newly introduced competitor, it is clear that selection is sometimes not powerful and/or fast enough to protect species from extinction due to introduced predators and pathogens (King 1984, Kaufman 1992, Fritts and Rodda 1998).

Evolutionary diversification

Although non-native species are typically characterized as having a negative impact on native species, some ecologists have emphasized that non-native species bring, not just threats to the native species, but opportunities for evolutionary diversification as well (Vellend *et al.* 2007). For example, although humans may lament the loss of historical native genotypes through hybridization with recently arrived species, from an evolutionary perspective, the creation of new genotypes from hybridization can increase species diversification (Roman and Darling 2007, Vellend *et al.* 2007, Fig. 5.4). That hybridization often can yield benefits, including hybridization involving non-native species, has long been common knowledge, as illustrated by the following observation by Ralph Waldo Emerson, 'A nation, like a tree, does not thrive well unless it is engrafted with a foreign stock.'

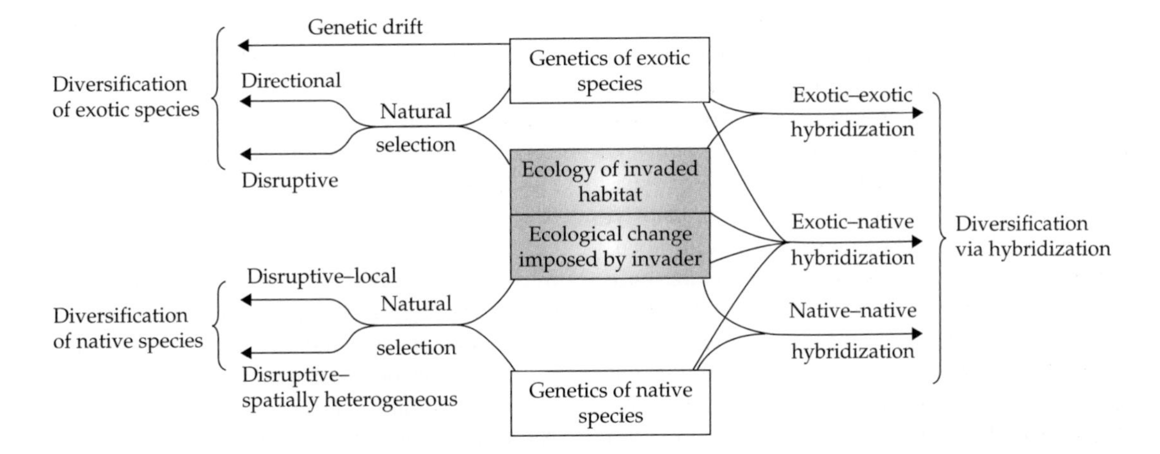

Fig. 5.4 Conceptual framework for understanding effects of exotic species invasions on evolutionary diversification. Characteristics of the genetics of exotic and native species (top and bottom boxes) and the ecology of their shared habitats (center boxes) interact via genetic drift, natural selection and hybridization to result in evolutionary diversification of exotic species, native species, or hybrid combinations of one or both of the exotic and native species. Redrawn and printed, with permission, from Vellend *et al.* (2007), copyright Elsevier Limited.

In addition to hybridizations between native and introduced species, hybridization may occur between non-native species that, prior to their introductions, had been allopatric in their distribution. Until recently, hybridization (other than polyploid hybrid speciation) was viewed primarily as a phenomenon that tended to homogenize gene pools, and hence was more likely to reduce, than increase, biodiversity (Mayr 1963). The field is actively reconsidering this perspective, and hybridization, including homoploid hybridization, is now being proposed as a potentially important mechanism for speciation (Howarth and Baum 2005, Mallet 2007), particularly when assortative mating is involved. (Mavárez *et al.* 2006). Rosenzweig (2001) also emphasized the evolutionary diversification opportunities associated with species introductions, arguing that, in the long run, the global mixing of species should not be expected to lead to a decline in global species diversity due to subsequent radiation and diversification.

Recent studies have suggested that evolutionary change may be more likely to occur in species-rich than species-poor communities (Emerson and Kolm 2005), and that different selection regimes among environments, such as presence or absence of a particular predator, can fuel divergence and speciation (Langerhans *et al.* 2007, Meyer and Kassen 2007), perspectives that support the notion that introduced species can provide evolutionary opportunities to the host community. Tella and Carrete (2008) argued that, although non-native parasites have caused some extinctions, novel parasites and pathogens may promote evolutionary diversification by eliciting an increase in genetic variation among their hosts. All these arguments and perspectives are consistent with Vermeij's (2005) conclusion that the fossil record generally shows that following the invasion of new species, the number of species resulting from adaptive radiations and evolutionary diversification exceeds the number of extinctions.

Due to the fact that most speciation events take place over longer time periods than the few decades of recent scientific study of non-native species and their impacts, there is little empirical evidence of full speciation events involving native species that have occurred as a result of the introduction of non-native species into their community. Nevertheless, there are examples of phenotypic diversification that have taken place. Some of the best examples involve phytophagous insects, many of which have evolved genetically novel strains that feed on one or more introduced non-native species (Strauss *et al.* 2006b). For example, the native apple maggot fly, *Rhagoletis pomonella*, which originally fed on native hawthorns (*Crataegus* spp.) has evolved an ecotype that feeds on the introduced apple trees (Filchak *et al.* 2000). Likewise, the Colorado potato beetle, *Leptinotarsa decemlineata*, which originally fed on native Solanaceous plants, now exhibits genetically distinct populations that feed on introduced potatoes and tomatoes (Horton *et al.* 1988). Similar diet shifts involving genetic change have been documented for other phytophagous insects (Tabashnik 1983, Sheldon and Jones 2001). Given that there is often very limited gene flow among strains feeding on different hosts (Filchak *et al.* 2000, Sheldon and Jones 2001), the evolution of distinct host preferences may represent an early stage of speciation (Vellend *et al.* 2007).

These many examples concerning evolutionary opportunities notwithstanding, it should be pointed out that while some evolutionary processes can take place quite rapidly (Thompson 1998, Hairston *et al.* 2005, Yoshida *et al.* 2007), speciation rates are not about to exceed extinction rates any time soon. Thus, while it is true that the redistribution of the world's biota does provide new opportunities for evolutionary diversification, speciation being just one of them, many negative impacts of non-native invasive species are more much more immediate, and they deservedly warrant our attention and concern.

Summary

One of the difficulties in studying the importance of evolution in the establishment and spread of non-native invasive species is that one never knows how many invasions failed due to a lack of evolutionary potential (genetic variability) (Vellend *et al.* 2005). Nevertheless, it is clear that evolutionary processes influence the invasion process beginning with the dispersal of the first arriving propagules. In many instances, new arrivals are able to adapt to new environmental conditions and biotic interactions through natural selection. At the same time, the new species impose new selection pressures on the native species. In most cases, there is likely ongoing reciprocal selection pressure being imposed by the recently introduced species and the long-term residents, meaning that the relative impacts of each side on the other is almost certainly to change over time. Although it has been more common to emphasize undesirable genetic and evolutionary impacts of recently introduced species, a fair appraisal must also acknowledge that species introductions can enhance diversity as well, through hybridization and the creation of new genotypes. In addition, by imposing new selection pressures, introduced species may provoke new paths of evolutionary change and diversification among the long-term residents.

The players and the conditions in an invasion episode are always changing. Various ecological processes and events continually alter the nature of the playing field, and evolution and phenotypic changes constantly amend the character of the players, both friends and enemies. With groups of continually morphing organisms interacting within a relentlessly changing world, it is no wonder that ecologists have found it a daunting challenge to understand and predict biological invasions. Nevertheless, ecologists have learned much in recent years, and more progress should be possible.

CHAPTER 6

Understanding and predicting invasions: an Integrated Approach

Despite confronting a very complex system, in which history plays a major role, invasion biologists have relentlessly tried to make the field a predictive science. There are likely two reasons for these persistent efforts. First, if the science is going to be able to inform management efforts, the field needs to provide knowledge that involves some reliable cause and effect relationships, i.e. predictability. Second, scientific knowledge and theories are commonly evaluated on their ability to make sound predictions, and thus predictions are needed so the field can utilize empirical data to evaluate competing ideas and hypotheses. One of the first efforts to impart some predictability to the invasion process was the tens rule (Holdgate 1986, Williamson and Brown 1986, Williamson 1996). Developed with British vertebrates, insects, and flowering plants in mind, this rule states that roughly 10% of the species that are introduced and escape cultivation or captivity actually establish, and that approximately 10% of those that establish, spread and become pests. The tens rule rests on the belief that there are statistical aspects to invasions that should yield some predictability to the process. Despite proposing the rule, Williamson (1996) noted that exceptions were not uncommon, and studies conducted since 1996 have indicated that it is unlikely that different organism types will exhibit similar probabilities of of escape, establishment, and of becoming pests (Richardson and Pyšek 2006). For example, a recent analysis of the reciprocal introductions of vertebrates (birds, mammals, and fish) between Europe and North America concluded that establishment and spread rates averaged higher than 50% (Jeschke and Strayer 2005).

Using traits to predict invasiveness

One way to try to predict invasions is to try to identify which traits are likely to contribute to invasive behavior, or at least are associated with invasive behavior. As described in Chapters 3 and 4, many traits have been shown to be associated with invasiveness. Heger and Trepl (2003) emphasized that different traits may enhance different portions of the invasion process, including the initial dispersal event, the establishment of the individual at the new site, the establishment of a viable new population, and subsequent spread. The fact that multiple traits are involved in invasion events means that as conditions vary from event to event, one would expect the relative importance of different traits to vary as well. This fact may explain part of the challenge in making predictions based on traits (Moyle and Marchetti 2006). For example, owing to particular conditions in the environment during one invasion episode, certain traits may be vital to success during the establishment phase. However, the same traits may not be so crucial in another invasion episode due to different conditions during the establishment period.

Trying to predict invasiveness on the basis of traits is an essential goal of risk analysis. Risk analysis is recognized as a crucial part of efforts to prevent introductions of invasive species (Kolar and Lodge 2001, Simberloff 2005, Keller *et al.* 2007), although the extent of enthusiasm for this approach varies (Ruiz and Carlton 2003c). One common type of risk analysis involves identifying traits of species that are indicative of invasive potential. For example, many studies have shown that rapid growth rate characterizes many invasive plant species (Grotkopp *et al.* 2002, Burns 2004). Thus, rapid growth rate, along with other traits associated with high growth rate, e.g. high foliar levels of N and P, may be good indicators of invasiveness in plants (Leishman *et al.* 2007). Based on a review of the literature, Kolar and Lodge (2001) identified vegetative reproduction and low variability in seed crops as additional good indicators of invasiveness in plants. However, while vegetative reproduction may provide benefits once a plant has been introduced into an area, this trait may impede dispersal ability if sexual reproduction (leading to seed dispersal) is compromised due to clonal growth (Pyšek and Richardson 2007). Rejmánek *et al.* (2005b) argued that Grime's functional strategies (Grime *et al.* 1988) could be a powerful tool for predicting eventual ranges and occupied habitats of European plant species. Shipley *et al.* (2006) showed that a model based on the functional traits of plants was highly accurate in predicting changes in relative abundances of the species during succession. They argued that similar models could be used to predict the invasive potential of plant species.

Inherently, risk analysis is based on information obtained from the past and present, e.g. based on generalizations drawn from comparisons of previously documented invasive vs non-invasive species. Risk analysis is, by virtual definition, a probabilistic venture. Like other enterprises based on risk analysis, e.g. insurance companies, analyses of species for invasive risk should be able to achieve some measure of success, providing the analyses are based on sound data. This means that risk analyses should be able to successfully identify many potentially harmful species prior to their introduction, information that is of enormous value. At the same time, the probabilistic nature of this approach means that predictions will not always be the right ones. In addition, as argued above, knowledge of individual traits only provides so much predictive ability as to the behavior of the entire organism. Every species must have been a 'demon' at one point, or else it would never have become established (Silvertown 2005). With respect to invasive species, this is not a new insight. Trying to determine specific characteristics that make some plants weeds, American botanist Asa Gray (1879) concluded that he 'could discern nothing in the plant itself that would give it an advantage.' Continuing, he wrote: 'the reasons for predominance may be almost as diverse as the weeds themselves.'

A common refrain in the law enforcement world is, 'the best indicator of future violence is past violenc.' The business and sports version of this adage is, 'the best indicator of success is prior success.' Consider the following analogous scenario involving two people trying to predict the winner of an upcoming sporting event, e.g. a running race or a golf tournament. The first individual carefully examines and observes each participant before the event. What is the ratio of fat to lean body mass of the respective runners? What are their lung capacities? How far does each golfer hit their drives on the practice range and how accurate is their putting? And so on. Using the empirically collected data, the first individual develops an algorithm or model to predict the likelihood of winning. If the right data are collected, and then analyzed appropriately, it is likely that this approach would be of real value, i.e. producing results with greater accuracy than would be accomplished by selecting the winner at random.

The other individual takes a much less time-consuming approach. This individual simply asks each competitor how many races or golf tournaments they have won in the previous two years, and chooses as the predicted winner the competitor with the most victories. For betting purposes, on whose prediction would the reader choose to rely?

The first individual is handicapped by the fact that the performance of the athlete is more than the sum of the athlete's individual abilities. Each of the athlete's abilities and dispositions function in the context of all the others, interacting in ways likely not fully understood. In fact, many of the relevant abilities may not even be known. No doubt this at least partly accounts for the fact that predicting outcomes based on individual traits has met with only moderate success, whether the field be business, sports, law enforcement, or invasion biology. On the other hand, prior success is an emergent property, the integrated outcome of all the traits and abilities, both those known and unknown. Whether in the physical, social, or biological realm, there is great predictability in momentum. Thus, it should not be surprising that a good predictor of invasiveness has often been the extent to which the species has been invasive in other places where it has been introduced (Kolar and Lodge 2001, Marchetti *et al.* 2004).

As described above, risk analysis often focuses primarily on the traits of a species. However, researchers have also developed more general decision-making strategies to predict the likely threat of invasiveness in a new region, ones that include information beyond organism traits. For example, a weed risk assessment tool, originally developed to screen plants being considered for introduction in Australia and New Zealand (Pheloung *et al.* 1999), and later amended with additional decision analysis and applied in other regions (Daehler and Carino 2000, Daehler *et al.* 2004), consists of questions involving climate range, geographic distribution, the extent to which the species has been cultivated, and the degree to which the species has been invasive elsewhere, as well as questions addressing specific life-history and other plant traits. Using a data set consisting of 180 non-native tree species that have been introduced into the Czech Republic, a subset of which have become invasive, Křivánek and Pyšek (2006) showed these risk assessment tools can identify invasive and non-invasive species with remarkable accuracy, providing the information needed to complete the evaluations is available.

It needs to be remembered that at the root of risk analysis is an understanding of potential harm, and that this understanding, i.e. what is considered to be harmful, is ultimately rooted in social values and is not scientific in nature (Andow 2005). This fact has enormous implications, the most important being that scientists have not been bequeathed the authority to make these decisions on their own. Scientists may have the knowledge and authority to describe the nature and extent of particular ecological impacts, but whether or not these impacts should be considered harmful is a social decision, in which the public needs to participate, and, along with the scientists, ultimately make.

Predicting invasibility

Considerable progress has been made in understanding the large number of factors that contribute to the invasibility of an environment (see Chapter 3). However, understanding what makes an environment invasible is not the same as being able to predict invasibility ahead of time. As previously emphasized, not only does the invasibility of an environment vary in space and time (Davis *et al.* 2000, Stachowicz and Byrnes 2006, Melbourne *et al.* 2007), but it varies from one species, or even one genotype, to another (Davis *et al.* 2005a). Thus, being able to predict invasibility really means being able to predict the spatio-temporal patterns and dynamics of an environment's ecological processes, as well as the suite of traits that the new arrivals will be bringing. It is hardly any wonder that ecologists have found it so difficult to transform the field of invasion biology into a predictive science.

Whittier *et al.* (2008) used calcium concentrations in North American rivers and streams to assess the risk of these environments to invasions by zebra and quagga mussels (*Dreissena* spp.) and, based on current distributions, concluded that calcium concentrations can be an effective broad-scale predictor of invasion risk for these species. Sax *et al.* (2007) suggested a new hypothesis to account for differences in invasibility among environments based on the relative abundance of specialist enemies, e.g. predators and pathogens, and specialist mutualists and facilitators (Fig. 6.1). Specifically, they proposed that in environments where predators and pathogens are more likely to be specialists, while the mutualists and facilitators are more likely to be generalists, invasibility should be high. This would be because the introduced species would be expected to encounter relatively little resistance from enemies, while receiving considerable assistant from the mutualists and facilitators. Conversely, environments with proportionately more specialist mutualists and facilitators and generalist enemies would be less invasible, since resistance would be expected to exceed facilitation. Of course, who are the enemies and who are the mutualists will vary depending on the introduced species. Thus, the same environment could be experienced as quite invasible by some species and quite resistant by others. This would make it difficult, if not impossible, to categorize an environment as inherently more invasible than another, unless one restricted the discussion to ecologically similar species.

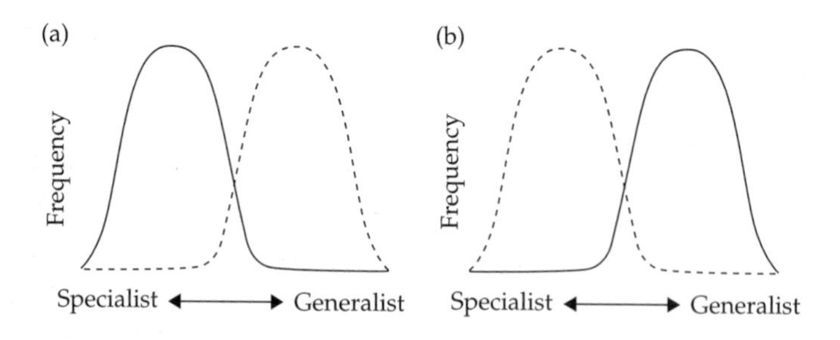

Fig. 6.1 Specialization in species' interactions and invasibility. Individual ecosystems can vary such that the frequency distribution of species along a continuum from absolute specialists to absolute generalists differs between (solid lines) predators and pathogens, and (dashed lines) mutualists and facilitators. (a) In a high invasibility system, predators and pathogens are more frequently specialists, whereas mutualists and facilitators are more frequently generalists. Such systems would be relatively easy to invade because few predators would be able to prey upon exotic species (for which they would not be specialized), whereas many mutualists would be able to assist exotic species. (b) A low invasibility system, with the opposite distribution and invasion outcome. The curves illustrated here are for heuristic purposes only; the actual shape of these curves is unknown empirically. Their impact on invasibility should operate as described here, however, as long as there is a difference in the mode of the two distributions, and as long as the frequency of interactions determines the average outcome of invasions. Redrawn and printed, with permission, from Sax *et al.* (2007), copyright Elsevier Limited.

As described in the previous section, past performance has been shown to be one of the best predictor's of invasiveness. This seems to be the case for invasibility as well. Just as past success represents the integrated effect of all of the traits possessed by a species or population, the extent to which an environment has been 'invaded' in the past represents the integrated consequence of all the environment's ecological conditions and processes. For example, one of the best predictor's of the invasibility of an environment is the species-richness of the native species, e.g. 'the rich get richer' concept (Stohlgren *et al.* 2003). At some point in the past, all the existing native species had to have colonized and established in the environment, and they needed to have been able to persist from that time until today. Although certainly propagule limitation can play a role, sites with low native diversity likely manifest conditions and processes that inhibit the successful establishment of most native species, while species-rich sites must be comparatively more invasible. Since native and non-native species generally are similarly influenced by basic ecological conditions and processes, then past invasibility, for which current native diversity can be regarded as a surrogate, should generally be a good predictor of current and future invasibility with respect to non-native species. In a comparative study of non-native plants in three regions in Europe, Chytrý *et al.* (2008) found that neophytes (plants introduced post-1500) tended to inhabit the same habitats as archaeophytes (introduced pre-1500). The researchers concluded that a good indicator for the susceptibility of a habitat to the introduction of neophytes would be the abundance of archaeophytes in the habitat, i.e. past invasibility should be a good indicator of future invasibility.

Although the invasion process has often appeared to be quite idiosyncratic, there is now in the field a detectable optimism that we should be able to improve our predictive abilities. The general assumption in the field has been that our predictive ability with respect to invasions will increase as we increase our understanding of the various complexities of the invasion process. This certainly seems like sound reasoning. Unfortunately, it may not so easy. In an evaluation of climate models, Roe and Baker (2007) concluded that reducing the uncertainties associated with individual processes will do little to increase the predictive power of the models. According to the authors, this was due to the considerable feedback dynamics of the climate system, particularly positive feedback, in which small uncertainties in the feedbacks can become highly amplified. In a somewhat discouraging final statement in which they concluded that 'we are constrained by the inevitable,' Roe and Baker explained that the more likely a large warming event is, the greater will be the uncertainty of the magnitude of the warming. Since it is abundantly clear that feedbacks of various types are often involved in the invasion process, including positive feedbacks, it is possible invasion biologists may face similar limitations in improving their predictive powers.

Despite this pessimistic perspective, there may be a way to untie this Gordian knot. The reader may be familiar with the following phenomenon: if asked to predict the number of jelly beans in a jar, few individuals will guess close to the correct number; however, the mean of everyone's guess will be quite close to the real number. Described in detail by James Surowiecki in his popular 2004 book, *The wisdom of crowds*, this phenomenon is rooted in mathematics. Essentially, each individual guess includes some information and some error. As described by Surowiecki, by combining the independent guesses of many people (and it is very important that the guesses be independent), the errors of individual guesses tend to cancel each other out, leaving the information. It is believed that the French mathematician P. Laplace first formally described this phenomenon in 1818 (Araújo and New 2007). The practical value of this phenomenon was recognized by Nobel Prize Economist Clive Granger, who, along with a colleague, JM Bates, published a seminal paper in the area of forecasting titled, *The combination of forecasts* (Bates and Granger 1969). Now known as 'ensemble forecasting', or sometimes 'consensus forecasting,' the practice of combining the results of multiple separate and independent forecasts in order to develop a best forecast has been used in many fields, including medicine, economics, management, meteorology, and climatology (as described and cited by Araújo and New 2007). Araújo and New present a compelling argument for the value of applying ensemble forecasting to prediction efforts in ecology. Although they addressed the value of ensemble forecasting for predicting species distributions in general, e.g. in response to climate change, this approach should be seriously considered by those interested in predicting future ranges, or potential ranges, of invasive species. As the authors emphasize, the chances of making poor or spurious forecasts are substantially reduced with an ensemble-forecasting approach.

Traits, invasibility, and propagule pressure: an integrated approach

The 1983 SCOPE scientific advisory committee on biological invasions posed three questions to guide the SCOPE invasion program:

1. What factors determine whether a species will be an invader or not?
2. What are the characteristics of the environment that make it either vulnerable to or resistant to invasions?
3. How can the knowledge gained from answering the first two questions be used to develop effective management strategies?

The scientific advisory committee was smart in delineating a few basic questions because it focused subsequent research. At the same time, as is the case with any paradigm, while providing a structure, the three questions also imposed some limitations on subsequent conceptual development. Characterizing the impacts of traits and environmental conditions separately, the questions precipitated two lines of research that proceeded rather independently: one that tried to identify traits associated with invasiveness, and one that tried to identify environmental factors influencing invasibility. This was unfortunate, since it is now recognized that the two cannot be adequately investigated independently. Traits are understandable as making invasion more likely only in the context of a particular environment (Burns 2006, Pyšek and Richardson 2007), and the invasibility is understandable only in the context of a particular species, or even a particular genotype (Davis *et al.* 2000, 2005a). The same environment, at a particular moment, may be quite invasible to one species and quite resistant to invasion by another. Richardson and Pyšek (2006) explicitly emphasized the interconnectedness of traits and invasibility, referring to species invasiveness and community invasibility as two sides of the same coin. Facon *et al.* (2006) also emphasized that invasion predictions will never succeed if traits of organisms are considered separately from the conditions of the environments of interest. Duncan (1997) found that the invasion success of the same bird species often differed substantially from one region to another, indicating that traits alone are not sufficient in predicting invasion success. Thuiller *et al.* (2006) examined the distribution of non-native invasive plants in South Africa in the context of both traits and environmental factors, and concluded that the distribution and spread of the species is best understood by a combination of life-history traits and environmental factors, as well as human uses of the plant. Specifically, they found that species that had successfully established and spread in certain environments, while typically taxonomically diverse, often shared common traits. For example, non-native species that thrived in warmer areas tended to have small seeds, to be succulent, and to have limited human uses (Thuiller *et al.* 2006).

Understanding that traits and invasibility cannot be studied independently is equivalent to the recognition that extinctions are ultimately due to an interaction between the traits of the species or population and the nature of the extrinsic extinction threats (Fréville *et al.* 2007). Concluding with a sentence that could easily have come from the invasion literature if the word 'extinction' were replaced with invasion, Fréville *et al.* stated, 'From a conservation perspective, our study strengthens the emerging idea that predictions about extinction risk cannot be made on the basis of species' traits alone.'

The integration of the traits and invasibility as a way to understand the invasion process is an important development, but it is not enough. Ultimately, invasion success can only be understood by taking into account propagule pressure as well (Lockwood *et al.* 2005, Rejmánek *et al.* 2005a, Barney and Whitlow 2008). Thus, the combined effects of three factors need to be considered—traits, invasibility of the new environment, and propagule pressure. In their development of a decision-making scheme to predict aquatic invasives, Ricciardi and Rasmussen (1998) came to the same conclusion, emphasizing the importance of considering all three factors—traits of the species, factors associated with the transport, and characteristics of the donor and recipient regions.

The concept of invasion pressure

Atmospheric pressure, often referred to as barometric pressure, describes the force exerted by the mass of the atmosphere over a given unit of area. It is a frequently used metric in the field of meteorology, and isobar maps, showing areas with similar atmospheric pressures, are a common visual in weather reports by the media. A number of factors contribute to changes in atmospheric pressure, including changes in altitude and changes in surface land and ocean temperatures.

By analogy, one could consider a concept of invasion pressure, defined as the probability that an environment will experience an invasion within a specified time period. Low probability would equate with light pressure, while high probability would equate with heavy pressure. As proposed, invasion pressure (*IP*) is not the same as propagule pressure, although propagule pressure would be one of the factors contributing to invasion pressure. In addition to propagule pressure, the invasibility of the environment and the traits of the arriving species would contribute to the invasion pressure of an area. An environment experiencing high invasion pressure could be the result of high levels of both propagule pressure and site invasibility, the latter matched with an organism possessing traits well-suited to the new environment. However, an environment could also experience high invasion pressure in the context of low invasibility, as long as the propagule pressure was exceedingly high. The fact that a high level of dispersal can compensate for low invasibility, even very low invasibility, likely explains why no environment is immune from invasion (Williamson 1996). Thus, invasion pressure is the integration of propagule pressure and invasibility, the latter defined in the context of a particular species or suite of traits (Davis *et al.* 2005a). Since both propagule pressure and invasibility can be quantified, it is possible to calculate, theoretically at least, the invasion pressure of an environment at a particular point in time.

The extent to which a dispersal event leads to establishment is the combined result of the successes and failures of individual propagules. In the end, there are ecological explanations for the success or failure of a propagule to establish in a new environment. Individuals do not die for no reason; they die because something ate them, or because they got burned up in a fire, or drowned, or died of desiccation, disease,and so on. Nevertheless, the success of groups of arriving individuals can be treated statistically (Drake 2004, Leung *et al.* 2004, Drake and Lodge 2006).

If invasibility is defined as the probability of establishment of an arriving propagule (Davis *et al.* 2000), then the invasion success of the dispersal event will be a function of the number of individuals that successfully dispersed to the site and the probability of individual establishment (with establishment of an individual defined as the individual persisting long enough in the new environment to reproduce). If a successful invasion event is defined as the successful establishment, of at least one individual from a single dispersal event, the probability of a successful invasion can be described as a simple function of the invasibility of the environment (the probability of establishment of an arriving propagule) and the number of propagules that arrive (Leung *et al.* 2004):

$$Y = 1-(1-P)^N \tag{6.1}$$

where Y = the probability that at least one individual in a dispersal event will successfully establish, P = the probability of establishment of individual arriving propagules, and N = the number of propagules that arrive at the site in the dispersal event.

Since invasibility only makes sense in the context of a particular species or population (Davis *et al.* 2000, 2005a), the invasibility term, P, can be viewed as an integration of the traits of the dispersers, and the biotic and abiotic conditions of the new environment. Although this simple equation does not include any specific ecological mechanisms other than propagule pressure (defined here as the number of arriving propagules), all three factors that determine the success of a single invasion episode are integrated into it—the traits of the arriving organism, the conditions of the new environment, and the number of arriving propagules. Integrated together in this simple equation, the three factors—propagule pressure, invasibility, and traits—combine to form the invasion pressure (*IP*) of an environment, the probability over a specified period of time that the environment will be successfully 'invaded' by a particular species.

Table 6.1 shows *IP* for a range of values of invasibility and the number of arriving propagules. As shown, success (defined in this figure as the likelihood that at least one propagule successfully establishes) is virtually certain if the number of propagules is at least one order of magnitude greater than the inverse of invasibility. For example, if invasibility (probability of establishment of an individual arriving propagule) equals 0.01, then any propagule number exceeding 1000 would virtually guarantee that at least one of the propagules successfully establishes. Conversely, if the number

Table 6.1 Values of invasion pressure (probability of a successful invasion) for different combined values of the number of propagules in an invasion event (N) and the probability of establishment by an individual propagule (P)—in this case, a successful invasion is defined as the establishment of at least one individual in the invasion event

		P					
		10^{-5}	10^{-4}	10^{-3}	10^{-2}	10^{-1}	10^{0}
N	10^5	0.632	1	1	1	1	1
	10^4	0.095	0.632	1	1	1	1
	10^3	0.010	0.095	0.632	1	1	1
	10^2	0.001	0.010	0.095	0.634	1	1
	10^1	0	0.001	0.010	0.096	0.651	1
	10^0	0	0	0.001	0.010	0.100	1

of propagules is at least three orders of magnitude less than the inverse of invasibility, the likelihood of invasion success, as defined, is very small. Thus, if $P = 0.0001$ and N is less than 100, there is only a 1% chance any individual will establish. If the dispersal pool only contains 10 individuals (and $P = 0.0001$), the likelihood is virtually zero.

Figure 6.2 shows the isobars for different values of invasion pressure, as a function of invasibility and the number of arriving propagules. Figure 6.2 and Table 6.1 show that, based on equation 6.1, invasibility and the number of arriving propagules both play a nearly equal role in determining invasion success; meaning that increasing the number of propagules by a certain factor has nearly the identical effect of reducing invasibility by the same factor. Figure 6.2 also shows the relative values for invasibility and propagule number that result in a 50% chance of invasion success, as currently defined. Specifically, when the number of propagules equals approximately 69% of the inverse of invasibility, there is a 50% chance that at least one arriving propagule will establish. For example, if the probability of establishment of a single propagule is one chance in a thousand, then 690 arriving propagules are required to achieve a 50% success rate (success being defined as at least one individual successfully establishing).

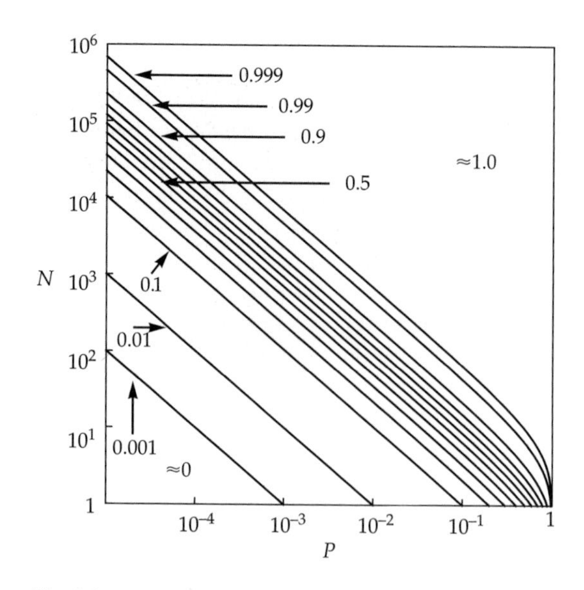

Fig. 6.2 Invasion pressure isobars shown for different combined values of the number of propagules in an invasion event (N) and the probability of establishment by an individual propagule (P).

Figure 6.3 shows a 3-D representation of Fig. 6.2. Perhaps the most important take-home message from Figs 6.2 and 6.3 is that relatively modest changes in either propagule pressure or invasibility, e.g. less than an order of magnitude, can have a very large impact on the likelihood of a successful invasion. Substantial changes in either invasibility or propagule number (even by changes of several orders of magnitude) will make little difference in the lower-left region of the graph, the lowlands in Fig. 6.3 (low invasibility and low number of arriving propagules), where establishment probability is basically zero. Similarly, substantial change in either or both the variables will normally matter little in the upper-right region of the graph, the high plateau (high invasibility and high propagule pressure), where the probability of establishment is virtually certain (Fig. 6.3). However, invasion success is much more sensitive to changes in invasibility and propagule pressure in the cliff area (Fig. 6.3), or the area surrounding the diagonal running from the upper left corner to the lower right corner in Fig. 6.2. This is also illustrated in Fig. 6.4, which shows the sigmoid curve describing invasion pressure as a function of propagule pressure or invasibility when the other variable is held constant. This region of sensitivity, as shown in Figs 6.2 and 6.3, is approximately defined as the area in which the number of propagules and the inverse of invasibility (as currently defined) differ from one another by less than two orders of magnitude (assuming that a successful invasion is defined as the successful establishment of at least one individual).

While the above equation describes the probability of the successful establishment of at least one individual during a single dispersal event of a species or population, it is not difficult to modify the equation using the binomial distribution to predict the probability that at least a given number of individuals successfully establish in any single dispersal event. Figure 6.5 shows the probability isobars if a successful invasion is defined as the establishment of at least 10 individuals. Naturally, the isobars in this adjusted IP map shift up and

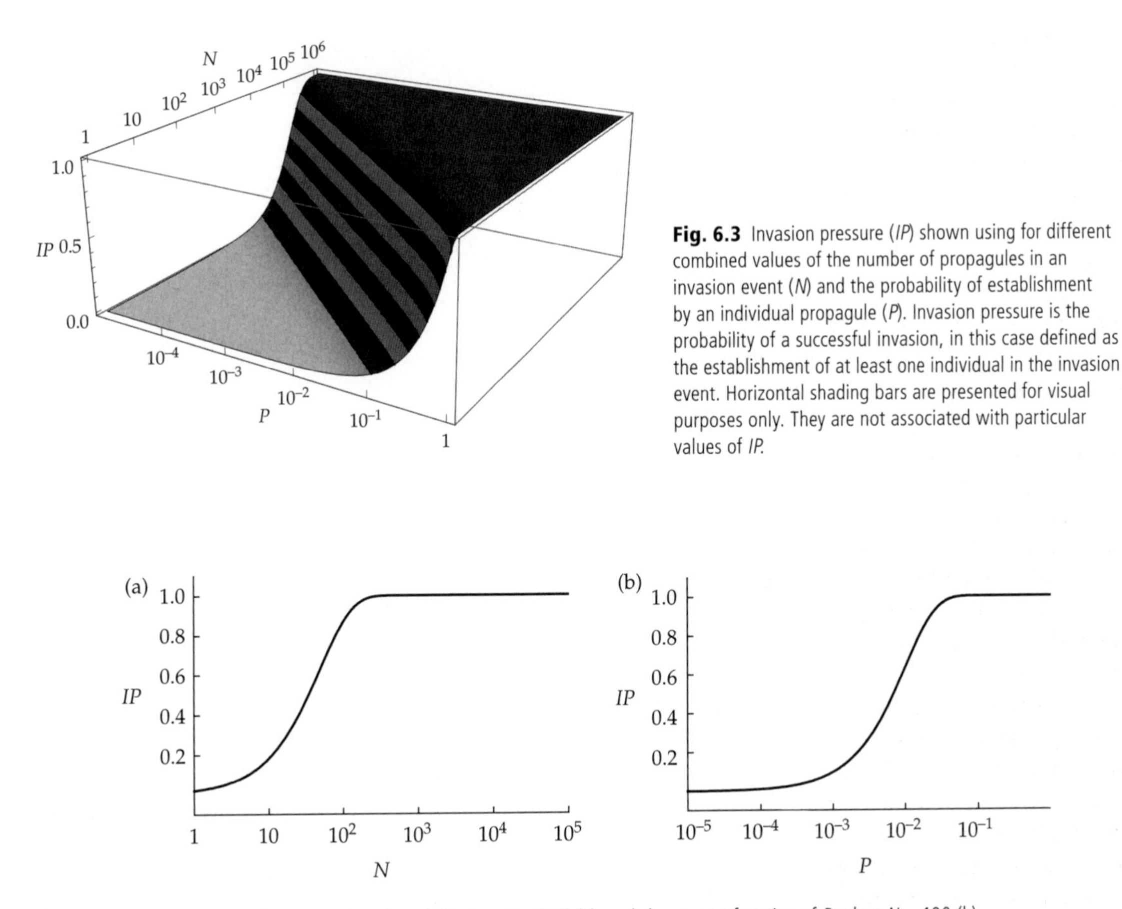

Fig. 6.3 Invasion pressure (*IP*) shown using for different combined values of the number of propagules in an invasion event (*N*) and the probability of establishment by an individual propagule (*P*). Invasion pressure is the probability of a successful invasion, in this case defined as the establishment of at least one individual in the invasion event. Horizontal shading bars are presented for visual purposes only. They are not associated with particular values of *IP*.

Fig. 6.4 Invasion pressure shown as a function of *N* when $P = 0.02$ (a); and shown as a function of *P*, when $N = 100$ (b).

to the right. With a minimum of 10 individuals required for successful establishment, the 50% success isobar occurs when the number of propagules equals approximately 9.7 times that of the inverse of the invasibility. Thus, if the likelihood of establishment for a single individual is 0.001, approximately 9700 would be needed for there to be a 50% chance that 10 or more of the dispersers would successfully establish. What has not changed is the compression of probability values that occurs along the diagonal region extending from the upper-left to the lower-right corners of the graph. In fact, the degree of compression has substantially increased (the cliff is much steeper), as illustrated in Fig. 6.6, a 3-D version of Fig. 6.5. As the number of establishing individuals required for a successful invasion increases, so does the steepness of the invasion cliff. Figure 6.7 shows the *IP* contours when at least 10, 100, and 1000 individuals must establish for the invasion episode to be considered successful.

The *IP* model does assume that the environmental conditions of the environment do not change during the dispersal event, as well as that the environment is homogeneous in space. Of course, neither of these two assumptions is realistic (Davis *et al.* 2000, Stachowicz and Byrnes 2006, Melbourne *et al.* 2007). However, it is not difficult to accommodate changing environmental conditions, or patchy invasibility, with this equation. One would simply consider the dispersers arriving to a different set of environmental conditions as part of a separate dispersal event, since each disperser now would

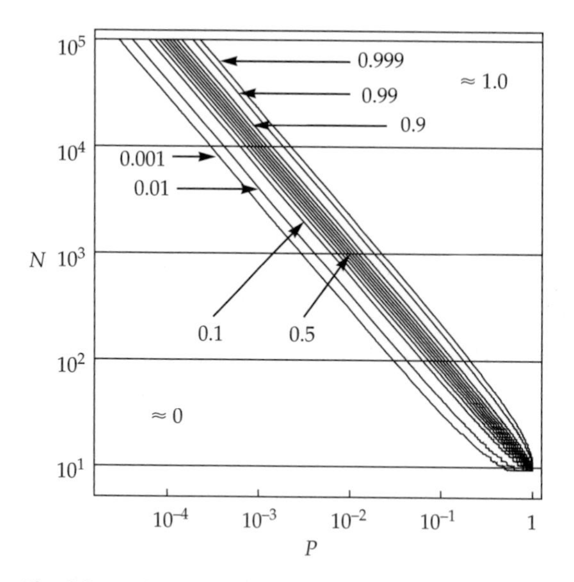

Fig. 6.5 Invasion pressure isobars shown for different combined values of the number of propagules in an invasion event (N) and the probability of establishment by an individual propagule (P), when a successful invasion is defined as the establishment of at least 10 individuals in the invasion event.

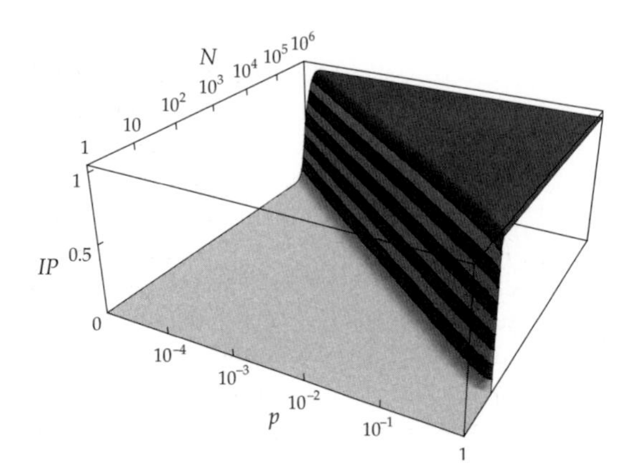

Fig. 6.6 A 3-D representation of Fig. 6.5.

be confronting a different probability of establishment success. The equation also assumes that all individuals in the arriving group share the same traits. Again, if one wanted to distinguish between individuals with different traits, one would simply consider each subgroup a separate dispersal

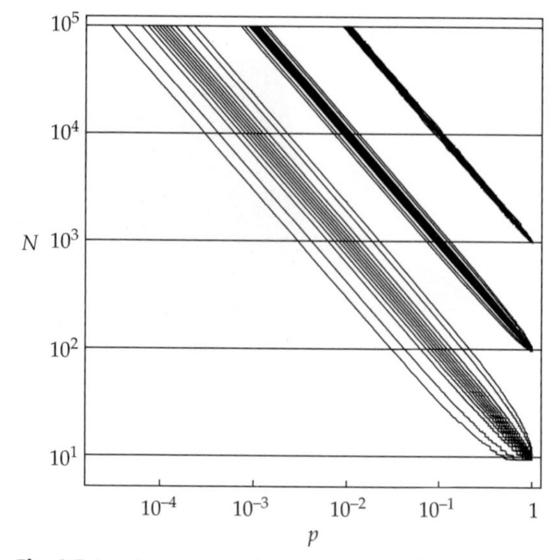

Fig. 6.7 Invasion pressure isobars when a successful invasion is defined as the establishment of at least 10 individuals (lower isobar set), at least 100 individuals (middle isobar set), and at least 1000 individuals (upper isobar set).

event. In any case, once one decides on the temporal and spatial boundaries of the dispersal event and the composition of traits of the dispersers, the combined effect of number of arriving propagules and integrated invasibility (a combination of the species' traits and the conditions of the new environment) is very predictable.

The *IP* maps shown in the preceding pages likely aptly characterize invasion episodes of small organisms, which are easily dispersed, or organisms that produce large numbers of dispersal-capable offspring. Invasion episodes of these organisms, e.g. most insects, microbes, plants, and many aquatic organisms, can easily involve propagule numbers that can vary by many orders of magnitude. In the case of animals that do not typically produce very large numbers of offspring, e.g. birds, lizards, amphibians, snakes, and mammals, a single invasion episode may often involve only a few individuals and numbers are seldom likely to exceed a thousand for a single invasion episode. An example of an *IP* map that might be more appropriate for these organisms is shown in Fig. 6.8. As shown, the zone of highly compressed isobars, the *IP* cliff, is also present with linear axes. Thus, for these

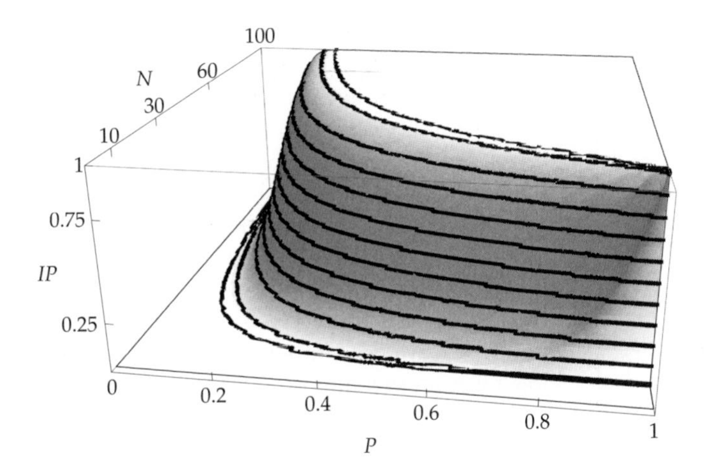

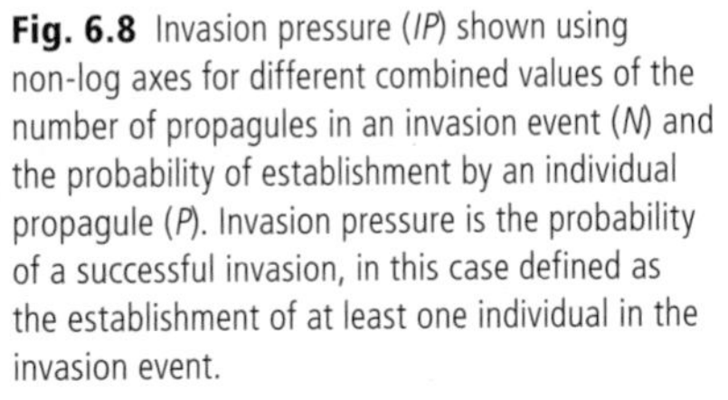
Fig. 6.8 Invasion pressure (*IP*) shown using non-log axes for different combined values of the number of propagules in an invasion event (*N*) and the probability of establishment by an individual propagule (*P*). Invasion pressure is the probability of a successful invasion, in this case defined as the establishment of at least one individual in the invasion event.

species as well, there are certain locations in the *IP* landscape where modest changes in invasibility and/or propagule number will be expected to have a significant impact on invasion success, while in other locations in the *IP* landscape, variations in the variables of similar magnitudes would not be expected to make much of a difference.

Implications of the invasion cliff

As shown above, whatever the number of establishing individuals required for a successful invasion event, one would expect that comparatively small changes in either invasibility or propagule number, or both, have the potential of substantially affecting the probability of a successful establishment. Similar, as well as more complex, probabilistic invasion models have documented this same non-linear behavior (e.g. Drake 2004, Leung *et al.* 2004, Drake and Lodge 2006), which is consistent with the view that certain critical thresholds must be exceeded before invasive spread can occur (Henderson *et al.* 2006). While establishment success or failure for each arriving propagule is ultimately due to particular biological and ecological mechanisms, the fact that a predictable and non-linear function can describe the likelihood of success for an entire invasion event has important implications for our understanding of invasions. For example, the compression of isobars that occurs in the diagonal regions of the *IP* maps helps to explains why it has been so difficult to predict invasions, e.g. why species introductions occur at one place and not another, and why introductions and spread may happen during a particular year and not during prior ones.

Even if propagule pressure does not change from one year to the next, a temporal change in the trait–invasibility complex could shift the system from a point where successful establishment is very unlikely to a point where it would even be expected. For example, if the likelihood of establishment per arriving propagule is 0.001 and the number of propagules is 200, then the probability that at least one of the arriving propagules will successfully establish is approximately 18%. If the propagule number remained at 200, but the establishment probability per propagule increased to 0.005, the probability of establishment by at least one individual increases to 63% (see *A*→*B* in Fig. 6.9). Invasibility could be increased by a variety of factors, such as increased resource availability due to a disturbance (Davis *et al.* 2000), introduction or increase in abundance of an important mutualist (Richardson *et al.* 2000b), and/or the fact that the arriving individuals possessed different traits than those exhibited by prior dispersers, better enabling them to overcome whatever biotic or abiotic resistance had been preventing the establishment of prior propagules. If invasibility does not change, an increase in propagule pressure could likewise shift the system from a state in which successful establishment is unlikely to

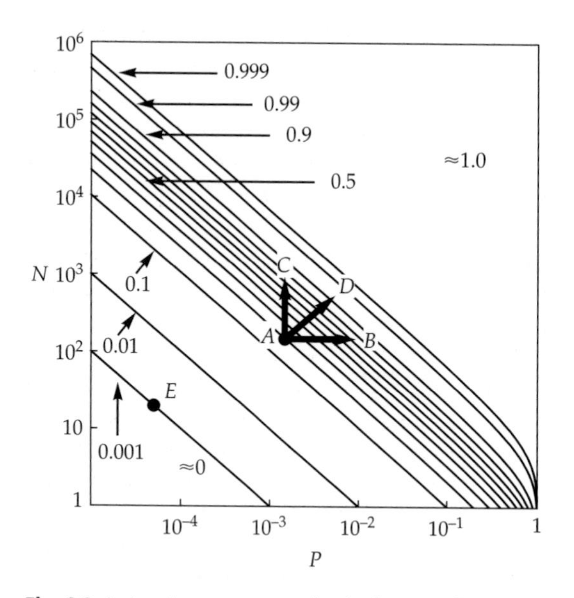

Fig. 6.9 An invasion pressure map showing how a modest increase in propagule pressure (*A*→*B*), invasibility (*A*→*C*), or both (*A*→*D*) could substantially increase the invasion pressure on an environment. If the system begins in position E, similar increases in either variable, will have little effect on invasion pressure. This phenomenon could help explain why there is often a lag period following an initial introduction prior to subsequent invasive spread. Invasion pressure is the probability of a successful invasion, in this case defined as the establishment of at least one individual in the invasion event.

one where it is probable. In the above example, if invasibility remains at 0.001 and the number of propagules increases from 200 to 1000, then the likelihood that at least one individual successfully establishes (Invasion Pressure) also increases to 63% (*A*→*C*, Fig. 6.9). If the number of propagules fortuitously increased five-fold during a time when invasibility also increased five-fold, then the probability of successful establishment by at least one individual would be nearly certain, i.e. 99%.(*A*→*D*, Fig. 6.9). Similar changes of this magnitude in either or both variables would have virtually no effect on invasion pressure if the system were at location *E* in Fig. 6.9.

Since the basic challenges and processes associated with subsequent spread are the same as those involved with the initial establishment, the *IP* maps also illustrates why there is often a lag between initial establishment and spread (Williamson 1996, Crooks 2005). A lag could occur for a variety of reasons. The established population, often small, may not be producing sufficient propagules to overcome the invasibility barrier in nearby environments. Or, given the existing local propagule pool, temporal fluctuations in the invasibility of these environments may seldom result in a sufficient enough increase to move the system out of the state in which the likelihood of a successful invasion event is essentially zero. In Fig. 6.9, an environment in location *E* would continue to experience negligible invasion pressure despite modest increases in either variable.

The sudden occurrence of spread may be due to a particularly large production of dispersing propagules that are able to overcome the invasibility barrier of the surrounding habitats. In Fig. 6.9, this would mean the system would move substantially vertically upward (*A*→*C*). Another common explanation for the lag between initial establishment and eventual spread is that the individuals and descendents of the initial established population adapt to the conditions in their new environment, either genetically or phenotypically, thereby providing the individuals with the traits needed successfully disperse and establish (Cox 2004, Crooks 2005, Dietz and Edwards 2006, Facon *et al.* 2006). Using the integrated trait–invasibility concept, this would be illustrated in Fig 6.9 by the system moving substantially from left to right (*A*→*B*). The system could also move left to right (reduction in the invasibility of the environment) by changes in the physical or biotic environment that increase the susceptibility of the environment to invasion by a particular species. Or, changes in the environment and traits of the species may both occur. Finally, spread, like the original establishment, may occur due to combination of changes in invasibility and number of propagules (*A*→*D*). For example, Jazdzewski and Konopacka (2002) suggested that the recent dramatic increase in the introduction of Ponto-Caspian species into the European river systems could be due to both an increase in propagule pressure and an increase in the salinity of the rivers, due to industrial and agricultural pollution, which may have reached a threshold permitting oligohaline species to enter the systems.

Population growth and spread, or decline and range contraction, following an introduction are the integrated results of subsequent establishment successes and failures of individual propagules, including those produced by the new population and those originating from the external dispersal pool. Thus, subsequent population dynamics can also be illustrated on an *IP* map. For example, a decline in the abundance of a non-native species could be due to a decline in the invasibility of the environment, which could be due to changes in abiotic conditions or to an increase in the biotic resistance of the other resident organisms, perhaps the result of an adaptive response (genetic and/or phenotypic) to the new species. If the population were located on or just above the cliff in the *IP* landscape, this reduction in invasibility would reduce the likelihood of successful reestablishment by the new set of propagules (see *A*→*B*, Fig. 6.10). Or, if the number of propagules being produced by the new population declined substantially, e.g. due to predation or other types of mortality of the adults, then one would likewise expect a reduction in the number of successfully established individuals in the next generation (*A*→*C*, Fig. 6.10). Of course if both invasibility and the number of new propagules decline, then one would expect an even more pronounced decline in the size of the population (*A*→*D*, Fig. 6.10). Again, if the population were located on the high plateau region of the *IP* landscape and some distance away from the cliff (point *E*, Fig. 6.10), comparable proportional changes in invasibility, and/or number of propagules, would not be expected to result in a decline in the establishment success of the next generation's set of propagules. However, the effect would be to move the system closer to the cliff, meaning that future declines in either or both variables might be sufficient to begin a substantial decline in population size and spread.

In recent years, considerable attention has been paid to the occurrence in nature of non-linear responses, alternative stable states, thresholds (sometimes referred to as tipping points), and hysteresis, the latter referring to the lag in return-time of a system to a prior state, even after some of the initial conditions have been restored (Beisner *et al.* 2003). The *IP* landscape manifests some, but not all, of these phenomena. Most clearly, it exhibits a non-linear response, specifically a threshold or tipping point (the invasion cliff). As the number of establishing individuals required for a successful invasion event increases, progressively smaller changes in invasibility and/or number of propagules have the potential to produce very large changes in *IP*, essentially either increasing the probability of a successful invasion event from zero to near certainty, or vice versa.

Although the *IP* lowlands and the *IP* plateau constitute alternative states, strictly speaking, they are not stable states since there are no feedback processes that tend to keep the system in either a lowland or plateau state. However, the invasion system may often appear as two alternative stables due to the substantial changes in invasibility and/or propagule numbers often needed to move the system from an uninvaded condition to an invaded one, or vice versa. The *IP* landscape does not exhibit

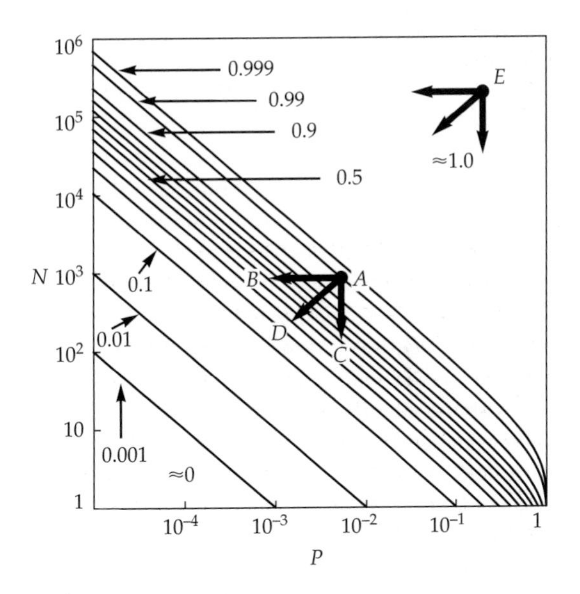

Fig. 6.10 Illustrated substantial declines in invasion pressure (*IP*) due to declines in invasibility (*A*→*B*), propagule pressure (*A*→*C*), or both (*A*→*D*) when the system is situated near the top of the invasion cliff. If the system begins further back on the invasion plateau, position *E*, comparable reductions in either or both variables will barely have any effect on invasion pressure. However, the reductions would move the system closer to the invasion cliff, making it more likely that any additional reductions in either variable could cause a significant decline in invasion pressure.

hysteresis. Changes to a particular combination of values for invasibility and propagule number will yield the same *IP* value irrespective of the direction of change, i.e. whether the system is moving up or down in elevation in the *IP* landscape.

There has been increasing recognition in recent years that substantial and unanticipated changes in species' abundances are very common events (Doak *et al.* 2008). Termed ecological surprises, these events have been attributed to a variety of factors, including complex community interaction webs, shifting abiotic conditions, variability in the composition of the interacting organisms (e.g., involving changes in their traits), and the fact that organisms affect one another in a myriad of ways, only a few of which are generally incorporated into models (Doak *et al.* 2008). This perspective suggests that ecological surprises are surprises primarily because of the complexity and dynamic nature of ecological systems, as well as our lack of detailed knowledge about them. While this is certainly partly true, the simple invasion pressure model, along with similar invasion models (Drake 2004, Leung *et al.* 2004, Drake and Lodge 2006), shows that unpredictability in population abundance, i.e. an ecological surprise, can also be an outcome of the underlying statistical properties of the system. Although shifts in invasibility and/or in the number of propagules are certainly due to ecological factors, such as those described by Doak *et al.*, the high degree of sensitivity to small shifts in these variables, the threshold response associated with the invasion cliff, is ultimately statistical in origin. This means that even were we able to fully understand all the interacting biotic and abiotic factors of a dynamic ecological system, the statistically-based threshold response of invasions means that invasions would still be extremely difficult to predict.

Summary

Understanding invasions is easier than predicting them, and the former is challenging enough. While many traits have been found to be associated with invasiveness, the connection is often not a strong one. This is undoubtedly partly due to the fact that invasiveness results from the interaction of particular traits with particular environment conditions. Thus, traits that may promote invasiveness in one environment may not do so in another. For the same reason, while certain environmental conditions, e.g. available resources, commonly facilitate introductions, not all species will respond similarly to the same set of conditions. Moreover, since invasibility is ultimately the integrated result of a myriad of biotic and abiotic factors, being able to predict changes in invasibility means being able to predict changes in these factors along with their interactions, and all of this at multiple spatial and temporal scales (Eppstein and Molofsky 2007). Finally, the likelihood of an invasion is always going to be greatly influenced by propagule pressure. Despite these realities, some progress is being made in the area of risk analysis, primarily by developing more general decision-making strategies that include information beyond traits. The extent of prior invasiveness by a species elsewhere, and of prior invasibility by an environment, continue to be among the best predictors of future invasiveness or invasibility.

Gilpin (1990) argued that the study of biological invasions should be 'self-consciously statistical, with an emphasis on characterizing the probability distribution of outcomes.' Gilpin's characterization describes the concept of invasion pressure (*IP*) as presented in this chapter. Defined as the probability that an individual dispersal event will result in a successful invasion, the notion of invasion pressure was introduced in a simple equation that integrated traits, invasibility, and propagule pressure (number of propagules in the dispersal event). Although this model is about as spare and simple as a model can be, simple models can sometimes be quite effective in yielding general ecological principles and conclusions (Levins 1966, Smith 1974). In this case, the *IP* model shows that propagule pressure and invasibility contribute approximately equally to invasion success. Significantly, and consistent with prior theoretical work on invasions, the model shows that changes in invasion pressure can alternatively be very sensitive or very insensitive to changes in invasibility and/or propagule numbers, depending on the magnitudes of the two variables, as well as on their relative values. In a three-dimensional graph of the three variables, this sensitivity is illustrated by a cliff-like feature (e.g. Figs 6.3, 6.6. 6.8). The invasion cliff, which connects the invasion pressure lowlands (where invasion is unlikely) to the invasion pressure high plateau (where invasion is virtually certain), graphically illustrates that the relationship between invasion pressure and its two primary driving variables is far from linear. The *IP* landscape shows that invasion pressure is best described as consisting of two relatively stable states, separated by a tipping point. This helps to explain many well-known aspects of invasion dynamics, including why invasions are often episodic, why there is often a lag in invasion spread, why some invasions experience a rapid collapse following a period of irruption, and, in general, why invasions have typically been so difficult to predict (and, unfortunately, why invasions are likely always going to be difficult to predict).

While understanding the patterns and mechanisms of the invasion process and being able to predict the likelihood of invasions are some of our primary objectives, they do not describe invasion biology's full agenda. There has been an enormous amount of research during recent years, as well as earlier, on the impacts of invasions. As the findings to date have shown, and as described in the next chapter, the impacts are many, often substantial, and sometimes transformative.

PART II

Impacts and management

CHAPTER 7

Impacts of invasions

Non-native species, like native species, can impact human health, national and local economies, and the ecosystems and ecological communities in which they reside. In fact, most non-native species do not have a large impact in any of these three areas. Some even have desirable effects. However, a small proportion of non-native species are considered harmful or undesirable owing to their impacts. In some instances, the harmful impacts can be dire. Introduced pathogens can threaten human health, crops, and livestock. Other introductions can seriously disrupt valuable ecosystem services, such as the provisioning of fresh water and timber, and some can cause extinctions of other species, as well as other undesirable ecological effects.

Impacts on human health and safety

Fire has always posed a threat to human safety and it is widely known that non-native plant species can modify fire regimes, often increasing the likelihood of fire due to the vegetation's flammability and phenology (e.g. high biomass during dry periods) (D'Antonio 2000, Brooks and Pyke 2001, Brooks *et al.* 2004). In the US southwest, some cities have passed fire ordinances restricting or banning the planting of certain flammable non-native species, e.g. pampas grass, *Cortaderia selloana*. In addition to increasing fire threats, some introduced plants can create new, or exacerbate, human health problems, including respiratory and skin allergies (McNeely 2005).

From a human perspective, it is difficult to dispute that the non-native species of greatest concern are those that threaten human health. As described by McMichael and Bouma (2000), non-native species can threaten human health in a variety of ways:

(1) an introduced pathogen can infect humans with a new disease;
(2) an introduced species may serve as a new and effective vector for the transmission of some already established diseases, increasing their infection rates;
(3) in some areas, an introduced species may kill crop plants to such an extent that the resident people experience caloric or other dietary deficits; and,
(4) in some cases, introduced species produce biotoxins, which can harm people, e.g. by contaminating potable water sources.

As described, non-native species can threaten human health in ways other than by the introduction of a new human disease. The extent of death and suffering caused by the Irish potato famine testifies to the human health dangers posed by introduced crop pathogens. Nevertheless, most would agree that the most serious health threats are posed by the non-native species threatening human health through the introduction and spread of infectious diseases.

With few exceptions (Chagas' disease being one of the few), most major human diseases originated in 'Old World' regions, perhaps because of the increased use of domesticated animals (that may have been the ancestral source of the human pathogens) and the fact that humans are phylogenetically more distant from 'New World' than 'Old World' monkeys, making it more likely that diseases would spread from monkeys to humans in the Old World (Wolfe *et al.* 2007). Of particular current concern is the ongoing spreading among the world's human populations of many viral diseases, as the respective viruses are successively transported into new regions as non-native species, e.g. severe acute respiratory syndrome (SARS), West-Nile encephalitis, Ebola hemorrhagic fever, dengue hemorrhagic fever, avian influenza, and AIDS. The global spread of these and other diseases is due to many factors, including the spread of humans into most of the earth's terrestrial environments, the high densities of many human societies, the increase in the number of domesticated animals and the increase in human contact with them, the global network of dispersal vectors created by humans traveling intra- and internationally, and a changing climate, which is permitting some pathogens and disease-carrying organisms to expand their ranges (Woolhouse and Gowtage-Sequeira 2005, Heeney 2006, Smith *et al.* 2007, Wolfe *et al.* 2007). An emerging disease may be the result of the introduction of a new pathogen to a region or it may be due to changes in the region that permitted a long-time resident pathogen to substantially increase in virulence, e.g. changes in the environment and/or in the genetic makeup of the pathogen or host (Storfer *et al.* 2007). As emphasized by Storfer *et al.*, it is very important to determine whether one is dealing with a new introduction or an emergence of a long-time resident pathogen, since many of the response strategies will differ dramatically depending on which is the case.

Taylor *et al.* (2001) concluded that approximately 61% of the more than 1400 infectious human diseases also infect animals (zoonotic diseases), and 75% of the emerging human diseases are zoonotic. A similar assessment in 2008 likewise emphasized the threat of zoonotic diseases, concluding that 60% of emerging infectious diseases had a non-human animal as its source, with 72% of these diseases originating in wild animals (Jones *et al.* 2008). The extent to which human specific infectious diseases have been spread throughout the world is demonstrated by the slope of the species–area relationship for these diseases. Whereas slopes for other groups, such as plants and animals, generally range from 0.15 to 0.35 (Rosenzweig 1995), Smith *et al.* (2007) calculated the slope for human-specific infectious diseases in general to range from 0.003 to 0.03, meaning that few new diseases are found as the area is increased, i.e. the diseases are mostly global in their distribution (Fig. 7.1). Smith *et al.* attributed this finding to two factors, the extent of human travel throughout the world, and the habitat homogeneity that humans have provided for the infectious agents. The global distribution of zoonotic diseases currently is more similar to that of plants and animals, with a species-area slope of 0.18 (Smith *et al.* 2007; Fig. 7.1). Although historically the distribution of zoonotic diseases has been more local and regional, intentional and unintentional introductions of non-native animal species throughout the world will inevitably globalize these diseases (Smith *et al.* 2007).

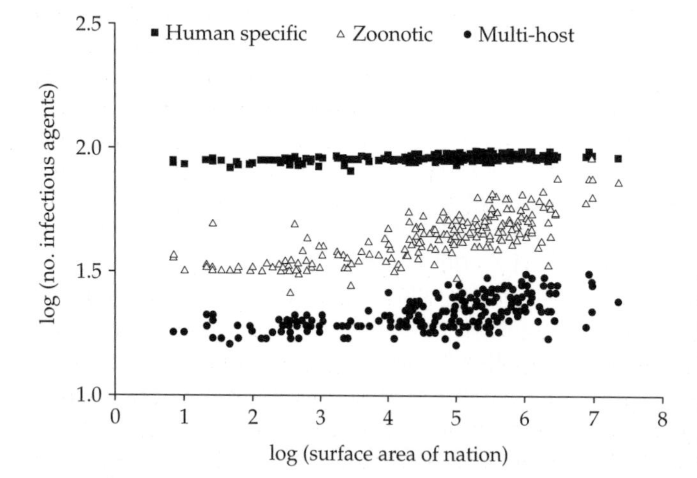

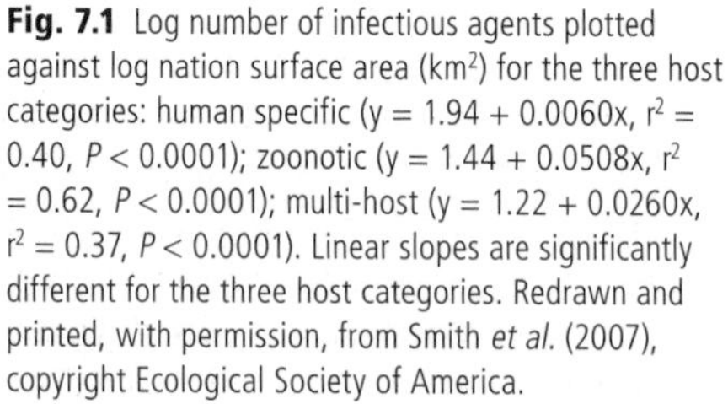

Fig. 7.1 Log number of infectious agents plotted against log nation surface area (km^2) for the three host categories: human specific (y = 1.94 + 0.0060x, r^2 = 0.40, $P < 0.0001$); zoonotic (y = 1.44 + 0.0508x, r^2 = 0.62, $P < 0.0001$); multi-host (y = 1.22 + 0.0260x, r^2 = 0.37, $P < 0.0001$). Linear slopes are significantly different for the three host categories. Redrawn and printed, with permission, from Smith *et al.* (2007), copyright Ecological Society of America.

Being able to predict the ability of pathogens and parasites to enter new communities and the ability of these organisms to shift hosts, thereby resulting in new diseases, is one of the primary challenges facing researchers and health professionals (Pederson and Fenton 2007). In their review of emerging infectious diseases, Jones *et al.* (2008) concluded that zoonotic diseases represent the most significant growing threat to global human health. A major challenge facing health professionals is that the majority of emerging human infectious diseases are originating in countries with few resources available, or allocated, to detecting the emergence of these diseases (Jones *et al.* 2008).

Some of the above diseases (e.g. West-Nile encephalitis, dengue hemorrhagic fever), as well as other prominent and deadly diseases (e.g. malaria, yellow fever), are mosquito-borne. Non-native mosquitoes may affect human health in three ways:

(1) by introducing a novel pathogen if the introduced mosquitoes are already infected;
(2) by providing a new transmission vector for a native pathogen; and/or
(3) by providing a new transmission vector for a novel pathogen independently introduced (Juliano and Lounibos 2005).

Thus, the introductions of new mosquito species into new regions, carry with them the threat of new human health threats. For example, in summer 2007, the tiger mosquito, *Aedes albopictus*, which has become established in southern Europe, infected more than 200 people with the viral disease chikungunya, the first documented instance of transmission of chikungunya outside the tropics (Enserink 2007). This is an example of an introduced mosquito species providing a new transmission vector for a novel pathogen, independently introduced.

Public interest in new and different pets has resulted in the transport of many animals from their native wild habitats to homes throughout the world, bringing with them potential serious human health risks (Brown 2008). In 2005, 210 million animals were legally imported into the US for the pet industry and an unknown number were imported illegally, both of which have the potential to introduce new zoonotic diseases (Brown 2008). In the case of the highly pathogenic avian influenza (HPAI H5N1) virus, although much attention has been given to the role that migrating birds may have played in its global dispersal, a careful examination of the evidence led Gauthier-Clerc *et al.* (2007) to conclude that commercial activity associated with poultry is the primary engine driving its dispersal.

Of recent concern is the regional and even global transport of bushmeat, which can serve as a vector of human disease, such as monkey pox and Ebola hemorrhagic fever. It has been estimated that in the Congo basin, nearly 300 g of bushmeat are eaten per day per person, with an estimated total of 4.5 million tons of bushmeat extracted annually (Fa *et al.* 2002). Although the bushmeat serves basic food needs, in many instances it has also acquired cultural meaning, e.g. being served as part of celebrations and religious rituals. Due to civil unrest in this and other regions of Africa, many people have emigrated to other parts of the world. While they may no longer need bushmeat to meet their dietary needs, they may still desire bushmeat for their cultural celebrations. In 2006, federal inspectors at JFK airport in New York City discovered a shipment of bushmeat hidden under smoked fish. According to court papers described by New York Times reporter Ellen Barry (2007), the bushmeat consisted of skulls, limbs, and torsos of non-human primate species, as well as the leg of a small antelope. The intended recipients of this shipment, immigrants from West Africa, argued that eating monkey meat was part of their religious rituals, including baptisms, Easter, Christmas, and weddings. This example illustrates how the global spread of some human diseases involves not only the spread of people but the spread of culture. It also demonstrates how prevention efforts may sometimes result in substantial conflicts between individuals with very different cultural beliefs and practices.

There do not appear to be many obvious instances of non-native species enhancing human health and safety. The one major exception, of course, are those non-native species that contribute positively to human nutrition, e.g. many of the introduced food crops planted throughout the world (e.g. wheat, originally from Southwestern Asia; corn, originally from southern North America or Mesoamerica; potatoes, originally from South America; cassava, originally from Central and South America; sorghum, originally from Africa; rice, originally from Southern Asia).

Economic impacts

Ecological, or ecosystem, services have been defined as 'the conditions and processes through which natural ecosystems, and the species that make them up, sustain and fulfill human life' (Daily 1997). Since replacing these services would normally exact an economic cost on society (Farber *et al.* 2006), for purposes of this discussion, impacts on ecological services are considered economic impacts. The Millenium Ecosystem Assessment (WRI 2005) used a four-category scheme to describe ecosystem services: supportive functions and structures (e.g. nutrient cycling and pollination and seed dispersal), regulating services (e.g. soil retention and disturbance regulation), provisioning services (e.g. provisioning of fresh water, timber, and food), and cultural services (e.g. opportunities for recreation and spiritual activities) (presented in Farber *et al.* 2006).

Economic costs to society of harmful non-native species, like the economic costs due to harmful native species, involve the costs associated with losses and damages, as well as the costs of efforts to control the species and their impacts. Pimentel *et al.* (2000) attempted to quantify the costs of non-native species in the United States and came up with a figure of $137 billion, which included damage and control costs. This is a very loose estimate, which, admittedly, is probably all that can be expected. It is not difficult to quibble with some of their estimates. For example, more than 10% of the estimated costs, $17 billion, were attributed to cat predation on birds, a figure arrived at by estimating that more than 500 million birds were killed annually by cats and assigning a $30 value figure to each bird. It stretches the imagination to imagine that the economic costs of cat predation on birds rivals that of crop weeds ($26.4 B), crop arthropod pests ($14.4 B), and crop pathogens ($21 B) (Pimentel *et al.* 2000). In addition, many of the 2000 estimates have been criticized for relying heavily on secondary sources (Reaser *et al.* 2003), some of which were reputed to have been incorrectly applied (Hoagland and Jin 2006).

Pimentel *et al.* (2005) provided an update on the environmental and economic costs associated with non-native invasive species in the US, with the 2005 estimate being $120 billion. However, there are a number of frustrating shortcomings with the 2005 paper. First, while the 2000 figure ($137 B) included estimated control costs, the 2005 figure left out control costs, including only costs associated with damage and loss. Thus, a comparison of the 2005 with the 2000 estimate suggests that the economic costs of invasive species in the US has declined by about 12% between 2000 and 2005. In fact, with the control cost estimates provided by Pimentel *et al.* in the 2005 paper, the comparable 2005 figure, i.e. with control costs included, is approximately $149 B, an increase of nearly 9%. Second, only 5 of the 131 cited references were published after 1999, 3 of which were on-line references. With less than 4% of its references published after the 2000 article, and the majority (81%) of the cost estimates provided in the 2005 article being identical to those listed in the 2000 version, it is difficult to view the 2005 publication as a useful update on the economic costs of non-native species in the United States.

While it is not difficult to take issue with the specifics of the Pimentel *et al.* (2000, 2005) cost estimates of non-native invasive species, it is clear that the economic costs of non-native species in the United States run into the many billions of dollars. It is even more difficult to quantify worldwide economic impacts on non-native species (Bright 1999), but it is not difficult to imagine that harmful effects on crops, domesticated animals, timber, waterways, human disease, and ecological services must easily run into the hundreds of billions of dollars per year.

In New Zealand, native tussock grasslands in upland regions have been found to play an important role in the provision of human water supplies (Mark and Dickinson 2008). Due to their morphology and physiology, the grasses are efficient at capturing water, whether in the form of rain, snow or fog, and exhibit comparatively low rates of transpiration. Thus, the upland native grassland environments typically release 64–80% of the annual precipitation to surface waters (Mark and Dickinson 2008). However, the amount of runoff can be dramatically reduced if the native grasses are replaced with non-native pasture grassland (Holsdworth and Mark 1990). A similar phenomenon has been documented in South Africa, where the encroachment of non-native woody plants (*Pinus*, *Eucalyptus*, and *Acacia*) into native upland environments has also reduced water yield (van Wilgen 2004). Restoration efforts in South Africa have shown that yield can be increased with the removal of the non-native species, but at considerable expense (van Wilgen 2004).

In 2007, a devastating insect-borne viral disease known as bluetongue (with 24 known serotypes) spread rapidly among domestic animals in Europe, particularly sheep. Belgium lost 15% of its sheep to this disease in 2007 and by early 2008 fears were that the disease could rival the 2001 outbreak of foot-and-mouth disease in the UK (Enserink 2008). In this case, the disease is not transmittable from animal to animal. Rather, it is transmitted by biting midges (*Culicoides*). Historically, this has been a disease occurring in tropical and subtropical regions. Its spread northward into Europe, including northern Europe, is believed to have been facilitated by recent temperature increases in Europe (Enserink 2008). Huanglongbing (HLB), a serious disease of citrus crops caused by a bacterium, *Candidatus* Liberibacter, and spread by citrus psyllids, has spread from China and is causing or threatening great economic harm throughout much of the world, including southern Asia, Africa, and South and North America. Tree mortality rates from this disease can exceed 50%, devastating local and regional economies dependent on the citrus industry (Callaway 2008).

Non-native plant pathogens may have economic effects beyond their impacts on their respective crops. Since plant pathogens are viewed as being a possible terrorist weapon, some countries are investing considerable additional resources in trying to prevent the introductions of particular pathogens. For example, in 2002, the United States implemented the Agricultural Bioterrorism Protection Act (ABPA), which listed particular non-native plant pathogens as 'select-agents.' By 2008, the US Department of Agriculture's biodefense budget had risen to $340 million. But, the economic impacts of the ABPA extended beyond federal budgets. US laboratories researching species listed as select agents are required to implement considerable added security measures, including video cameras, biometric security devices for the doors, and security-checks on all personnel, which can cost individual labs tens of thousands of dollars (Callaway 2008).

In many instances, the economic impacts of non-native species are greatest on the world's poor. Although farmers in a developed country may have to pay a price to try to reduce the impact of agricultural diseases or pests, at least they are usually able to pay the price and reduce the impact. Poor farmers in other parts of the world often do not have the ability to pay these costs. For example, the tomato yellow leaf curl virus (TYLCV), native to Egypt and introduced into the Caribbean and Central America in the mid-nineties, has been spreading north through Central America using the white fly, *Bemisia tabaci*, itself an introduced species, as its vector (Dalton 2006). There are currently no effective anti-viral control measures; however, pesticides can be used to control the dispersal vector, the white fly. Unfortunately, in some regions in Mexico where the disease has spread, many tomato farmers cannot afford the cost of pesticides that can control the white flies. As a result, in winter 2006, some regions experienced losses approaching 100% of their plants (Dalton 2006).

Of course, not all the economic impacts of non-native species, even non-native invasive species, are negative. The introduction of several non-native crayfish species has revived local fisheries in some places in Europe where the native populations had been previously decimated by the crayfish plague, *Aphanomyces astaci* (Gutiérrez-Yurrita *et al.* 1999). Great Lakes fisheries, which had crashed due to the introduction of the sea lamprey, *Petromyzon marinus*, earlier in the century, similarly improved following the introduction of non-native salmon species in the 1970s and 1980s. Economic benefits are also provided by non-native and non-invasive horticultural species. Non-native food crops, which constitute the bulk of many people's diet, are an important sector of many economies. Additional reviews of economic impacts of non-native species can be found in Perrings *et al.* (2000) and Pimental *et al.* (2000 and 2005).

Ecological impacts

Although one could argue that almost any ecological impact has the potential to affect some ecological service, the impacts of many non-native species are mostly restricted to other species and to the structures and processes of their new ecosystems, with little known impact on ecological services. Parker *et al.* (1999) identified five types of ecological impacts caused by non-native invasive species: genetic impacts, impacts affecting individual organisms, impacts affecting population dynamics, impacts altering community structure, and impacts affecting ecosystem processes. Vitousek (1990) identified three types of ecosystem impacts by non-native species: changes in nutrient availability by altering biogeochemical cycles, trophic changes in food webs, and physical or structural alterations of the environment. In fact, the impacts are not so cleanly segregated. Impacts on populations are normally the summed effects of impacts on individuals, and any substantial impact on the population of a particular species is also going to have a community impact. And, changes in biogeochemical cycles can differentially affect species resulting in changes in food webs. Although no scheme of ecological impacts by non-native species can avoid overlap between categories, for current purposes, four types of ecological impacts will be presented: impacts on populations, and biodiversity impacts on food webs and communities, impacts on biogeochemical processes, and impacts altering the physical structure of the environment. Genetic impacts were discussed as part of the evolutionary processes taking place during dispersal and establishment in Chapter 5.

Impacts on populations and biodiversity

Introduced pathogens

In many instances, new species cause reductions in the size of long-term resident populations. This is especially true in the case of introduced predators and pathogens (Daszak *et al.* 2000), which can dramatically reduce population levels of prey and host species, particularly on islands and other geographically isolated environments, such as lakes, in which the predators and pathogens can drive prey populations and species to extinction (Warner 1968, Jehl and Everett 1985, Keitt *et al.* 2002, Steadman 2006). As illustrated by the Irish potato famine, and the North American chestnut blight, some introduced pathogens can devastate host species even on continents and large islands. One of the best-documented cases of an introduced pathogen's effect on animals is the introduction of the rinderpest virus into Africa in the late-1800s with the introduction of Indian cattle by Italian troops stationed in east Africa. Within a decade, the virus had decimated wildebeest and other antelope populations, as well as domesticated livestock (Plowright 1982). It is estimated that the disease killed 90% of the cattle in sub-Saharan Africa and that as many as one-third of the population of Ethiopia and two-thirds of the Maasai of Tanzania died of starvation as a result (Normile 2008). Some have claimed the rinderpest epidemic to be 'the greatest natural calamity ever to befall the African continent' (Reader 1999, cited in Normile 2008).

A combination of genetic-based resistance among the surviving animals and a concentrated effort to immunize livestock enabled the wild populations to rebound in the twentieth century (Spinage 2003). However, due to a decline in vaccination and surveillance programs in the 1970s, the disease rebounded in the early 1980s, causing massive livestock deaths throughout much of Africa, the Middle East, and southern Asia. A more recent example of a devastating disease on a wild population is the introduction of a new strain of the Ebola virus to western gorilla, *Gorilla gorilla*, populations in Gabon and Congo, which have experienced mortality rates of 90–95%, resulting in the deaths of thousands of gorillas from 2002 to 2005. (Bermejo *et al.* 2006)

The introduction of the West-Nile virus into North America in the late-1990s is believed to have caused declines in several bird species, including the American Crow (*Corvus brachyrhynchos*), Blue Jay (*Cyanocitta cristata*), American Robin (*Turdus migratorius*), House Wren (*Troglodytes aedon*), Black-Capped and Carolina Chickadees (*Poecile atricapilla* and *P. carolinensis*), and Eastern Bluebird (*Sialia sialis*)

(LaDeau *et al.* 2007). In many areas, the population changes in North American songbirds attributed to West-Nile virus have significantly altered the composition of many North American avian communities (LaDeau *et al.* 2007). In their study, LaDeau *et al.* documented considerable spatial heterogeneity in the disease impacts, which, they suggested, could be due to regional differences in the relationships between vector abundance and land use, regional differences in mosquito feeding preferences, and regional differences in the dominant disease vectors. They also documented substantial temporal fluctuations in the disease impact, with some species, e.g. Blue Jay and House Wren, showing population recoveries by 2005 to pre-disease levels (LaDeau *et al.* 2007).

There is currently no evidence that any North American bird species are threatened with extinction due to West-Nile virus, or any other pathogen for that matter. However, introduced pathogens and parasites have caused avian extinctions on islands (Warner 1968, Van Riper *et al.* 1986). For example, in Hawaii, introduced diseases (avian malaria and avian pox virus) along with their introduced vectors (mosquitoes), are believed to have been the primary causes of extinctions of many Hawaiian native bird species (Lafferty *et al.* 2005). Although the mosquito-borne diseases also infect non-native bird species, native species are normally much more susceptible (Van Riper *et al.* 2002). This means that the non-native species serve as reservoirs for the disease, increasing the risk of infection of native birds (Lafferty *et al.* 2005). There is also evidence suggesting that introduced pathogens have caused extinctions of some island mammals (Pickering and Norris 1996).

In some cases, pathogen-caused extinctions of island species may be facilitated by other factors, such as a changing climate. In Hawaii, because most low-elevation forests have been cut for agriculture, native forest bird species are confined to higher elevation forested areas, whereas introduced species are most abundant in the low-elevation anthropogenic environments (van Riper *et al.* 1986). Due to the low density of hosts at low elevations and to cold temperatures at high elevations (the latter which inhibits high altitude spread of the dispersal vector, the introduced mosquito *Culex quinquefasciatus*), the malarial parasite, *Plasmodium relictum*, peaks in abundance at mid-elevations (Fig. 7.2). This leaves native birds at high elevations substantially protected from avian malaria. However, models of the impacts of warmer temperatures showed that *C. quinquefasciatus* will be able to persist at higher elevations, thereby transmitting the disease to the remaining high-elevation populations of native birds, and possibly causing some of them to go extinct (Benning *et al.* 2002).

Similar to the effect of West-Nile virus on North American birds, i.e. causing declines but no extinctions, the introduction of the parapoxvirus into Great Britain through the introduction of one of its hosts, the gray squirrel, *Sciurus carolinensis*, is believed to be a major cause of the decline of the native red squirrel, *S. vulgaris* (Tompkins 2003). The introduction of the protist *Bonamia ostreae* into European Atlantic coastal waters in the 1970s decimated many of the native oyster populations (Goulletquer *et al.* 2002). Several recently introduced fish viruses, including hemorrhagic septicemia virus, are currently threatening fish in the Great Lakes and surrounding inland lakes in North America. And there is evidence that a primary contributing factor in the collapse of many honeybee populations in North America (the honeybee itself being an introduced species in NA) is an introduced virus, the Israeli acute paralysis

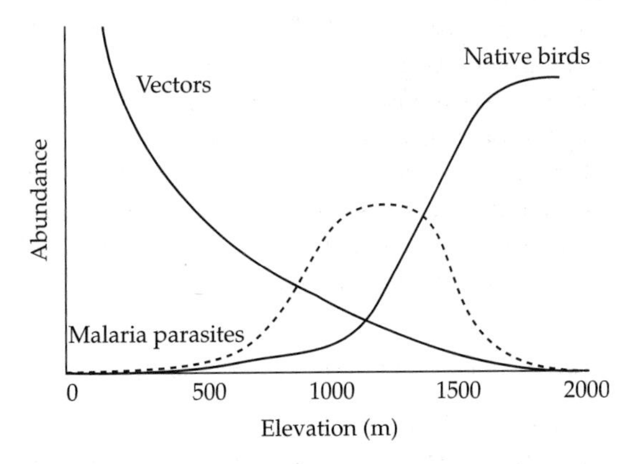

Fig. 7.2 A general characterization of the distribution along an elevation gradient of active Hawaiian birds, the malaria parasite (*Plasmodium relictum*), and its mosquito vectors. Redrawn and printed, with permission, from van Riper *et al.* (1986), copyright Ecological Society of America.

virus (IAPV) (Cox-Foster *et al.* 2007). It is possible many hives are particularly vulnerable to the effects of the IAPV because the bees are already in a weakened state due to heavy parasitism by two species of non-native mites, *Varroa destructor* and *Acarapis woodi* (Cox-Foster *et al.* 2007). The introduction of the crayfish plague, caused by the fungus *Aphanomyces astaci*, into Europe in the mid-1800s has resulted in sharp declines in the populations of native European crayfish species (Westman 2002). Besides directly causing mortality on wildlife, new diseases are likely to impact populations by increasing the susceptibility of animals to other sources of mortality. For example, birds with higher rates of blood parasites have been found to experience higher predation rates from avian predators (Møller and Nielsen 2007). This finding suggests that predation on prey recently infected by an introduced disease may influence the virulence and transmission rate of the disease (Møller and Nielsen 2007).

As illustrated by the North American chestnut blight, *Cryphonectria parasitica*, and many crop diseases, introduced pathogens can be a major cause of plant mortality. The introduction into North America of the white-pine blister rust, *Cronartium ribicola*, is causing mortality rates in some pine populations as high as 90% (Kendall and Arno 1990). The introduced pine pitch canker fungus, *Fusarium circinatum*, is currently threatening California stands of Pinus radiate (Richardson *et al.* 2007). Dutch elm disease, *Ophiostoma* spp., has similarly devastated American elms, Ulmus americana, throughout North America. Introduced *Phytophthora* pathogens are currently causing high mortality in a number of North American and European tree species (Brasier 2000, Rizzo *et al.* 2002).

As described above, the spread of humans into more and more natural environments has increased our risk of zoonotic diseases. However, it must be remembered that this can be a two-way street. While we are experiencing increasing health risks due to increased interactions and proximity between humans and animals, we are also exposing the animals to our pathogens. A study of a chimpanzee population in Côte d'Ivoire found that some recent deaths from respiratory disease have been caused by human viruses, most likely spread through ecotourism and by the researchers themselves, both of which brings humans and these animals, which have become habituated to humans interaction, into close proximity with one another (Köndgen *et al.* 2008).

Introduced predators

Introduced predators, like pathogens, have consistently had larger population impacts on island than mainland fauna (Blackburn *et al.* 2004, Cox and Lima 2006, Sax and Gaines 2008). Introductions of predatory snails to Pacific island have caused the extinctions of several species of native island snail species (Hadfield *et al.* 1993, Cowie 2002). It is believed that human predation was primarily responsible for the extinction of large flightless birds in New Zealand, e.g. exceeding 3.75 kg, while introduced small mammals, particularly rats, were responsible for most extinctions of the smaller island flightless species (Roff and Roff 2003). Seabirds often utilize islands for breeding and numerous studies have shown that nesting populations have suffered greatly due to predation by introduced predators, such as cats (McChesney and Tershy 1998, Keitt *et al.* 2002), rats (Kepler 1967, Grant *et al.* 1981, Roff and Roff 2003), mongoose (Hays and Conant 2007), and snakes (Savidge 1987). Some of these predators can also decimate native mammal, lizard, and turtle populations (Seaman and Randall 1962, Nellis and Small 1983, Hays and Conant 2007).

In cases where introduced predators do not actually cause extinctions in birds, they may be contributing to an increased skew in the sex ratio of wild populations (Donald 2007). Wild populations of birds have consistently been found to contain more males than females (Mayr 1939, Donald 2007), and since birth ratios are consistently not skewed (Donald 2007), the difference in adult sex ratios must involve higher mortality rates among the females. There are many possible reasons why female birds may experience higher mortality rates, one of them being that females experience increased predation rates during the breeding season (Donald 2007). A review of the adult sex ratios of globally endangered bird species showed that the skew is greatest in those species for which introduced predators have been listed by the IUCN as a severe threat (Fig. 7.3).

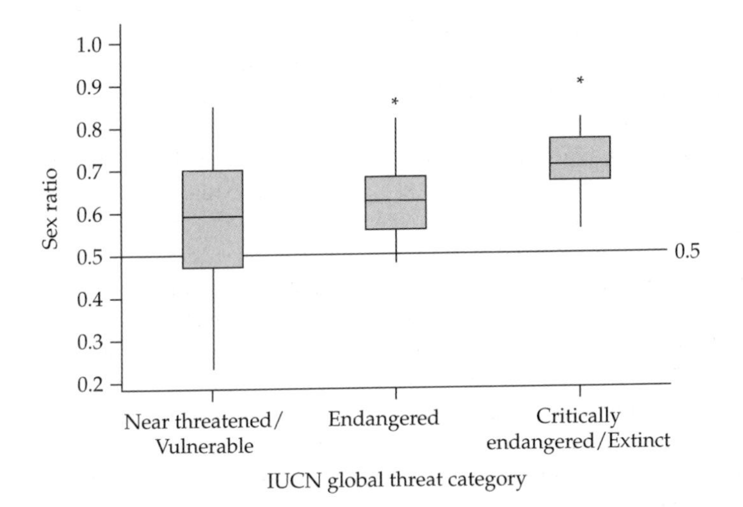

Fig. 7.3 Adult sex ratios (ASR) of birds in different IUCN threat status categories. There was a tendency for ASR skew to increase with increasing threat category (Kruskal–Wallis test, $P < 0.05$). The horizontal line represents the median, the box the interquartile range and the vertical lines span the range of the values lying between the interquartile and 1.5 times the interquartile range. Outliers beyond this are represented by asterisks. Redrawn and printed, with permission, from Donald (2007), copyright Blackwell Publishing.

Of course, it matters what type of a predator is introduced into an environment, but even when multiple non-native predator species appear quite similar, their impacts on their prey may differ substantially. Along the northeast coast of North America, a previously introduced predatory crab, the European green crab (*Carcinus maenas*), is being replaced by a more recently introduced predatory crab, the Asian shore crab (*Hemigrapsus sanguineus*). A recent study of the two species found that *C. maenas* exhibited much higher levels of intra-specific aggression than did *H. sanguineus*, which translated into an average of a six-fold greater density of the latter in sites sampled (Griffen and Delaney 2007). Griffen and Delaney concluded that the substantially higher densities already exhibited by *H. sanguineus* (and the species is still establishing and expanding in these coastal areas) may destabilize the predator–prey system, possibly resulting in oscillatory cycles of increasing amplitude.

Introduced fish predators have been found to elicit a number of phenotypic changes among its prey. For example, Lippert *et al.* (2007) showed that the introduction into Ontario lakes of several large predatory fish (smallmouth bass, *Micropterus dolomieu*, walleye *Sander vitreus*, and northern pike, *Esox lucius*) impacted yellow perch, *Perca flavescens*, populations in a number of ways, including earlier ontogenetic diet shifts from zooplankton to benthos, poorer growth during their first growing season, and reduced energy investment (egg lipid content) in eggs by mature females. In addition, the body shape of the perch in lakes with recently introduced predators showed a shift from a typical streamlined, pelagic body form towards a deeper-bodied benthic body form, a change, the authors suggested, that may increase foraging efficiency in a benthic environment, or may represent an anti-predator strategy.

Introduced predators seemed to have negatively impacted freshwater and inland sea populations, more so than oceanic ones (Moyle 1986, Vermeij 1991). For example, in several Russian lakes, a number of native amphipod species are believed to have been almost completely replaced by an introduced Baikalian amphiod, *Gmelinoides fasciatus*, the most likely mechanism being predation on juveniles of the native species (Panov and Berezina 2002). By altering age-dependent survival probabilities, particularly by decreasing survival probabilities of larger fish prey, non-native fish predators may select for smaller size and earlier maturation in the prey populations, a process known as stunting, which can also reduce the economic value of the prey fish (Lehtonen 2002).

Cox and Lima (2006) and Salo *et al.* (2007) suggest that prey naïveté among island animals, particularly among birds, probably contributed to their extinctions by introduced predators, both human and non-human. Blumstein (2002) showed that Tammar wallabies, *Macropus eugenii*, which had been introduced onto Kawau Island, NZ, which was free of large wallaby predators, had lost some of their predator-recognition abilities. Long-term isolation from certain predatory archetypes, e.g. snakes and ground mammals, is believed to be the cause of prey naïveté for many of these islands species. The type of naïveté discussed by Cox and Lima and Salo *et al.* is evolutionary naïveté, in which the species has not evolved recognition abilities for certain predator types, as opposed to ontogenetic naïveté, which refers to the lack of individual exposure to a particular predator type during the prey's lifetime.

Continental terrestrial prey are generally not as likely to exhibit naïveté to an introduced predator, since it is unlikely the new predator represents a new predatory archetype. However, this is not the case for continental aquatic systems, in which the insularity of many freshwater systems is believed to have similar effects as the insularity of oceanic islands (Cox and Lima 2006). Although it is difficult to assess the extent to which prey naïveté is responsible for a dramatic decline in an aquatic prey species following introduction of a novel predator, Cox and Lima (2006) present several examples in which they believe this to be the case. A stunning example is the case of the introduction of Nile perch, *Lates niloticus*, into Lake Victoria. These lakes had lacked a large pike predator and it is estimated that more than 100 species of haplochromine cichlids have gone extinct due to predation by the Nile perch (Ogutu-Ohwayo 1990). Introduced European brown trout, *Salmo trutta*, into South America and New Zealand (Leveque 1997, Townsend 2003), and introduced mosquitofish, *Gambusia holbrooki*, into Australia (Hamer *et al.* 2002) have caused major reductions in native fish and amphibians, respectively, and prey naïveté is believed to have played a role in each of these situations as well. Although a number of freshwater extinctions due to the introduction of a predator have been documented, no recent extinctions of any marine species due to an introduced predator have been recorded (Vermeij 1991), a finding Cox and Lima argue is consistent with their hypothesis of increased prey naïveté in freshwater systems.

Native North American moose that have lived for multiple generations in the absence of predators, such as wolves and grizzly bears, have also been shown to exhibit prey naïveté when these predators have been reintroduced (Berger *et al.* 2001). However, Berger *et al.* also found that predator recognition and avoidance behavior in the moose developed quite quickly through learning, leading them to conclude that it was highly unlikely that the moose would experience a predation 'blitzkrieg.' Given the lifespan of the moose, and the rapidity with which they regained their predator-avoidance behavior, the change was almost certainly phenotypic. However, in other instances, the acquisition of anti-predator responses to a novel predator may involve natural selection and genetic changes. For example, Kiesecker and Blaustein (1997) showed that the red-legged frogs, *Rana aurora*, had developed recognition abilities (chemical cues) and anti-predator responses to the introduced bull frog, *Rana catesbeiana*, changes that were believe to have a genetic component to them. Predator aversion in mice has been shown to have both a learned and genetic component (Kobayakawa *et al.* 2007), suggesting that natural selection could play a role in reducing prey naïveté. While these findings provide some hope for prey species threatened by introduced predators, other examples, particularly involving island or other insular populations, clearly show that no behavioral responses were sufficient to ward off predation-induced extinction.

Even if a new predator does not represent a new predatory archetype, and hence the prey does not suffer from naïveté, this does not mean the new predator cannot drastically reduce the size of the prey population, or even cause its extinction. If the predator is very efficient, the prey populations can be substantially reduced, despite the fact that the prey recognizes the predator as a predator and tries to take evasive action. Predator–prey examples of this phenomenon include the very heavy predation on the European native water vole, *Arvicola terrestris*, by the introduced American mink, *Mustela vision* (Macdonald and Harrington 2003), the predatory impact of the red fox, *Vulpes vulpes*, on eastern grey kangaroos, *Macropus giganteus* (Banks *et al.* 2000), and human hunters using modern technology on just about any species.

Introduced herbivores

Introduced herbivores can significantly impact the vegetation in the new region. The hemlock woolly adelgid, *Adelges tsugae*, has caused a decline in many populations of eastern hemlock, *Tsuga canadensis*, in the eastern United States (McClure and Cheah 1999, Lovett *et al.* 2006; Fig. 7.4). Herbivory in European wetlands by the North American muskrat, *Ondatra zibethicus*, and the South American coypu, *Myocastor coypus*, has altered the relative abundance of wetland plant species (Toivonen and Meriläinen 1980, Gosling 1989). In some cases, herbivory from the NA muskrat creates patches of open water in otherwise dense stands of aquatic macrophytes, and by increasing habitat heterogeneity, increases plant diversity at the site (Nummi 2002). Extensive grazing by introduced herbivores (cattle and rabbits) is believed to have played a major role in converting historical sclerophyllus Chilean forests to savannas (Holmgren 2002). In addition to their effects on vegetation due to their herbivory, introduced terrestrial herbivores often impact the vegation and other animals that feed on or live in the vegetation, through trampling and soil disturbances created during foraging (Mueller-Dombois and Spatz 1975, Cox 1999, Beever *et al.* 2003), which in turn can facilitate the establishment of non-native vegetation (Aplet *et al.* 1991, D'Antonio *et al.* 1999, Cushman *et al.* 2004).

Introduced competitors

There are very few examples of introduced species causing extinctions of native species through competition, whether on continents or islands (Davis 2003, Sax and Gaines 2008). Nevertheless, native

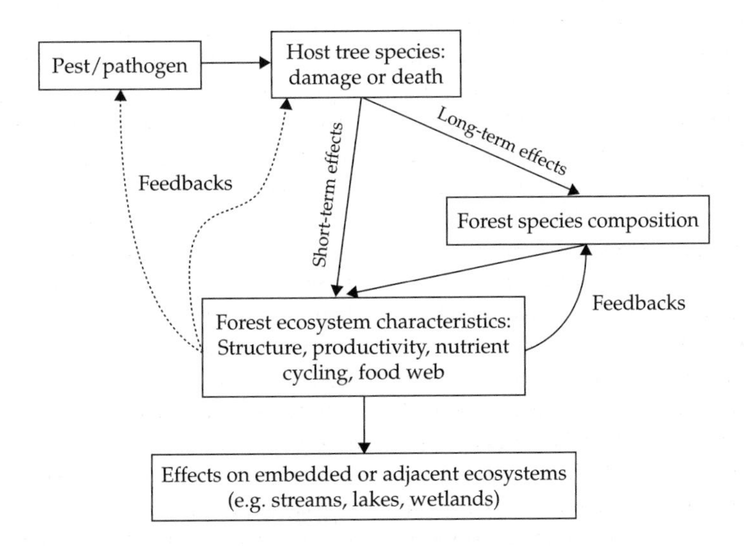

Fig. 7.4 Pathways of impact of pests and pathogens on forest ecosystem processes. Ecosystem characteristics can be affected by the direct, short-term action of the pest or pathogen on the tree—for instance defoliation or mortality. Longer term effects are caused by pest-induced changes in forest species composition, which then produce changes in ecosystem processes. These ecosystem characteristics can feed back to affect the pests (e.g. increased nitrogen availability can increase the survival of phytophagous insects), the trees (e.g. increased light availability from tree death may improve the condition of the survivors), or the forest composition (e.g. increased light, water, and nutrients may change the relative competitiveness of different tree species). Redrawn and printed, with permission, from Lovett *et al.* (2006), copyright American Institute of Biological Sciences.

population declines due to competition from non-native species are commonly documented (Bruno *et al.* 2005). Introduced North American crayfish are believed to be displacing native European crayfish species via competition, although the North American species are also likely to be contributing to the decline in the native species by carrying the crayfish plague, *Aphanomyces astaci*, to which the native species are much more vulnerable than the non-native crayfish (Westman 2002). In addition to habitat loss and overhunting, competition from the American mink, *Mustela vision*, is believed to be one of the causes for the decline of its close relative European mink, *Mustela lutreola* (Maran and Henttonen 1995, Maran *et al.* 1998). An experimental study of the impacts of the introduced black rat, *Rattus rattus*, on the endemic Santiago rice rat, *Nesoryzomys swarthi*, on Santiago Island in the Galápagos Islands concluded that interference competition through aggressive encounters by *R. rattus* were primarily responsible for the decline of *N. swarthy* (Harris and Macdonald 2007). Harris and Macdonald predicted that an increase in the frequency of El Niño episodes, which tend to increase the numbers of black rats due to higher resource levels, would further intensify the competitive effects on the rice rats, possibly leading to complete competitive exclusion.

Liu *et al.* (2007b) documented a case in which a non-native species has displaced a native species by disrupting the mating behavior of the native individuals. The case involved an introduced strain of the white fly, *Bemisia tabaci*. Because the mating dynamics of the introduced strain were so similar to those of the native strain, non-native males frequently courted native females, thereby obstructing mating attempts by the native males, resulting in reduced reproductive success in the native strain. However, as Liu *et al.* documented, this behavior was asymmetrical in that native males did not court non-native females. Liu *et al.* concluded that this behavior has been an important mechanism behind the substantial declines, and even regional extirpations, in native white fly strains endemic to China and Australia.

Although competition from non-native plant species is seldom likely to result in extinctions of long-term resident species (Sax and Gaines 2008), it can substantially impact community composition and regenerative processes. Kueffer *et al.* (2007) studied the effects of an abundant non-native tree, *Cinnamomum verum*, on tree regeneration in the Seychelles Islands. In their field experiment, Kueffer *et al.* found that in nutrient-poor soils, *C. verum* suppressed the growth of tree seedlings due to below-ground competition for nutrients. Because this effect was impacting tree regeneration and successional processes in these forests to such a great extent, the researchers considered *C. verum* to be a 'transformer species' (*sensu* Richardson *et al.* 2000a). Using models parameterized with field data, Williams and Crone (2006) predicted that populations of the native North American grassland forb, *Anemone patens*, would exhibit a gradual decline in numbers in areas dominated by *Bromus inermis*, whereas populations growing with native grasses were expected to remain stable. An experimental study involving the planting of seedlings from 10 native tree species under the canopies of common buckthorn, an invasive shrub/small tree in North America, found that the seedlings actually did better (measured by cover) under than outside the buckthorn canopy (Knight *et al.* 2007), perhaps due to increased soil nitrogen, which is known to occur underneath buckthorn canopies (Knight *et al.* 2007). Introduced seaweeds commonly negatively impact native seaweed species (Williams and Smith 2007), presumably mostly through competition.

In some instances, factors associated with climate change may increase the competitiveness of a species. Baruch and Jackson (2005) found that increased levels of CO_2 improved the competitive ability of two non-native invasive grass species in tropical savannas. Although documented extinctions due to competition are very rare to date, one would expect that, should they occur, it is most likely that they would happen on islands or other geographically isolated and confined environments, in which populations of the native species are likely to be small and less able to avoid the new competitor.

Transgenic crops and biofuels

Transgenic crops represent a new kind of non-native species threat, specifically the creation of new species through hybridization with native wild species (Ellstrand 2003, Chapman and Burke 2006). In some cases, the hybrids have exhibited greater survival and/or fecundity than the wild types (Campbell *et al.* 2006). Mercer *et al.* (2006, 2007) found that the success of transgenic crop × wild type hybrids depended on the particular environmental conditions. An interesting, though frustrating, issue regarding efforts to prevent and monitor the escape of transgenes into wild populations is that the genetic information associated with the transgenic crops may be proprietary, and thus developers and regulators of transgenic organisms may not be legally obligated to provide the genetic information that managers might need to be able to detect the introduction of transgenes into wild populations (Schoen *et al.* 2008).

It has been pointed out that the development of biofuels could also create invasion problems (Raghu *et al.* 2006). Some of the species being considered for biofuel production in the US are non-native grasses that are already invasive in some environments (Raghu *et al.* 2006). The use of native grasses for biofuel production does not necessarily eliminate the problem, since while the native grasses may not be invasive in their native environment, they may become invasive if planted outside its native range (Raghu *et al.* 2006). Barney and DiTomaso (2008) pointed out that many of the attributes being selected for in biofuel stocks are those often associated with invasive plant species, e.g. susceptibility to few pests, tolerance of poor growing conditions, and the ability to produce highly competitive monospecific stands.

Indirect impacts

When the impact of a non-native species on a native species is mediated by the involvement of a third species, the effects of the third species are sometimes referred to as 'indirect effects' (White *et al.* 2006). In a more general discussion of the impacts of one plant species on another, Jones and Callaway (2007) emphasized that the mediating factor may often be an abiotic one. In either case, the third-party effect underscores the context-dependency of the impacts of non-native species.

In addition to directly competing with native plants, non-native plants may also negatively affect native plants via allelopathy and interactions with other organisms. Studies have shown that plant exudates of some non-native species are toxic to native species, thereby enabling the non-native species to spread and dominate in some habitats (Ridenour and Callaway 2001, Rodgers *et al.* 2008). This effect has been termed the 'novel weapons' effect (Callaway and Ridenour (2004). In other instances, introduced plant species may alter the soil microbial community (Kourtev *et al.* 2002, Ehrenfeld 2003), which may alter the plant–microbe relationship of native species (Callaway *et al.* 2004, 2005, 2008). In the case of garlic mustard, *Alliaria petiolata*, its ability to inhibit the activity of many soil organisms through its chemical exudates is believed to be so great that the species has been used as a natural soil fumigator in agriculture (i.e. they are planted in alternation with the crop species and then tilled into the soil to biofumigate it) (Brown and Morra 1997).

In an attempt to control the spread and impact of musk thistle, *Carduus nutans*, in the Great Plains, NA, *Rhinocyllus conicus*, a weevil that feeds on musk thistle in its native range, was introduced (Gassman and Louda 2001). Although the weevil is not a feeding specialist and feeds on both native and non-native thistles, *C. nutans* is regarded as its preferred host plant (Russell *et al.* 2007). Nevertheless, the susceptibility of native thistle plants to the weevil can be influenced by its proximity to patches of *C. nutans*. Russell *et al.* (2007) found that the number of weevil eggs laid on the native wavyleaf thistle, *Cirsium undulatum*, was associated with the native thistle's co-occurrence with the non-native thistle, a phenomenon known as associated susceptibility (Brown and Ewel 1987).

A particularly interesting example of a non-native plant species impacting native plants through the actions of another organism involves the community-wide effects of tall fescue, *Lolium arundinaceum*. Introduced from Europe to North America, *L. arundinacum* commonly hosts the fungal endophyte, *Neotyphodium coenophialum*, also introduced from Europe. *Neotyphodium coenophialum* is toxic to most herbivores. In an experimental field study, Rudgers *et al.* (2007) found that tree seedling establishment was much lower in endophyte-infected plots and evidence indicated that much of the effect was due to increased tree seedling herbivory/

predation by voles (*Microtus* spp.), which exhibited increased preference for the tree seedlings in these plots due to the presence of the unpalatable tall fescue. The authors argued that the effect was great enough to impact successional processes in many anthropogenic grasslands, specifically inhibiting the transition from a grass-dominated habitat to one dominated by woody plants. Meiners (2007) showed that non-native shrubs can negatively impact tree regeneration in both old fields and successional forest. He found significantly higher seed predation rates underneath the canopies of *Rosa multiflora* (in old fields) and *Lonicera maackii* (in successional forests). A similar finding was made by Orrock *et al.* (2008), who found that the presence of a non-native annual, *Brassica nigra*, resulted in an increase of seed predation by small mammals on a native bunchgrass, *Nasella pulchra*, in a California grassland.

In some instances, non-native shrubs and small trees, including *Rhamnus*, have been found to create what has been termed an 'ecological trap' for some songbird species, in which the birds preferentially select to nest in the non-native species but then end up experiencing a higher rate of predation (Schmidt and Whelan 1999). However, the impact of these species on the lifetime fitness of the birds is not clear. For example, the American Robin, *Turdus migratorius*, one of the species studied by Schmidt and Whelan and which experienced higher predation when nesting in buckthorn, *Rhamnus* spp., also regularly feeds on the buckthorn fruits, which remain on the trees for several months after their late summer and fall production. During the winter in Minnesota, where robins have only recently been overwintering in significant numbers, one of the most reliable places to view flocks of robins are large buckthorn stands that are still bearing the fall fruit (personal observation).

Another way that non-native species may impact populations is by altering existing mutualisms involving native species. Traveset and Richardson (2006) reviewed studies examining the effects of introduced species on the reproduction biology of native plants and concluded that non-native invasive species commonly disrupt important plant reproductive mutualisms. Disruptions can occur through the introductions of pollinators that reduce flower visitation by native species (Gross and Mackay 1998, Hansen *et al.* 2002), introductions of predators that can reduce the populations of pollinators (Kelly *et al.* 2007), and introductions of non-native plants that compete with the native species for pollinators (Brown and Mitchell 2001, Brown *et al.* 2002).

Interaction of non-native species and habitat change

Like all ecological events and processes, impacts by non-native species do not occur in a vacuum. As Didham *et al.* (2007) emphasized, non-native species and habitat modification can interact to produce declines in native species. Didham *et al.* proposed a binary framework for understanding the interactive effects of habitat change and non-native species on native species (Fig. 7.5). In some instances, the two effects occur sequentially, a process the authors refer to as 'interaction chain effects.' In these cases, the habitat modification occurs first, which results in an increase in the numbers of non-natives. (The increase in non-natives following the habitat change could be due to an increase in propagule pressure, an increase in invasibility, or both.) The habitat modification may negatively impact the native species directly, but, in addition, the introduction and/or increase of non-natives into the environment then adversely impacts the native species, in ways such as those described earlier in this chapter. Key in the interactive chain effects is the fact that the per capita impact of the non-native species on native species remains constant and is not affected by the nature or extent of habitat modification. Didham *et al.* (2007) argue that this sort of interaction can be effectively quantified using path analysis.

Didham *et al.* referred to the second type of interaction as 'interaction modification effects.' In these instances, the habitat modification directly impacts the native species and also facilitates the introduction and establishment of the non-native species, as was the case with the interactive chain effects. However, in this case, the per capita impacts of the non-native species on the native species are not constant but are affected by the nature and extent of the habitat modification, a process the authors call 'functional moderation.' An example of the latter is a reversal in competitive dominance that may

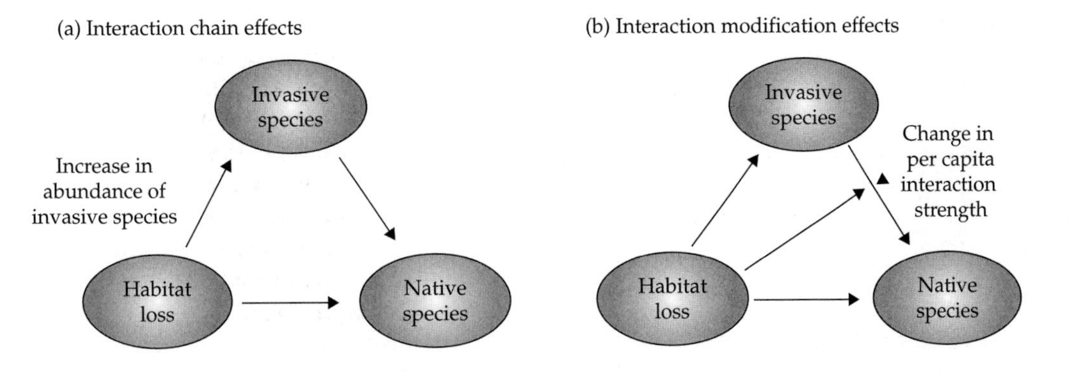

Fig. 7.5 Two major pathways by which habitat modification can interact with species invasion to increase the total impact on native species. (a) In an interaction chain effect, total invasive impact on native species is increased by the indirect effect of habitat modification on invader abundance, but the per capita invader impact (i.e. the slope of the abundance/impact relationship) remains constant at all levels of habitat modification. (b) In an interaction modification effect, total invasive impact is dependent not only on invasive abundance, but also on the degree to which habitat modification alters ecological interactions between invasive and native species (indicated by the opposing arrow-head symbols). For clarity, no feedback effects are shown between species invasion and habitat modification. Redrawn and printed, with permission, from Didham *et al.* (2007), copyright Elsevier Limited.

occur as a result of the habitat modification. Didham *et al.* cited a study of non-native earthworms in California (Winsom *et al.* 2006), which showed that the non-native species, *Aporrectodea trapezoides*, was outcompeted in semi-natural grasslands by the native species, *Argilophilus marmoratus*, because of the greater ability of the latter to acquire resources. However, in disturbed grasslands, the non-native species dominated because of its increased relative growth rates in the higher productivity system (Winsom *et al.* 2006). In the case of interactive modification effects, Didham *et al.* concluded that experimentation would be required to quantify the interactive effects on native species of habitat modification and non-native species.

Biodiversity impacts

The biodiversity impacts of non-native species have been an issue of contention. Claims that non-native species consistently, or even usually, threaten biodiversity have been challenged in recent years (Sax *et al.* 2002, Davis 2003, Gurevitch and Padilla 2004, Stohlgren *et al.* 2008a). With respect to terrestrial species, in many instances, the non-native species have been found to be more symptomatic of land-use change than causes of reduced native biodiversity, the latter declining in response to the human disturbances in the landscape (Maskell *et al.* 2006). MacDougall and Turkington (2005) effectively made this point with their passenger vs driver metaphor. In a comprehensive study of the serpentine flora in California, Harrison *et al.* (2006) found no evidence that non-native vegetation reduced any component of native herb richness. Rather, variation in herb richness could be fully explained by soil chemistry, disturbance, overstory cover, elevation, and precipitation.

As mentioned earlier, there is abundant evidence that introduced predators and pathogens can cause extinctions and hence reduce biodiversity on islands and in freshwater systems (Blackburn *et al.* 2004, Cox and Lima 2006, Sax and Gaines 2008); however, they have seldom done so on continental or in marine systems (Vermeij 1991, Davis 2003, Lotze *et al.* 2006, Reise *et al.* 2006; Fig. 7.6). For example, although more than 80 non-native marine species are believed to have established in the North Sea during the past 200 years, with respect to species-richness, their impact has been primarily additive, with no evidence that they have driven any native species to extinction (Reise *et al.* 2002, Reise personal communication). This may be the case with inland seas as well. Although more than 100 non-native species are believed to have been introduced into the Baltic Sea during the past two centuries, at least 70 of which have become established, no

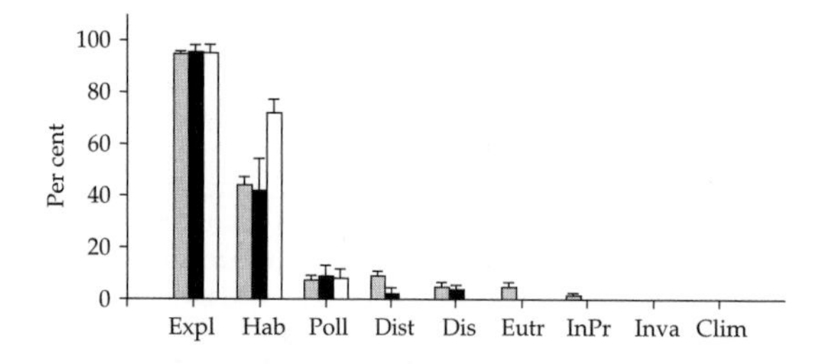

Fig. 7.6 Causes and consequences of change in 12 marine study systems (means + SEM). Per cent of species depletions (light gray) and extinctions (black) caused by different human impacts (Expl, exploitation; Hab, habitat loss; Poll, pollution; Dist, human disturbance; Dis, disease; Eutr, eutrophication; InPr, introduced land predators; Inva, invasive species; and Cli, climate change), and per cent of recoveries (white) resulting from impact reduction. Redrawn and printed, with permission, from Lotze *et al.* (2006), copyright American Association for the Advancement of Science.

extinctions of native species had been recorded as of 2002 (Leppäkoski *et al.* 2002, Ojaveer *et al.* 2002), and this was still the case at the end of 2007 (Leppäkoski, personal communication). In their characterization of the fauna in the Caspian Sea, Aladin *et al.* (2002) concluded that, while some of the introduced species certainly produced some undesirable effects, they primarily contributed to the Caspian rich biodiversity. In a study of the impacts of non-native species on coastal marine environments, Reise *et al.* (2006) stated that they 'found no evidence that they [non-native species] generally impair biodiversity.' On the contrary, they concluded that, more often than not, the new species expand ecosystem functioning by adding new ecological traits, intensifying existing ones and increasing functional redundancy.

The opening of the Suez Canal in 1869 enabled many residents of the Red Sea and the Indo-Pacific to move into the Mediterranean Sea, a phenomenon often referred to as the Lessepsian migration, named after the French engineer who supervised the construction of the canal, Ferdinand de Lesseps. Although there have been some local extinctions of some native species, the primary biodiversity impact on a regional scale has been a substantial increase in species-richness (Galil and Zenetos 2002). Likewise the species-richness of European aquatic coastal communities has been enhanced by the introductions of non-native species, particularly in the historically biodiversity-poor estuaries (Paavola *et al.* 2005, Wolff 2005). Reise *et al.* (2006) concluded that in coastal aquatic ecosystems, there is no support for an equilibrium perspective of community assembly, i.e. that if new species come in, others have to leave. This seems to be the case with plants on islands as well. Although the number of non-native plant species has increased steadily over the last few centuries (Fig. 7.7), very few native plant extinctions have been documented (Sax and Gaines 2008). While it is possible that there is an extinction lag, much evidence suggests that non-native, and even non-native invasive, species tend to become integrated into their new communities and environments, and that their negative impacts on native species usually diminish over time, due to evolutionary changes in both the native biota and the introduced species (Cox 2004). Nevertheless, while the introduction of non-native species may not always be causing extinctions, this does not mean they are not reducing biodiversity. They may still be causing declines in the abundance of some native species (Stinson *et al.* 2007), as well as the extirpations of some populations, which may result in the loss of genetic diversity (Galil 2006).

Despite causing extinctions and local declines in some environments, non-native species have generally increased regional biodiversity throughout the world, and in many instances have increased biodiversity at small local scales as well (Davis 2003, Sax and Gaines 2003). Sax *et al.* (2005d) described four ways that species-richness and species assemblages in native habitats might compare with those in the non-native environments (Fig. 7.8). First, species richness of particular taxa, e.g. birds, mammals, trees, may be similar in both habitat types (equivalency). Or, while the species-richness of one taxon might be lower in the non-native habitat, this decline might be compensated by an increase in richness in another taxon (compensation). In both of these instances, species-richness would be approximately the same in both habitats. However, if a decline in species-richness in one or more taxa are not compensated, then species-richness would be lower in the non-native environment (inhibition). Finally, if one or more species exhibit an increase in species-richness without a comparable

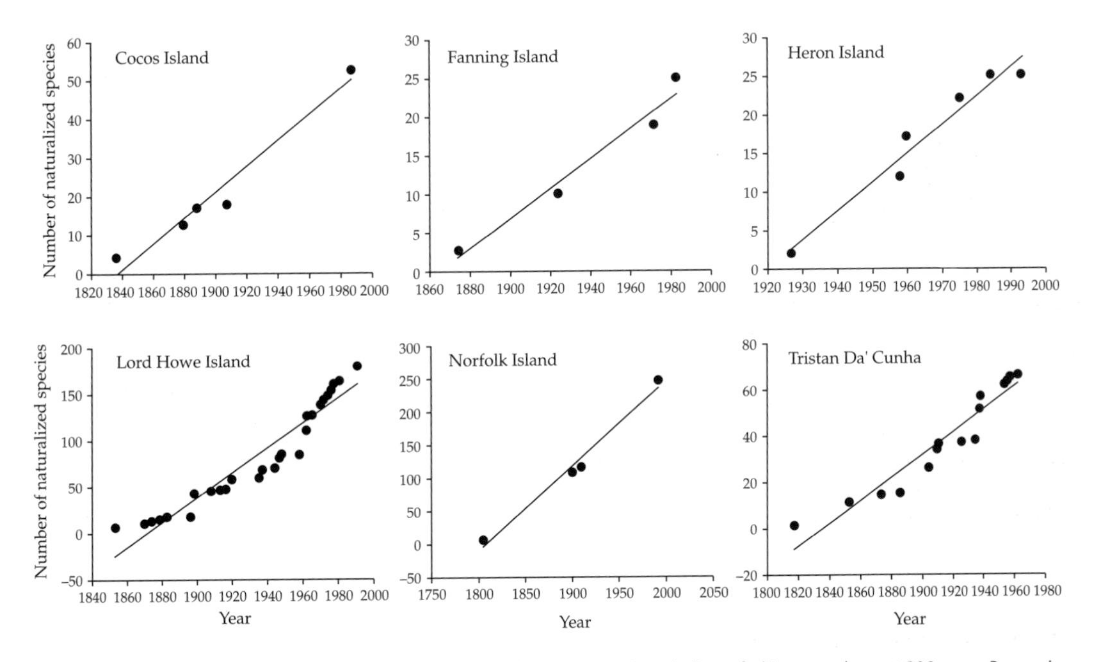

Fig. 7.7 Naturalized plant richness has increased on oceanic islands in an approximately linear fashion over the past 200 years. Regression lines are all highly significant. None of these islands show evidence of an asymptote in cumulative richness of naturalized species over time. Redrawn and printed, with permission, from Sax and Gaines (2008), copyright (awaiting approval).

decline in richness in other taxa, species-richness would increase (facilitation). Reviewing relevant literature, they found examples to support three of the four alternatives, the first, third, and fourth, although they felt that examples to support the second alternative will likely be found as more data sets are accumulated. Sax *et al.* emphasized that this simple graphical model could be used to compare single sites over time, as their species composition changes, as well as multiple sites in different locations.

Non-native plants may influence biodiversity by altering succession patterns. A number of cases have been documented in which historical old field succession dynamics are stalled due to the introduction and dominance of non-native species (Cramer *et al.* 2008). Through positive-feedback loops, the non-native species seem to be altering the environment to favor their own persistence and impede, or even prevent, the establishment of typically later successional species, such as woody plants. One way this can be accomplished is by altering the fire regime (Grigulis *et al.* 2005). Another is for the non-native species to produce dense stands that make it difficult for the later successional species to colonize (Cramer *et al.* 2008). The non-native species may also alter the soil-microbial community in a way that favors them, while making it more difficult for the woody species to establish (Davis *et al.* 2005d). Drake *et al.* (2008) studied the impact of non-native plant species on productivity–diversity relationships. Comparing the relationship in six different North American biomes, they found that the extent of invasion of an ecosystem did not necessarily affect productivity–diversity relationships, and concluded that the reason for this finding was the basic functional similarity between native and non-native species.

Biotic homogenization

Although the redistribution of species around the world has often resulted in regional and local increases in species-richness, there is no question that the world's biota has become more homogenized (Lockwood and McKinney 2001, Rahel 2002, 2007, Olden *et al.* 2006). Ultimately, biotic

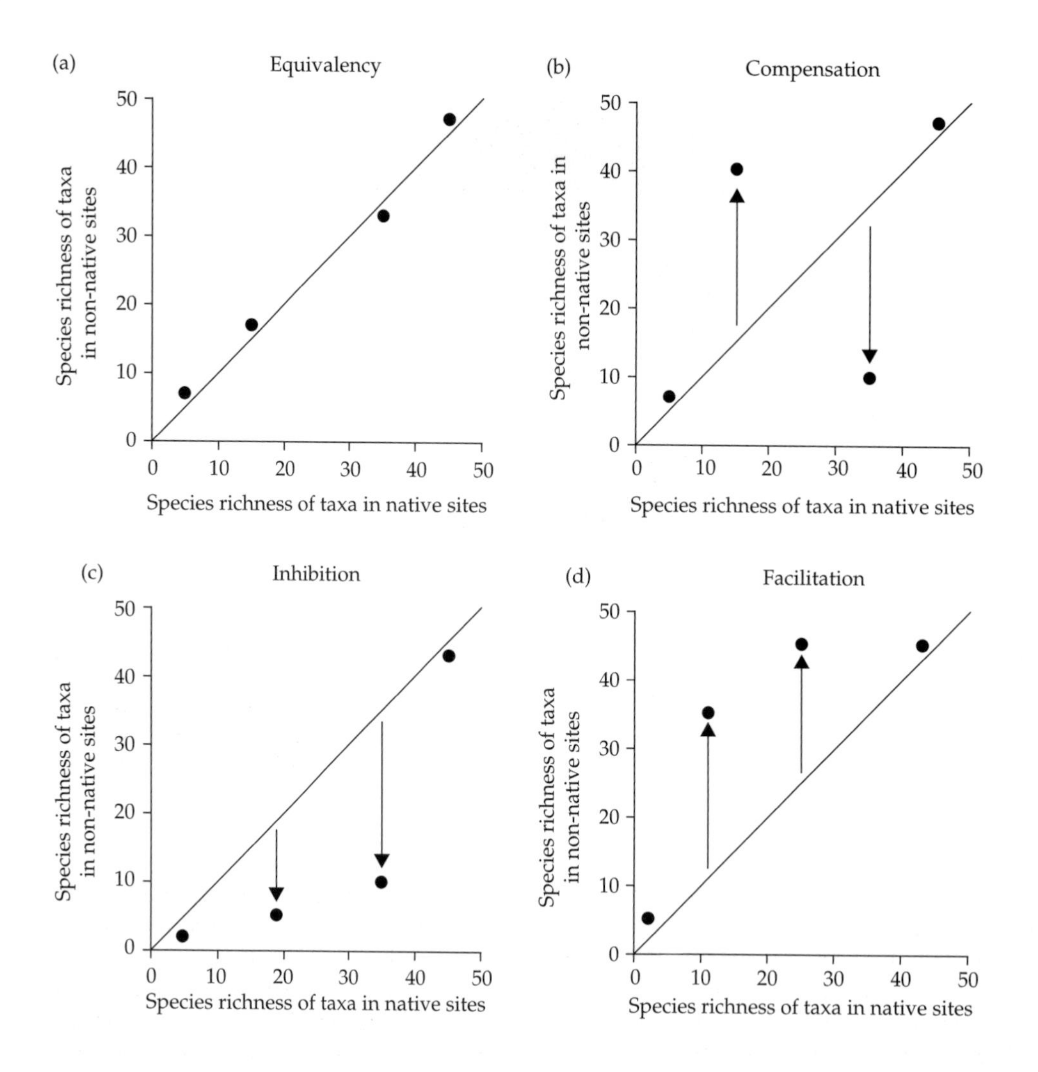

Fig. 7.8 Conceptual framework for comparing multiple taxonomic components of species assemblages in different (native and exotic) habitats. Taxonomic components (or taxa), which are represented as points in this figure, could include any natural division of organisms, such as birds, mammals, plants and insects. Any quantitative measure of a taxonomic component could be used, but here this framework is illustrated, with species richness values. (a Equivalency occurs when the (species richness) value of each taxonomic component that is compared between habitat types is roughly equal. (b) Compensation occurs when a decrease in the value of one component is compensated for by an increase in the value of another. (c) Inhibition occurs when one or more components decrease in value, without compensated increases in other components. (d) Facilitation occurs when one or more components increase in value, without compensated decreases in other components. Redrawn and printed, with permission, from Sax *et al.* (2005d), copyright Blackwell Publishing.

homogenization is the result of three interacting processes: introductions of non-native species, extirpation of native species, and habitat change that facilitates the first two processes (Rahel 2000, 2002, Olden and Poff 2003; Fig. 7.9). In the United States, the similarity in the fish faunas of the 50 states has increased dramatically since European settlement, a finding that was determined to be primarily due to widespread introductions of game fish, with extinctions of native species having less of an impact (Rahel 2000, Olden and Poff 2004). Rahel (2000) reported that 89 pairs of states

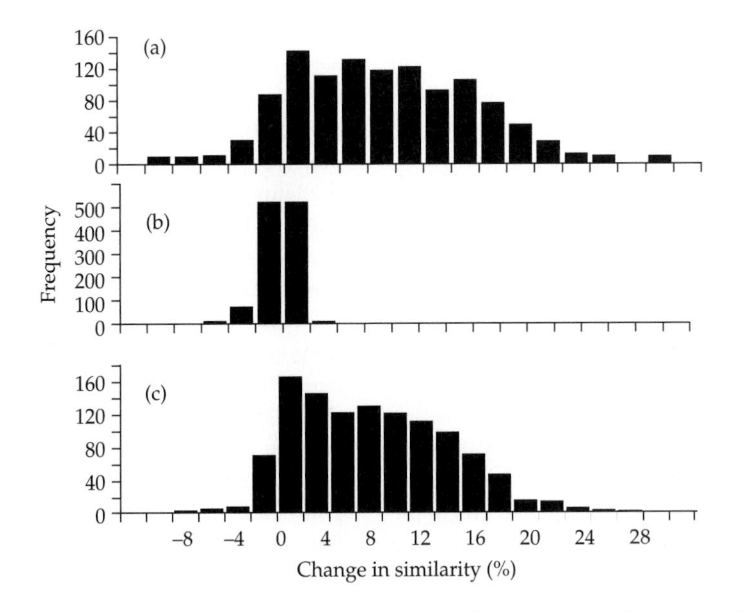

Fig. 7.9 Changes in similarity of fish faunas among 1128 pairwise combinations of the 48 coterminous United States. (a) Change in similarity based on combined effects of species extirpations and introductions. Distribution is skewed toward positive values, indicating fish faunas have become more similar by an average of 7.2%. (b) Change in similarity based on species extirpations only. Extirpations have caused negligible change in the similarity among state fish faunas. (c) Change in similarity based on introductions only. Distribution resembles than in (a), indicating most of the increased similarity in fish faunas is due to introduction of a group of cosmopolitan species. Redrawn and printed, with permission, from Rahel (2000), copyright American Association for the Advancement of Science.

in the US that had no species in common prior to European settlement shared, by the end of the twentieth century, an average overlap of 25 species. The effect on individual lakes due to introductions of this type is generally an increase in fish species-richness (Radomski and Goeman 1995). Vascular plants exhibit the same patterns. The introductions of large number of non-native species into the US have made the state floras more similar, but at the same time the species-richness of the respective state floras has also increased substantially as a result of the introductions (Sax *et al.* 2005e).

As pointed out by both Rahel (2000) and Olden and Poff (2003, 2004), it is possible that various combinations of introductions and extinctions can actually reduce the biotic similarity between sites. For example, the introduction of different species to environments that originally shared very similar biotas, and/or the extinction of different native species from the same environments, would result in reduced biotic similarity among the sites. The introduction of different aquarium fish in different California watersheds, which resulted in an increase of biotic heterogeneity, is an example of the former (Marchetti *et al.* 2001).

In some instances, although species-richness at local sites may be low, the species composition differs among the sites, meaning that the number of species residing in the region is still large. For example, in a multiple-scale analysis of plant communities in Great Britain, Smart *et al.* (2006) found that between 1978 and 1998, many individual sites experienced a decline in species-richness. However, during this same time, differentiation among sites in a locale increased. Smart *et al.* concluded that this relationship was due to human impacts on the environment, which resulted in the loss of many subordinate species at the site level, but that due to site-specific historical and environmental factors, it also resulted in species-poor communities that differed from one another in their species composition. Smart *et al.* referred to this landscape as a mosaic of species-poor communities. The findings of Rahel (2002), Davis (2003), Sax and Gaines (2003), Bruno *et al.* (2005), and Smart *et al.* (2006) are contrary to dire forecasts of a world eventually inhabited by only a small number of super-tramps.

Positive impacts on native species

Naturally, not all species introductions negatively impact native populations. A review of the literature published from 1993 through 2004 documented many instances in which the native species were facilitated by non-native species, including by non-native invasive species (Rodriguez 2006; Fig. 7.10). Rodriguez described five ways in which this facilitation occurred: habitat modification, trophic subsidy, pollination, competitive release, and predatory release. Rodriguez described the first three as direct mechanisms and the last two as indirect mechanisms. According to Rodriguez, habitat modification was the most frequently reported way in which the facilitation occurred, and included the creation of novel habitats and the increase in structural diversity of habitats, as well as the replacement of lost habitats due to the decline or disappearance of native species. An example of the former is the creation of hard substrate, a new habitat in some environments, by introduced bivalves (Crooks 1998, Bially and MacIsaac 2000). An example of the latter is the use of non-native plant species in restoration efforts in devegetated environments. Non-native grasses and forbs have been used in mine-reclamation projects in Illinois and Indiana, USA, and the resulting grasslands have created persistent refuges for many native grassland birds, partly compensating for the lost of prairie habitat in these regions (Scott *et al.* 2002). The importance of anthropogenic grasslands, dominated primarily by non-native grasses and forbs, in providing important habitat for native grassland birds was also documented in a study of avian use of old fields in east-central Minnesota (Goldsmith 2007).

In Rodriguez's scheme, trophic subsidy consists of nutrient enrichment and food augmentation and diversification, which can benefit native species (Quinos *et al.* 1998, Harding 2003). Although introduced green crabs, *Carcinus maenas*, are known to significantly reduce the populations of native prey species, they also may be responsible for the increase in other invertebrate species, which may benefit from the decline of their competitors, which happened to be the crabs' preferred prey (Grosholz *et al.* 2000), an example of facilitation through competitive release. An example of facilitation via predatory release involves the predation of native anurans by the cane toad, *Bufo marinus*, resulting in an increase in some of the preferred native prey species of the native anurans (Crossland 2000). While non-native pollinators can negatively impact some native species, e.g. the displacement of some native bee species by honeybees, *Apis mellifera*, (Thomson 2004), they can also supplement the pollination of plants by native pollinators, providing a particularly valuable function when the native pollinators have declined in abundance (Cox 1983, Gross 2001).

In some environments, fish are important seed dispersers (Correa *et al.* 2007), suggesting that the introduction of certain fish species could either enhance or suppress seed dispersal, depending on their impact. For example, the introduction of grass carp, *Ctenopharyngodon idella*, into reservoirs in Israel has increased seed germination of two native aquatic plants, and given the long-range movements of *C. idella*, it may increase long-distance dispersal (Agami and Waisel 1988). On the other hand, should introductions of non-seed dispersing fish result in a decline in the native seed-dispersing species, then the dispersal of some native plants would be compromised. For the most part, the impacts of fish introductions on seed dispersal of riparian plants have not yet been studied (Correa *et al.* 2007). Rodriguez concluded that facilitative effects of non-native species were most likely to occur when the non-native species provide a limiting resource, increase habitat complexity, replace a native species and fulfill their functional role in the ecosystem, or provide significant escape from enemies or competitors.

Additional studies since Rodriguez's review have documented similar instances where the impact of non-native species on native species was positive. The introduction of the Manila clam, *Tapes philippinarum* into British waters has apparently benefited the Eurasian oystercatcher, *Haematopus ostralegus ostralegus*, which now feeds extensively on the new species (Caldow *et al.* 2007). Simulations have

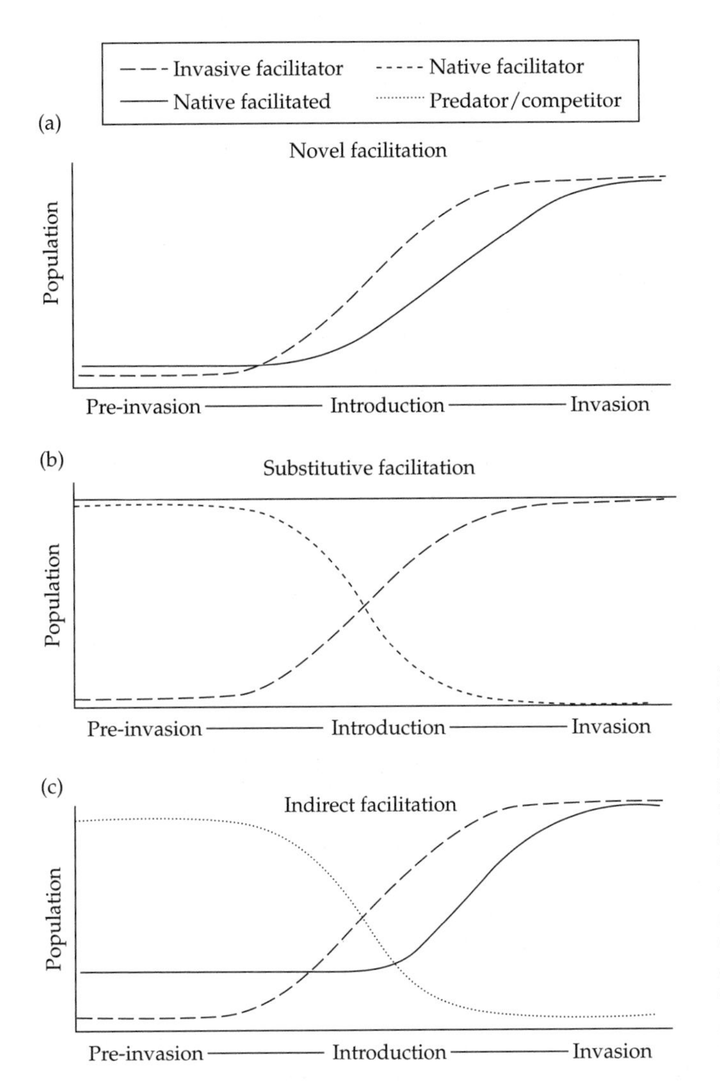

Fig. 7.10 Conceptual models for three scenarios that define why invasive species can facilitate native species. Depicted along a timeline of invasion events is the relative population size of different interacting species: invasive facilitator, native facilitated, native facilitator, and predator/competitor. Scenarios are: (a) novel facilitation, which occurs when no native facilitator existed; (b) substitutive facilitation, which occurs when an invader functionally replaces a native facilitator; (c) indirect facilitation, which occurs when the reduction of a predator or dominant competitor indirectly results in the facilitation of a native. Redrawn and printed, with permission, from Rodriguez (2006), copyright Springer.

predicted that the oystercatchers should experience reduced winter mortality due to this additional food source (Caldow *et al.* 2007). In the coastal waters of Washington, USA, the Asian hornsnail, *Batillaria attramentaria*, has been found to increase the densities of four other mudflat invertebrates, two native and two non-native (Wonham *et al.* 2005). Wonham *et al.* concluded that the primary facilitative effect on these four species by *B. attramentaria* was by providing hard substrate habitat for the species. In harsh environments, the establishment of many plants is facilitated by nurse plants, often shrubs or trees, which ameliorate the stringent physical conditions beneath and surrounding them (Padilla and Pugnaire 2006). Although invasive non-native plants often negatively impact resident plant species, they can also serve as nurse plants. For example, forest succession involving native tree species in Puerto Rico has been found to be facilitated by non-native tree species that established first (Lugo 2004) or were planted (Lugo 1997). A study of the association of soil mite assemblages with both native and non-native grasses found no effect of the grass type (native vs non-native) on the mites (St. John *et al.* 2006).

Impacts on food webs and communities

The introduction of the North American comb jelly, *Mnemiopsis leidyi,* into the Black Sea in the late 1980s rapidly and dramatically altered the aquatic food web. A predator with a broad diet (copepods, cladocerans, mollusks, fish eggs and larvae), it substantially decreased populations of zooplankton, and several economically valuable fish species (Kideys 2002). In turn, when another ctenophore, *Beroe ovata,* was introduced into the lake, it reduced the population of *M. leidyi* to extremely low numbers (Kideys 2002). *Cercopagis pengoi* (a predatory cladoceran) was introduced into portions of the Baltic Sea in the early 1990s and modeling efforts have indicated that, owing to its dietary overlap with several native fish species, it has the potential of causing a decline in the populations of these fish, including several commercially valuable species (Telesh *et al.* 2001). However, studies have also showed that *C. pengoi* is a common part of the diet of these fish (Antsulevich and Välipakka 2000), suggesting that the negative impacts of the introduced cladoceran on the fish populations might not be as substantial as feared.

Other examples of food web impacts include the invasion of the Asian clam, *Corbula amurensis,* into Northern San Francisco Bay, which has dramatically reduced phytoplankton abundance through its filter-feeding, in turn affecting, via a cascade effect, other trophic levels, e.g. herbivorous zooplankton and fish that feed on the zooplankton (Thompson 2005a). The introduction of the mongoose, *Herpestes javanicus,* on the Japanese island Amami-Oshima has been found to have substantially altered the island's food web by reducing the number of large native predators, which is believed to be the cause of an increase in the number of smaller animals (Watari *et al.* 2008). And, in Kenya, the introduction of the Louisiana crayfish, *Procambarus clarkii,* has reduced the abundance of native clams, a primary part of the diet of the African clawless otter, *Aonyx capensis,* and is believed to be a major cause for the decline of the otter (Ogada 2005).

The sea lamprey, *Petromyzon marinus,* an external parasite of fish, caused the collapse of several populations of large native fish species following its introduction into the Great Lakes (Christie 1974). The decline in top predators enabled the populations of several prey species to explode to high numbers, including the introduced alewife, *Alosa pseudoharengus.* Effects on this trophic level then cascaded down to the next one as several planktivorous species, the prey of the alewife and similar fish, declined precipitously (Christie 1974). The alewife invasion even disrupted courtship for some terrestrial mammals. I grew up on the western shore of Lake Michigan in the 1950s and 1960s, and painfully recall the alewife explosion in the 1960s, during which our beautiful sandy beaches became the final repository for millions of dead and rotting alewife. Their rancid smell and sharp fins and bones cut short many a moonlit and barefoot walk along the beach, substantially reducing the number and quality of romantic opportunities for an entire generation of Great Lakes teenagers. The author can attest that, for those romantically challenged teenagers who needed all the help they could get from moonlit beach ambiance, the alewife invasion was a tragic development.

Human harvesting of marine fish often involves taking the top predators, a process sometimes referred to 'fishing down the food web' (Pauly *et al.* 1998, Essington *et al.* 2006). Human impacts on freshwater food webs are often in the opposite direction due to the frequent introductions of non-native game fish that are top predators, e.g. largemouth bass, *Micropterus salmoides,* smallmouth bass, *M. dolomieu,* striped bass, *Morone saxatilis,* northern pike, *Esox lucius,* walleye, *Stizostedion vitreum,* rainbow trout, *Oncorhynchus mykiss,* brown trout, *Salmo trutta,* and brook trout, *Salvelinus fontinalis,* which have been stocked in lakes throughout North America and Europe (Eby *et al.* 2006). These introductions, a process that has been termed 'stocking up freshwater food webs' can increase the top-down control effects, thereby changing the abundance of species in lower trophic levels (Eby *et al.* 2006). In some cases, the introductions have increased the diversity of top predators in freshwater systems, with subsequent food web impacts, including a decline in littoral prey-fish diversity

(Vander Zanden *et al.* 1999). In other instances, the introductions have resulted in an overall decline in piscivorous fish diversity and simplification in the food web (Goldschmidt *et al.* 1993). In estuarine and marine coastal systems, loss of top predators due to overharvesting and habitat degradation has skewed dominance of these systems toward lower trophic levels (Stachowicz *et al.* 2007), a phenomenon that has been accelerated by the fact the majority of introduced species into these systems have been lower trophic-level organisms (Byrnes *et al.* 2007; Fig. 7.11).

Weasels (*Mustela* spp.) typically prey on both birds and small mammals. Previous studies of

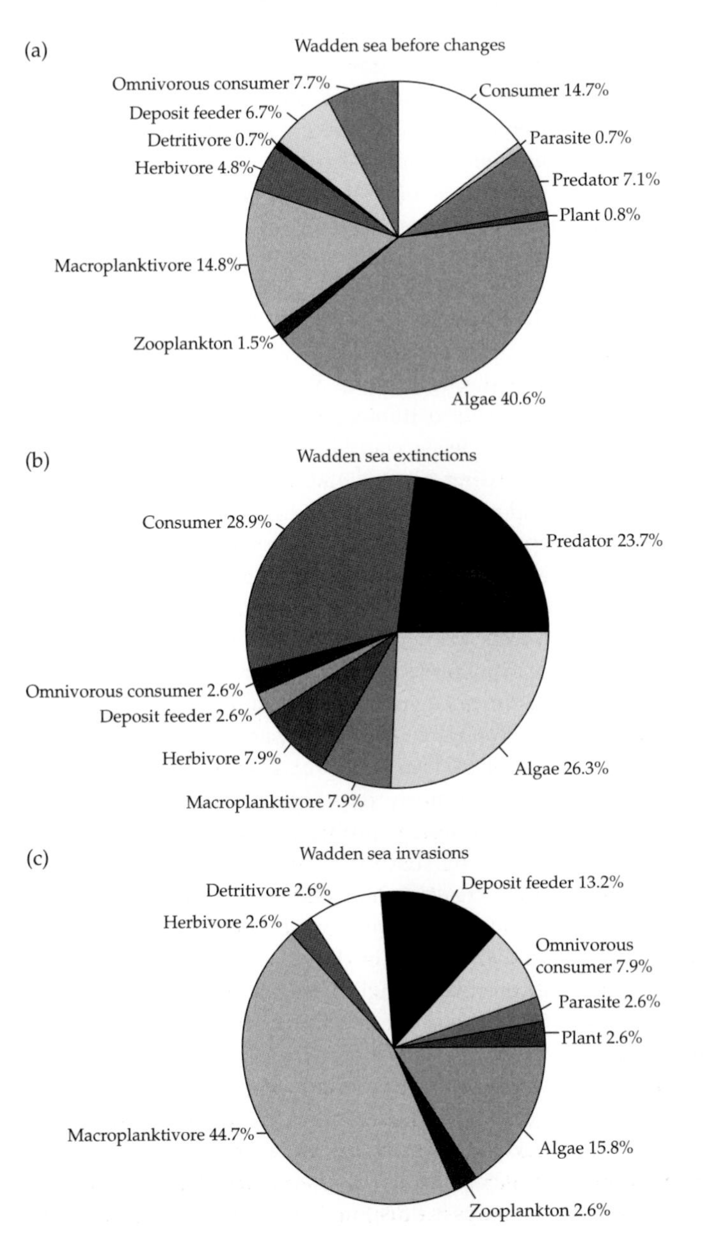

Fig. 7.11 An illustration of how invasions and extinctions can alter the species richness of different trophic levels. Shown are data from the Wadden Sea: (a) species-richness by trophic levels of communities prior to invasion and recent extinctions; (b) species-richness by trophic levels of the organisms that have gone extinct in these marine communities; (c) species-richness by trophic levels of the marine organisms that have been introduced into these communities. Shading levels indicate trophic level: white = 1 (bottom), light grey = 2, dark grey = 3, black = 4 (top). Species at higher trophic levels have been more likely to go extinct, while introduced species are more likely to be lower trophic organisms. Redrawn and printed from Byrnes *et al.* (2007).

weasels in their native environments have found that the weasels' impact on birds is less during years of high-density small mammal populations (Dunn 1977, Tapper 1979), meaning that the resident birds get periodic respite from high weasel predation. However, such predation respites may not occur, or may occur less frequently, in regions where the weasels have been introduced, due to the composition and population dynamics of the small mammal species in the new environment. This was determined to be the case in New Zealand beech forests, in which the most abundant small mammal, the introduced house mouse, *Mus musculus*, seldom reached high enough densities to reduce bird predation by *Mustela erminea* to the extent found in Europe during irruptions of the European small mammal populations (White and King 2006).

Introductions of plant species can also impact food webs. The spread of the *Spartina* hybrid (*Spartina alternifolia* × *S. foliosa*) into San Francisco Bay has shifted the food web from one that was algae-based to one that is detritus-based (Levin *et al.* 2006). It is expected that this trophic shift will produce future declines in populations of some of the resident fishes, as well as migratory birds that rely more on the algae-based food web. The introduction and spread of non-native forage grasses in the southern US was found to alter the abundance and composition of ground invertebrates, which is believed to be at least partly responsible for the reduced abundance of grassland birds in these non-native grasslands, particularly those species that forage on or near the ground (Flanders *et al.* 2006). The spread of Canada thistle, *Cirsium arvense*, in Yellowstone National Park, USA, is reported to be influencing the feeding habitats of both gophers and grizzly bears (Robbins 2008). Attracted to the starchy tubers, the gophers feed extensively on the plant, stockpiling some of the tubers for future consumption. Being very adaptive and flexible foragers, the bears, in recent years, have begun to seek out and raid the gopher caches. During their excavations, the bears often also come upon gophers and their young, which they then eat as well.

In urban areas, non-native plant species may provide crucial nectar resources for native butterflies (Shapiro 2002), and many native insect species have been found to oviposit and feed on non-native plants (Graves and Shapiro 2003, Cox 2004). On the other hand, there is concern that garlic mustard, *Alliaria petiolata*, is causing a decline in some native butterfly species (Renwick *et al.* 2001). In this case, the butterflies that oviposit on *A. petiolata* exhibit lower larval survivorship than those that oviposit on native species. Bukovinszky *et al.* (2008) found that variation in the food quality of plants, even between populations of the same species, can significantly affect the resulting food web. In their study, Bukovinszky *et al.* found that a herbivore–parasitoid–secondary parasitoid food web differed markedly depending on whether the herbivores (aphids) fed on feral or domesticated population of *Brassica oleracea* (Brussel sprouts, var. *gemmifera*). Specifically, the researchers concluded that feral *Brassica* was a better host for the aphids, and hence also benefited the parasitoids, than was the domesticated strain, as evidenced by the increased size and fitness of both the aphids and the parasitoids. Bukovinszky *et al.* believed that this difference in response by the higher trophic levels was due to differences in plant traits, including plant metabolites, defense chemicals, and plant architecture. In some cases, non-native species may influence the food web by facilitating the transfer of environmental contaminants. Studies in both the Great Lakes and the Rhine–Meuse basin have shown that *Dreissena polymorpha* is likely contributing to the intake of heavy metals and toxic organics by waterfowl, which now ingest *D. polymporha* as a major part of their diet (Mazak *et al.* 1997, Hendriks *et al.* 1998).

Garlic mustard is believed to be altering the tree species composition of many North American forests by disrupting mycorrhizal relationships (Rodgers *et al.* 2008). In soils conditioned by garlic mustard, Stinson *et al.* (2006) found a decline, almost to zero, in the colonization by arbuscular mycorrhizal fungi of the fine roots of sugar maple (*Acer saccharum*), red maple (*Acer rubrum*), and white ash (*Fraxinus americana*) in soils, which was associated with a dramatic decline in the growth rates of the seedlings (Fig. 7.12). That garlic mustard is impacting other plant species, both woody and herbaceous, by disrupting mycorrhizal relationships is illustrated by the fact that the negative effects of garlic mustard on growth rates of the other plant species is positively associated with the mycorrhizal dependency of the other species (Fig. 7.13).

Concern by ecologists over these food web and community impacts is similar to the concern of newly structured communities expected as a result of changing climate conditions (Williams *et al.* 2007). Referred to as no-analog communities (Williams and Jackson 2007), these consist of new combinations of species, a sort of reshuffling of the species occurring as a result of different migration rates

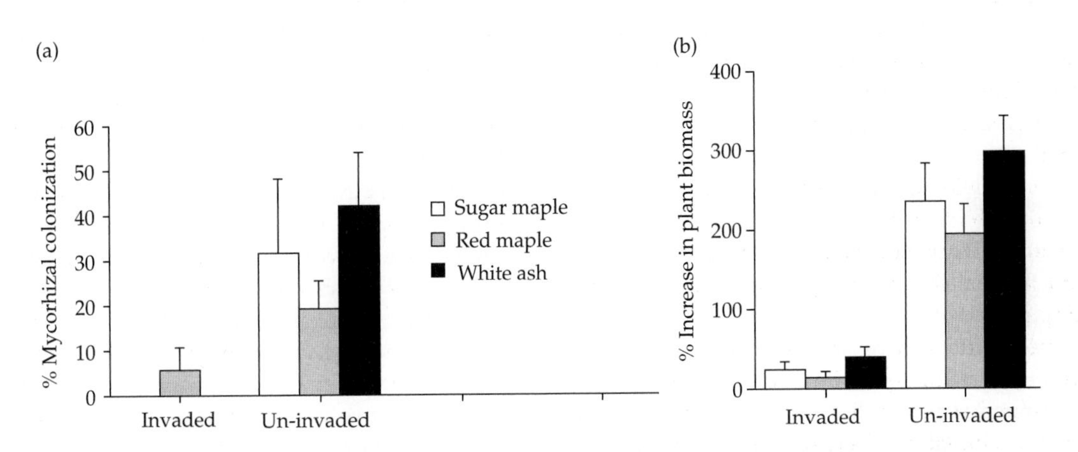

Fig. 7.12 The effect of garlic mustard invasion on (a) mycorrhizal colonization and (b) growth of native tree seedlings (mean + 1 SE). Redrawn and printed from Stinson *et al.* (2006).

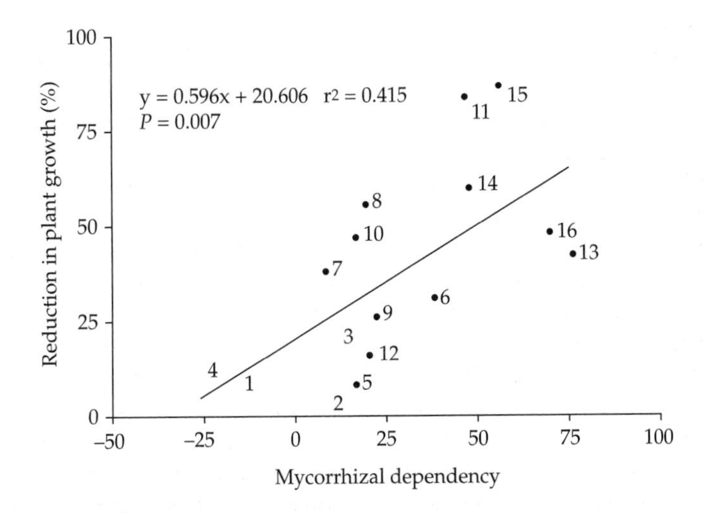

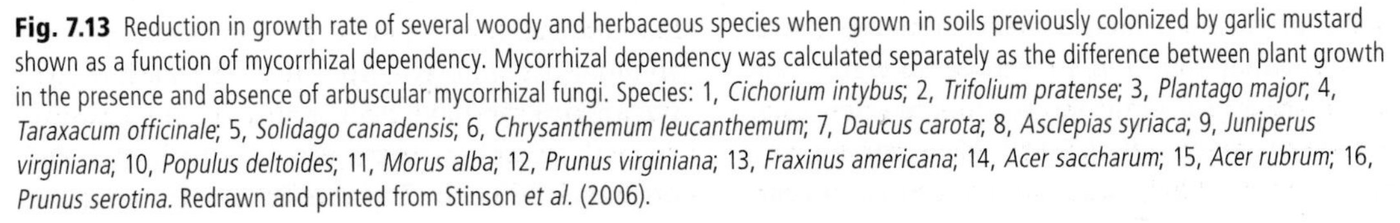

Fig. 7.13 Reduction in growth rate of several woody and herbaceous species when grown in soils previously colonized by garlic mustard shown as a function of mycorrhizal dependency. Mycorrhizal dependency was calculated separately as the difference between plant growth in the presence and absence of arbuscular mycorrhizal fungi. Species: 1, *Cichorium intybus*; 2, *Trifolium pratense*; 3, *Plantago major*; 4, *Taraxacum officinale*; 5, *Solidago canadensis*; 6, *Chrysanthemum leucanthemum*; 7, *Daucus carota*; 8, *Asclepias syriaca*; 9, *Juniperus virginiana*; 10, *Populus deltoides*; 11, *Morus alba*; 12, *Prunus virginiana*; 13, *Fraxinus americana*; 14, *Acer saccharum*; 15, *Acer rubrum*; 16, *Prunus serotina*. Redrawn and printed from Stinson *et al.* (2006).

and new combinations of climate variables. The widespread introductions of new species similarly provide more raw materials for this reshuffling. Of course, the comparison communities, the analogs, are simply the ones with which we are familiar. The future reshuffled communities will be neither more nor less ecologically correct or appropriate than the current communities, or the communities of the past, the composition of which often differed substantially from that of the present communities, due to different climate regimes (Williams *et al.* 2001) or different dispersal rates (Davis *et al.* 1986).

Impacts on biogeochemical processes

One of the first reported cases of ecosystem impacts of a non-native species was by Vitousek *et al.* (1987), who documented that the nitrogen-fixing introduced shrub, *Myrica maya*, increased by four-fold the nitrogen input of the historically nitrogen-poor volcanic soil in which it was establishing. Similar impacts by nitrogen-fixing introduced plant species have been since documented elsewhere (Witkowski 1991, Haubensak 2001). Nitrogen levels in the leaf litter of common buckthorn, *Rhamnus cathartica*, an invasive shrub/small tree in North America, are high compared to most native trees in the same environments, facilitating faster decomposition and increased soil nitrogen under buckthorn stands (Knight *et al.* 2007). In turn, the high-nitrogen litter may facilitate the establishment and spread of non-native earthworms (Knight *et al.* 2007). In some instances the changes in N cycling can be so substantial, affecting not only species composition but disturbance regimes, that the systems affected by these changes can be considered to have been moved into an alternative stable state (D'Antonio and Hobbie 2005).

Recent studies have shown that changes in nitrogen cycling caused by non-native plants can be due to changes in the abundance and composition of ammonia-oxidizing bacteria in the soil (Hawkes *et al.* 2005). This suggests that, even if the non-native plants can be eradicated from an area, the altered soil fauna may represent an ecological legacy having a lasting impact on the ecosystem (Heneghan *et al.* 2002, Hawkes *et al.* 2005). Non-native plants have also been found to alter the carbon and water cycles in their new environments. Studies on North American prairies concluded that the grasslands store less carbon when dominated by crested wheatgrass, *Agropyron cristatum*, compared to when the systems are dominated by native grasses (Christian and Wilson 1999, Curtin *et al.* 2000). Litton *et al.* (2006) found that replacement of native forests with non-native grasslands in Hawaii resulted in a 93% reduction in above-ground live biomass, indicating the dramatic reduction in the size of the above-ground carbon pool. Some non-native species have been found to increase evapotranspiration rates, thereby drying the soils and reducing runoff into nearby waterways (Mack *et al.* 2000, Dye *et al.* 2001, van Wilgen 2004, Mark and Dickinson 2008).

In a study of off-shore islands in New Zealand, Fukami *et al.* (2006) found that the soil nutrient levels and soil fauna differed significantly between rat-populated and rat-free islands. Soil basal respiration and litter decomposition also differed between the two island types, with rat-populated islands exhibiting lower levels of the former and higher levels of the latter. Fukami *et al.* provided evidence that the effects were substantially due to the rats' impact on seabird usage of the islands for breeding, the birds largely avoiding islands inhabited by the predatory rats. The birds' absence resulted in lower nutrient inputs (via guano) and also less trampling of the vegetation; the latter has been found to reduce tree-seedling generation (Maesako 1999). This example illustrates the range of cascading effects that can result from the introduction of a single predatory species into an island environment. In this case, Fukami *et al.* concluded that, from the point of view of the vegetation, the cascading sequence of events produced negative effects below ground (reduced nutrient supply) but positive effects above ground (reduced physical disturbance by trampling).

The introduction of game fish into freshwater systems is believed to significantly impact biogeochemical cycles (Eby *et al.* 2006). Introductions of top predators are known to produce substantial changes in the food web, and changes in the abundance of planktivorous fish, zooplankton, and phytoplankton have been found to impact nitrogen and phosphorous cycling (Schindler *et al.* 1993, Elser *et al.* 1998, Findlay *et al.* 2005). Zebra mussels,

Dreissena spp. were found to have significantly altered the cycling of phosphorus, ammonia, and nitrate in Lake Erie (Strayer *et al.* 1999). Non-native forest insect pests have been shown to substantially alter energy and nutrient fluxes in the impacted forests (Lovett *et al.* 2006). These effects can be both short-term (weeks to years), involving processes effected by the damage to or death of trees, and long-term (decades to centuries), involving changes in the species composition of the forests (Lovett *et al.* 2006). In some cases, the effects on nutrient fluxes of forest insects have been found to change over time. For example, Stadler *et al.* (2006) found that in forests 'invaded' by the hemlock wooly adelgid, *Adelges tsugae*, nitrogen fluxes decline early in the infestation but increase later as the hemlocks are gradually replaced with deciduous trees.

Except in cases where fungal species have caused substantial harm, e.g. chestnut blight, Dutch elm disease, and extensive frog mortality (chytrid fungus), little attention has been given to the spread and potential ecosystem impacts of non-native fungi or other soil microbes until recently (Schwartz *et al.* 2006, Desprez-Loustau *et al.* 2007, van der Putten *et al.* 2007b). The spread of fungi is expected to increase as more efforts are made to introduce mycorrhizal fungi in an attempt to improve agricultural and forest productivity (Gianinazzi and Vosátka 2004, Duponnois *et al.* 2005), and success in restoration and bioremediation (Miller and Jastrow 1992, Leyval *et al.* 2002). Currently, it is very difficult, verging on the impossible, to predict ecosystem impacts of non-pathogenic fungal species to a great extent because even base-line data on native fungal communities do not exist (Desprez-Loustau *et al.* 2007). In order for mycologists and ecologists to discover the ecosystem impacts of non-pathogenic introduced fungi, progress needs to be made in several areas, including the development of protocols for rapid identification of fungal species, increasing basic life-history knowledge of fungal taxa, and increasing understanding of community-wide interactions among fungal species (Schwartz *et al.* 2006).

Halpern *et al.* (2008) conducted a comprehensive evaluation of human impact on marine ecosystems. As shown in Fig. 7.14a, species invasions are believed to impact a very small proportion of marine area. However, they are believed to affect a proportionately larger area of coastal environments (Fig. 7.14b). It should be noted that the projections presented by Halpern *et al.* were based primarily on expert judgment and not on actual data. Obtaining basic data on ecosystem impacts of human activities in these marine environments, including the introduction of non-native species, was identified by the authors as the important next step in this research.

Ultimately, the impact of an introduced species on ecosystem processes is going to be context-dependent, meaning that the impacts will vary between habitats and also with changing environmental conditions within a habitat (D'Antonio and Hobbie 2005, Reise *et al.* 2006). Factors that will affect the nature and extent of the impact include the extent to which traits of the new species are already present among the native residents, whether or not the native residents respond to the new species in a way that would influence the abundance and/or per-individual impact of the new arrival, and whether or not any abiotic factors and/or processes might change in such a way as to buffer the impacts of the new species (D'Antonio and Corbin 2003). It needs also to be remembered that negative ecosystem impacts by non-native species are not inevitable. While individual non-native species may produce a particular effect, this effect may be buffered, not only by native species but non-native ones as well. Thus, as more species are added to a community, while there is the possibility that the singularity of one of the species may cause major changes in the environment, it is perhaps more likely that the singularities of a large number of species may cancel out the effects of one another, thereby giving rise to some stability and regularity (Pueyo *et al.* 2007). For example, Reise *et al.* (2006) concluded that the addition of non-native species to European coastal waters had not produced any observable directional impact on the coastal ecosystems, nor was there any evidence that the non-native species had impaired biodiversity or ecosystem functioning. Instead, the new species more often added new ecological traits and increased functional redundancy, thereby enhancing ecosystem functioning (Reise *et al.* 2006).

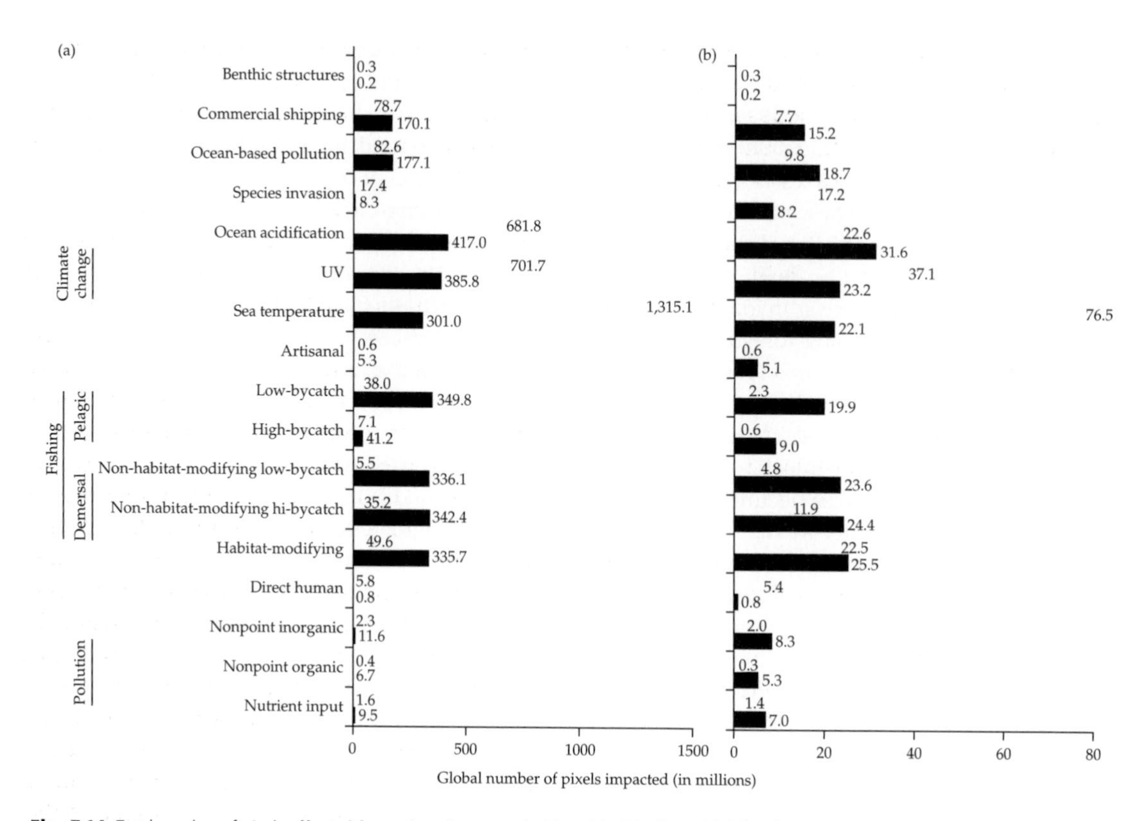

Fig. 7.14 Total number of pixels affected for each anthropogenic driver (a) globally and (b) for all coastal regions <200 m in depth. Values for each bar are reported in millions. Each pixel in the analysis represented 1 km² of ocean area. Redrawn and printed, with permission, from Halpern *et al.* 2008, copyright American Association for the Advancement of Science.

From an ecosystem perspective, it may matter whether species-richness of an environment increases or decreases as a result of species introductions. Hector and Bagchi (2007) argued that in grassland environments, high species-richness may be necessary to maintain multiple ecosystem processes at desired levels. It is important to note that Hector and Bagchi's argument was not based on a diversity impact *per se*; rather, they argued that because different processes are affected differently by different species, high species-richness will increase the probability that key species will be present. This emphasizes the point that, from an ecosystem perspective, it may not be as important that introduced species increase or decrease the system's species-richness, as it is whether species that substantially affect important ecosystem processes are added or lost as a result of the introduction.

Impacts altering the physical structure of the environment

In some cases, the ecological impact of non-native species, like that from many native species, substantially affects the physical environment, which in turn influences the success of other species. In these instances, the species producing them are often referred to as 'ecosystem engineers' (Jones *et al.* 1994, Crooks 2002, Hastings *et al.* 2007), a term that is applied to both native and non-native species, or 'transformer species', which was proposed to describe such effects by non-native species (Richardson *et al.* 2000a). (In addition to describing impacts on the physical structure of the environment, the term 'transformer species' was intended to also describe species producing other types of major impacts, such as substantially affecting resource availability.)

The foraging behavior of non-native herbivores can physically alter the host environment, in ways that affect the native species. One example of this phenomenon is the impact of non-native rabbits on the breeding seabirds of Macquarie Island, Tasmania (Scott and Kirkpatrick 2008). The recent explosion of the rabbit population (suspected due to a moderating climate, the decline of a primary predator, feral housecats, and/or the reduced effectiveness of an introduced myxoma virus) has resulted in herbivory levels that are so extensive that the ability of the vegetation to stabilize the soils is being compromised. This is resulting in landslides and the loss of breeding habitat for the seabirds, as well as direct mortality of eggs and hatchlings.

Grazing by sheep has been found to reduce structural diversity and plant biodiversity (Van Vuren and Coblentz 1987), and rooting of wild pigs can reduce above-ground plant biomass and reduce tree-seedling establishment (Sweitzer and Van Vuren 2002). Introduced Asian carp species have drastically altered the physical structure of many of their new aquatic environments, virtually eliminating aquatic macrophytes and substantially increasing the turbidity of their waters (Lehtonen 2002). In some instances, introduced species may add to the physical complexity or heterogeneity of an environment. As described earlier, the shells of zebra mussels, *Dreissena polymorpha*, often create additional hard surface habitat (Ricciardi *et al.* 1997), and foraging of feral pigs in a wetland environment was found to increase microhabitat diversity (Arrington *et al.* 1999).

The introduction and spread of European earthworms (Lumbricidae) into the cold-temperate and mixed deciduous-coniferous forests of North America has dramatically altered the detritivore communities and processes in these environments, which are having major impacts on the resident plant communities (Bohlen *et al.* 2004, Frelich *et al.* 2006, Hale *et al.* 2006, Holdsworth *et al.* 2007). Through their foraging and movements, the earthworms reduce the thickness of the litter layer, processing and distributing the litter and humus material into deeper soil layers. In some cases, the earthworm activity leads to reduced nitrogen and phosphorus availability (Frelich *et al.* 2006). These, and possibly other impacts, e.g. effects on fungal communities (Scheu and Parkinson 1994) are believed to be responsible for declines in many of the native forest herbs and tree seedlings (Hale *et al.* 2006).

The ability of non-native plant species to alter fuel properties and hence modify natural fire regimes has been well-documented (D'Antonio 2000, Brooks *et al.* 2004). The fuel properties of an environment can be altered if the non-native species produce a greater fuel load and/or increased horizontal fuel continuity, and if the plant tissues of the non-native species differ in moisture content and/or chemical composition of their tissues (Brooks *et al.* 2004). For example, the introduction of non-native grasses often increases both fuel loads and horizontal fuel continuity, resulting in hotter and more frequent fires (Brooks 1999, Rossiter *et al.* 2003). When altered fire regimes change the composition of the vegetation, the resident animal populations are often affected as well (Knick *et al.* 2003). In many cases, a positive-feedback loop is created, in which the non-native species alters the fire regime, which in turn reinforces the persistence of the non-native species (Mack and D'Antonio 1998). It has been suggested that changes in fire regimes caused by the introduction and establishment of non-native invasive species may actually alter regional climate patterns. For example, in areas of the Amazon basin, the spread of non-native grasses has created a fire regime that perpetuates the grassland and prevents the tropical forest from reestablishing, which reduces evapotranspiration, a primary input of atmospheric water in this region (Mack *et al.* 2000).

Non-native plant species can alter the physical environment in other important ways. In a study of Hawaiian montane flora, Daehler (2005) found that a large number of woody species have naturalized above 2000 m, substantially altering the physical environment. Some, like the non-native *Pinus*, add an entirely new structural form to the environment. Others form dense stands restricting native plant growth and recruitment. Sometimes, the planting or natural establishment of non-native tree species in substantially human-disturbed environments can dramatically impact the physical structure of the environment in ways that are desirable.

Pine and eucalyptus plantations planted on steep Ecuadorian slopes, previously cleared of vegetation by humans, were found to reduce erosion back to near the levels before the slopes were cleared (Vanacker *et al.* 2007). In a review of non-native ecosystem engineers, Crooks (2002) concluded that species that increased habitat complexity or heterogeneity tended to increase species-richness in the new environment, while species that reduced those habitat attributes usually caused a decline in species-richness.

In some instances, ecologists have proposed using non-native species in habitat restoration plans precisely because of their engineering impacts. In suggesting ways to restore salinized lands in Australia, Byers *et al.* (2006) proposed the planting of salt-tolerant trees and shrubs (including non-natives) with a variety of rooting depths in order to promote the downward movement of salt through the soil. Byers *et al.* indicated that the non-native species could then be removed once the salt levels had been sufficiently reduced.

The creation of novel ecosystems

As described in the preceding pages, the introduction of non-native species can change the nature and processes of an environment in many ways. While the impacts may be relatively minor in many cases, in others the biotic community and the ecosystem as a whole may be altered substantially. Sometimes the changes may result in new combinations of ecosystem patterns and processes, prompting some to refer to these as 'adventive ecosystems' (Seastedt 2005) or 'novel ecosystems' (Hobbs *et al.* 2006). In a similar vein, Williams and Jackson (2007) referred to 'no-analog communities' in their discussion of the new communities that are expected to result from climate change. These communities would experience novel climatic regimes, e.g. new seasonal combinations of temperature and precipitation, as well as new combinations of species. Williams and Jackson emphasized that these no-analog climatic-induced communities would also interact with new species introduced intentionally or accidentally by humans, i.e. biological invasions, to produce yet more 'ecological surprises.' It is important to recognize that novel, or no-analog, communities are being created not only by introduced species. For example, Nowacki and Abrams (2008) argued that the decline of fire in eastern US forests has created a positive-feedback cycle, which they refer to as 'mesophication,' that continually favors shade-tolerant and fire-intolerant species. The result, they conclude, is the development of ecosystems and plant communities with no ecological antecedent

The challenge posed by these future 'ecological surprises' and 'no-analog systems,' according to Williams and Jackson (2007), is that, for modeling purposes, these represent uncharted space, for which we have no empirical data to parameterize and validate model forecasts. While Williams and Jackson make a good point, it is not clear that current models are as severely handicapped as W & J suggest. In a fanciful comparison, W & J likened the impending novel climate regimes with the uncharted regions of the world during the era of European exploration, regions which on maps were sometimes inscribed with the foreboding words 'here there be dragons.' One problem with terms like 'no analog communities' and 'novel ecosystems' is that they are suggestive of qualitative differences between these and current ecosystems and communities. While the Europeans may have viewed different regions and peoples of the world as qualitatively different, there is no reason to believe that novel climate regimes, ecosystems, and communities will represent fundamentally different systems. Organisms will be influenced by the same sorts of events and processes as they are now (Lugo 1994). The players in a particular ecosystem may be different, but the fundamental nature of the processes should not be. While the challenges and pitfalls of ecological modeling and forecasting should never be underestimated (Pilkey and Pilkey-Jarvis 2007), if the differences between the dynamics of current and future systems will mainly be quantitative in nature, then current models should have something worthwhile to say with respect to future changes, whether they involve climate change or species introductions, or both. However, not everyone agrees with this perspective. Ricciardi (2007) argued that current invasions do represent a unique ecological and evolutionary phenomenon, for which understanding of prior environments can provide only limited insights.

Summary

Since our health, safety, economies, and environments are all impacted by various native species, often in ways we deem harmful, it is to be expected that non-native species would likewise produce these sorts of effects. The impacts of most non-native species are minor, as far as we can tell at this point, and some have been distinctly beneficial in their effects. Nevertheless, a significant number have caused, are causing, or are threatening to cause great harm. For humans, the species of greatest concern are those non-native species that threaten our health, primarily pathogens and their vectors. Most would probably agree that next in importance are those species causing, or threatening to cause, substantial economic harm, including species that are damaging ecological services.

Although societies have not normally been as concerned about other sorts of ecological impacts, non-native species have been found to affect their new environments in a variety of ways, sometimes substantially. The types of impacts include effects on individual organisms, populations, food webs and communities, disturbance regimes, biogeochemical processes, and the physical structure of the environment. In some cases, their introduction has even resulted in the creation of novel ecosystems. The new species sometimes cause a reduction in the species-richness of the new environment, and in insular habitats like islands and freshwater systems, introduced predators and pathogens have driven native species to extinction. However, more commonly, species introductions have increased species-richness at regional levels, and often at local levels as well. At the same time, introductions have typically increased the biotic similarity of different regions, although this is not always the case.

Of course, even without human remediation and restoration efforts, few of these impacts—health, economic, and ecological—would be permanent. The extent and nature of the impacts of introduced species would inevitably be altered due to changes in both the the biotic and abiotic environment, engineered by natural selection and other biological and physical processes constantly occurring at local and regional scales. However, in many cases, patience is not an option for us, or at least it is not an option we want to take. In these instances, we need to intercede, an endeavor we often refer to as management.

CHAPTER 8

Management of invasive species

Real and substantial threats to human health, great economic harm, and other undesirable ecological impacts have fueled efforts to manage the spread and impact of non-native species posing these threats. As is the case with most problems, there are two general approaches one can take to deal with these threats—prevention and mitigation. With respect to non-native invasive species, de Poorter *et al.* (2005) divided these two broad categories into four management strategies: prevention, early detection, eradication, and control.

In a simple sense, the role of invasive species management can be viewed as an effort to minimize the invasion pressure of an area, the probability of a successful invasion for a particular dispersal episode. This can be done by reducing the invasibility of the environment and/or the number of propagules arriving in the environment. For example, by killing or removing all organisms from the site of interest, eradication of a population effectively reduces the propagule pressure from within the site to zero, although the site may still experience non-zero propagule pressure if propagules continue to arrive from outside the site. In the context of a map of invasion pressure (*IP*), one type of management could be viewed as an attempt to maintain a system in a low level of pressure (Fig. 8.1, point *F*) by preventing the system from experiencing any increase in invasion pressure. Another type of management could involve an effort to move the system from a state of high invasion pressure to a lower one (down and/or to the left on the map), i.e. reducing the number of new propagules (both from outside the site and within the site) (Fig. 8.1, *A*→*C*), reducing the invasibility of the environment (Fig. 8.1, *A*→*B*), or both (*A*→*D*). If the system begins further back on the invasion plateau (position *E*), then management efforts may appear to have little impact, since invasion pressure will not be visibly reduced. Nevertheless, as shown, the efforts would have moved the system closer to the cliff, meaning that visible results may be seen from additional management efforts.

This non-linear response of invasion pressure to management efforts is a very important phenomenon for managers to keep in mind. Unfortunately, this phenomenon does not make the life of a manager easier. If a particular management strategy is showing little effect, it could be because, even though the efforts have substantially reduced invasibility and/or the number of propagules, they have not yet shifted the system off the plateau and into the cliff area, where results would begin to be seen; or it could mean that the management efforts are ineffective and not reducing either variable. Managers will have the difficult task of deciding which of these two alternatives is most likely. If they are able to determine that they are reducing invasibility and/or propagule input, they should continue with their efforts, with the understanding that they have simply not yet moved the system to the tipping point

Efforts to prevent introductions

Management efforts can be implemented throughout the invasion process, from the initial dispersal and introduction stages, through the naturalization and spread stages (Wittenberg and Cock 2005, Fowler *et al.* 2007; Fig. 8.2). Preventing the introduction of harmful non-native organisms is normally considered the most cost-effective approach to managing these species (McNeely *et al.* 2001, Carlton and Ruiz 2005), and while this is likely true

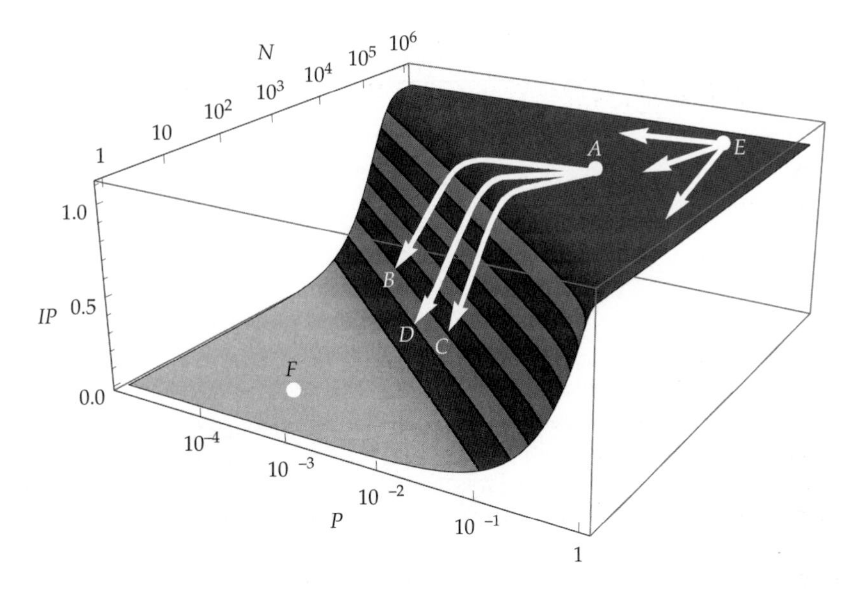

Fig. 8.1 Illustrated substantial declines in invasion pressure (*IP*) when management efforts reduce invasibility (*A*⟶*B*), propagule pressure (*A*⟶*C*), or both (*A*⟶*D*). If the system begins in position *E*, comparable reductions in either or both variables will not visibly reduce invasion pressure, suggesting that the management efforts have failed. However, as shown, the efforts have moved the system much closer to the invasion cliff, making it more likely that additional management efforts may successfully push the system over the edge. If the system begins in position *F*, the purpose of any management efforts would be to prevent increases in either or both invasibility and the number of arriving propagules in order to keep the invasion pressure from increasing from the current negligible levels.

in most instances, the costs of truly comprehensive and effective efforts to prevent introductions should not be underestimated (Keller *et al.* 2008), given the wide variety of dispersal pathways and entry points that must be monitored (Fig. 8.3). In any case, prevention efforts represent an attempt to reduce invasion pressure by reducing propagule pressure. Preventing initial introductions is extremely important in cases where rates of establishment and spread of introduced are quite high. Jeschke and Strayer (2005) stressed this point, with respect to vertebrate introductions in Europe and North American, having found that the establishment and spread rates of vertebrates introduced into Europe from North America and vice versa averaged higher than 50%.

As shown in Fig. 8.4, and discussed by Wittenberg and Cock (2005) and McNeely *et al.* (2001), prevention efforts consist of a combination of laws and regulations, risk analyses, and border control. The creation of laws restricting the introduction and transport of harmful non-native species is often an important step in the management process. However, in order to be effective, the laws need to be enforced and their implementation adequately funded. Thus, although well-intentioned laws may be passed, this is certainly no guarantee that they will be able to accomplish their intended objectives (Fowler *et al.* 2007).

Ruiz and Carlton (2003c) proposed a four-part framework to be used to try to prevent new introductions. They framed this approach as 'vector management.' The first step in the process is vector analysis. As presented by Ruiz and Carlton, vector analysis involves identifying the transfer mechanism(s) and assessing how and when they operate to deliver the new organisms, and from where the new organisms originate. Vector analysis is essentially an assessment of propagule supply. Since many organisms are transported via multiple vectors, this assessment needs to be undertaken for each vector. In addition, for each vector,

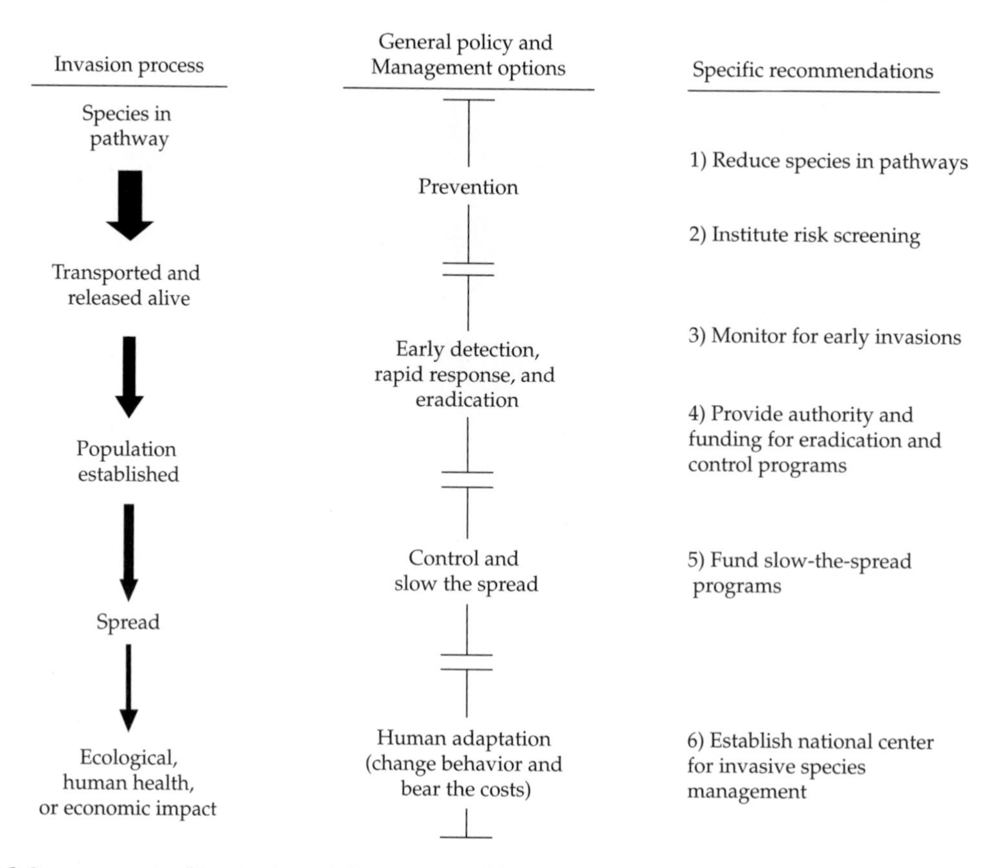

Fig. 8.2 Stages common to all invasions by non-indigenous species (left column), major policy and management options (middle column), and major recommendations (right column) associated with each stage of invasion. From the top to the bottom of the left column, each arrow is thinner than the preceding one because the proportion of species that proceeds from one step to the next is less than the previous one. Nevertheless, because the number of species entering pathways is increasing as global trade increases, the number of species causing harmful impacts is increasing with time. In the right column, recommendations do not correspond exactly with each stage of invasion; in particular, recommendation 6 underpins all policy and management options. Redrawn and printed, with permission, from Lodge *et al.* (2006), copyright Ecological Society of America.

an assessment needs to be made for each recipient region of interest. The second part of the management scheme (Ruiz and Carlton 2003c) involves an assessment of vector strength, an evaluation of the relative importance of the different vectors in actually causing new introductions. From a management perspective, there is great value in identifying the comparative strengths of different vectors, since this enables managers to focus their efforts on the key vectors, thereby promoting an efficient use of resources. In the Ruiz and Carlton scheme, the active management begins with the third part, vector interruption. As the words suggest, this involves efforts to disrupt the supply line. As described by Ruiz and Carlton, these efforts can involve education, voluntary guidelines, and regulations and laws. The final step involves ongoing evaluation of the efficacy of the vector interruption efforts. This includes assessing to what extent the propagule supply has been reduced, as well as the effectiveness in reducing actual introductions, the ultimate goal.

While many harmful non-native species are introduced by accident, many have been introduced intentionally, e.g. as part of the horticultural and pet trade (Reichard and White 2001, Cassey

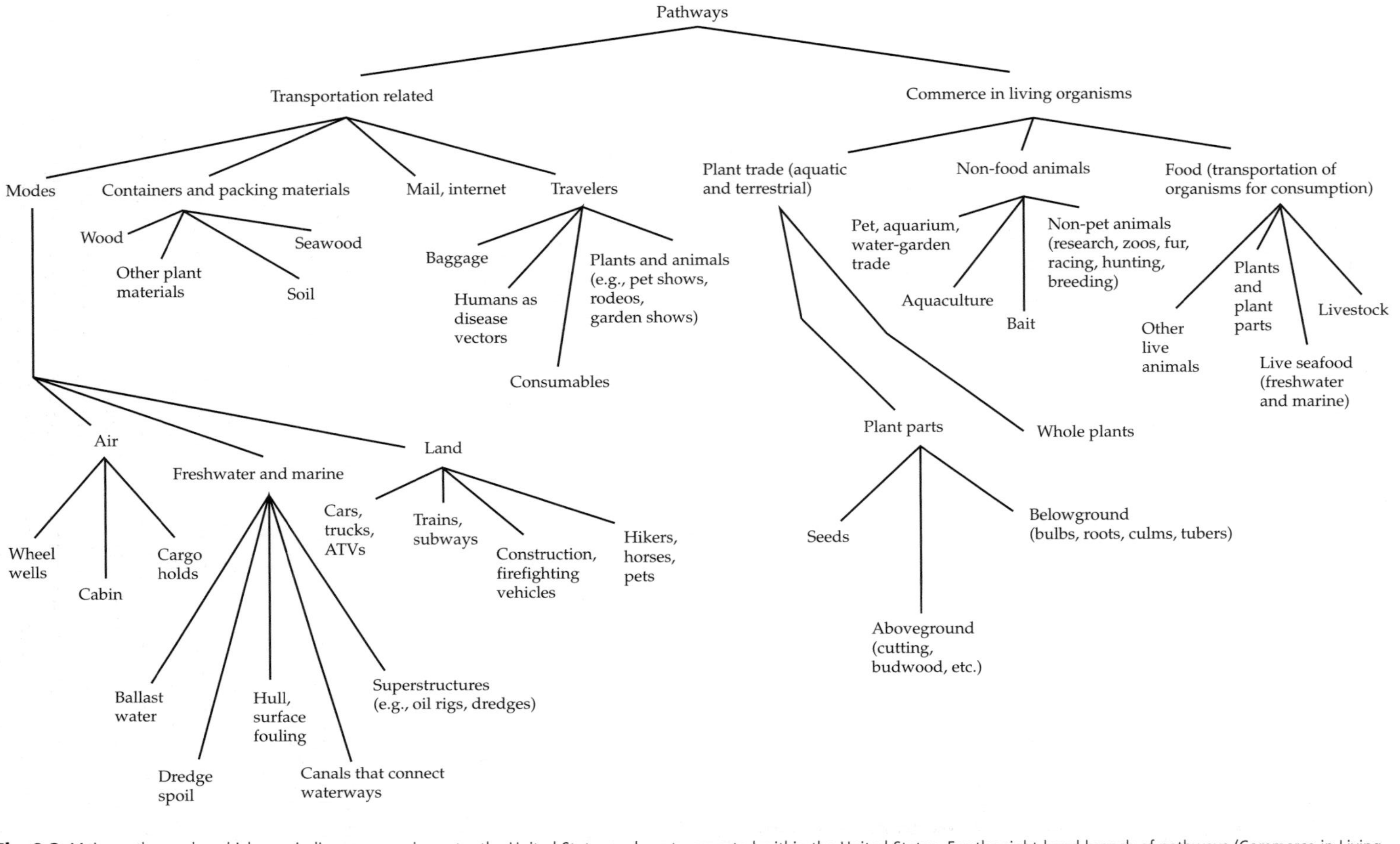

Fig. 8.3 Major pathways by which non-indigenous species enter the United States and are transported within the United States. For the right-hand branch of pathways (Commerce in Living Organisms), each pathway also entails the possibility of other species hitchhiking on or in the species that is the focus of trade, or in the medium (e.g. water, soil, nesting material) or food of the focal species. Hitchhiking organisms could include parasites and pathogens of the species in trade. Redrawn and printed, with permission, from Lodge *et al.* (2006), copyright Ecological Society of America.

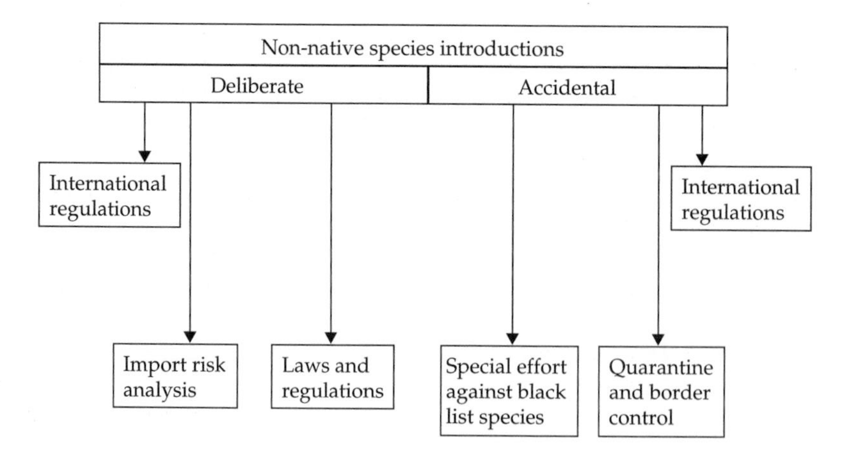

Fig. 8.4 A variety of types of initiatives are available and have been undertaken to try to prevent the introductions of undesired non-native species. Redrawn and printed with permission from Wittenberg and Cock (2005), copyright Scientific Committee on Problems of the Environment.

et al. 2004, Semmens *et al.* 2004). A major obstacle in trying to control the global spread of invasive species introduced intentionally is that the economic costs of the invasive species are almost never borne by those who introduced them. Perrings *et al.* (2005) emphasize that these costs are almost never reflected in the market price of the species in question (i.e. they are an externality), and argue that an effective management approach would be to internalize the costs.

The use of tariffs could be an effective way to internalize the costs associated with introduced species of this type (Costello and McAusland 2003, Margolis *et al.* 2005). However, making this sensible recommendation is a lot easier than implementing it, given constraints and obstacles associated with current world trade policies (Margolis *et al.* 2005, Perrings *et al.* 2005). It has proven challenge enough for individual nations to take the steps necessary to internalize environmental costs associated with their own products and industries. It seems difficult to imagine that a comprehensive and effective international plan to internalize the costs associated with intentionally introduced species will be implemented any time soon. Nevertheless, with strong leadership, it should be possible for particular commercial sectors to make some positive movement in this direction. For example, a small surcharge could be added to the retail end of the pet trade, to be used to fund various prevention and mitigation efforts associated with the introductions. Costello and McAusland (2003) showed how tariffs could be effective at reducing the import of these species to begin with, but many object to this approach because it tends to negatively impact the economies of developing countries, the common source for these animals. Thus, a more realistic approach may be to target the retail end in the process. For example, buyers in many developed countries are already now required to pay a recycling or disposal surcharge when buying certain items, such as tires and electronics. A conservation or invasive species surcharge applied to the sale of introduced pets would not present buyers with a radically new concept.

Without question, efforts to prevent harmful species from entering a region will be more efficient if attention can be focused on likely suspects. While profiling of people raises ethical questions, sometimes resulting in very inefficient use of border control resources, fortunately the same concerns are seldom raised with respect to the profiling of other species. For example, a recent analysis of the captive bird industry in Spain showed that the only birds that pose a serious threat of escape and naturalization are birds that were originally caught in the wild (Carrete and Tella 2008). Captive bred birds were generally found to have lost the ability to survive on their own in the wild. Since escaped birds from the pet trade were found to be the primary source of avian invasions in Spain (Martí and del Moral 2003), the findings by Carrete and Tella indicate that banning wild birds from the pet trade would dramatically reduce the invasion risks associated with this industry, while still permitting the industry to meet the desires of pet owners.

As described in Chapter 6, trying to identify likely invasion suspects is commonly referred to as risk analysis. A fundamental challenge of risk analysis approaches is that people differ as to whether the goal should be to minimize the likelihood of permitting the entry of harmful species or of minimizing the likelihood of excluding harmless and potentially economically valuable species. Ecologists have generally supported the first goal and, in making their case, have commonly invoked the precautionary principle, which, in this case, means that species should not be introduced unless they have been proven harmless. For example, Gollasch and Leppäkoski (1999) argued that treating all non-indigenous species as guilty until proven guilty is the only environmentally sound approach. However, some have criticized the use of the precautionary principle as being unscientific and untenable, at least when strictly applied (O'Riordan and Cameron 1994, van den Belt 2003). Simberloff (2005) argued for a more nuanced use of the precautionary principle, in which we should act to prevent introductions if there is a good scientific reason to expect that the species will cause serious harm, even if definite knowledge is lacking.

Of course, what constitutes good enough reason, and how much harm is serious harm, are the details in which the devil dwells. In 2008, the IUCN (International Union for Conservation of Nature) Council adopted a set of guidelines for applying the precautionary principle to biodiversity conservation and natural resource management, which also underscored the importance of a context-dependent approach (http://www.pprinciple.net/PP_guidelines_brochure.pdf). For example, the guidelines emphasized that 'a balance should be struck between the stringency of the precautionary measures, which may have associated costs, and the seriousness and irreversibility of the potential threat.' The guidelines also emphasized the importance of attending to equitability, i.e. that 'attention should be directed to who benefits and who loses from any decisions, and particular attention should be paid to the consequences of decisions for groups which are already poor or vulnerable.' In any case, it is important to remember that the decision to invoke, or not to invoke, the precautionary principle, i.e. to presume either harm or safety, is not a scientific one, even if scientists are making it; rather it is a social decision reflecting on how risk averse is society (Andow 2005).

Leung *et al.* (2002) and van den Belt (2003) have argued that any efforts that try to balance the risk of letting in a harmful species with that of keeping out valuable species should be based on our assessment of the costs, economic and other, of the two alternatives. For example, a cost-benefit analysis of a screening program would include the costs of developing and implementing the program, the cost-savings of reduced harm due to preventing the introduction of invasive species, as well as the costs associated with errors in the screening process, including letting in some invasive species and keeping out some non-invasive and economically beneficial species. Keller *et al.* (2007) took this approach in the development of a bioeconomic framework to quantify the net benefits from a prescreening program designed to identify invasive plants. They concluded that there were positive net benefits from their model screening program when the accuracy of the risk analysis protocol exceeded 69%.

In some instances, the new species are not arriving at our borders but instead are being created from within, e.g. involving the development of GMOs (Genetically Modified Organisms). In this case, screening programs will need to be implemented, not at border entry locations but within the laboratories where the modified species are being created. An example involves the development of biofuel feedstocks, which often are being selected, bred, and engineered from non-native plants (Barney and DiTomaso 2008). Given that these strains are generally being developed to produce dense monospecific stands, to tolerate poor growing conditions, and to escape from pests, there is concern that these strains may have great invasive potential. Barney and DiTomaso (2008) recommended genotype-specific pre-introduction screening for these new strains, which would consist of risk analysis, climate-matching modeling, and studies of fitness responses to different environmental scenarios.

Efforts to detect nascent invasions

In the face of considerable propagule pressure, efforts to prevent all introductions of a species over a large area are going to be very expensive, and may very well be impossible. While a societal investment of this magnitude may be easily justifiable when the harmful species is a virulent human pathogen, in other instances it may often not be possible to secure sufficient resources to effectively secure all borders. In these cases, alternative approaches need to be explored. One that is already being implemented is 'early detection' or 'early warning,' an approach that seeks to identify invasions very early in the process, when effective control measures, even including eradication, may be both economically and ecologically possible (de Poorter *et al.* 2005, McNeely *et al.* 2001, Wittenberg and Cock 2005, Lodge *et al.* 2006).

Early detection efforts can involve the regular monitoring of specified sites. This approach has been used for decades to monitor movements and establishment of pest insects (Carey 1996, Hadwen *et al.* 1998). Australia has several decades of experience with this approach, using what the Australian Commonwealth Scientific and Industrial Research Organization (CSIRO) refers to as 'sentinel sites,' locations, or sometimes herds (e.g. of cattle) that are believed to have a high potential of early establishment. In Australia, the sentinel approach has been used to monitor the spread of pest insects, fungi, viruses, and plants, both native and non-native. In North America, sentinel site monitoring has been conducted as part of management of invasive insects, e.g. the cactus moth, *Cactoblastis cactorum* (Westbrooks *et al.* 2006). There is increasing interest in the use of remote reconnaissance for early detection efforts. For example, as of 2008, the NASA Office of Earth Science and the US Geological Survey were developing a National Invasive Species Forecasting System (ISFS) based on NASA's remote sensing capabilities. When fully developed and implemented, the ISFS will be used for the early detection and monitoring of invasive species on Department of Interior and adjacent lands (http://invasivespecies.gsfc.nasa.gov/). Early detection of some species may also be possible using molecular techniques. Ficetola *et al.* (2008) described their ability to detect the present of bull frogs, *Rana catesbeiana*, a non-native invasive species in many European lakes and wetlands, by amplifying DNA fragments that persist in the environment for a period of time. This technique may hold great promise for early detection efforts for other species in other environments as well.

Sentinel site monitoring will be much more efficient if observers are able to concentrate their searches on particular species, i.e. those potentially harmful species that are most likely to arrive in a region. By identifying likely candidates for newly arriving species, good databases have much to offer to these efforts. Many such databases have emerged in recent years. The Global Invasive Species Database (GISD) (http://www.issg.org/database/) was developed by the Invasive Species Specialist Group (part of the World Conservation Union) to provide global information on non-native invasive species of all types to agencies, resource managers, decision-makers, and other interested individuals. The database contains information on the biology and ecology, distribution (native and alien ranges), and impacts of invasive species world-wide. It also provides general and management references, links, and contacts for each species. In addition, using a simple habitat-matching model, the database provides predictions as to those regions that are potentially at risk of invasion from a particular species. This database was developed specifically to assist sentinel and other early warning efforts. There are many other outstanding databases that have been developed, focusing on particular regions of the world and/or particular taxa. Many of these are listed on the National Invasive Species Information Center's website maintained by the US National Invasive Species Council (http://www.invasivespeciesinfo.gov/resources/databases.shtml).

The value of early detection ultimately depends on the ability and commitment of society to respond rapidly with an effective containment or eradication measure. McNeely (2005) described a situation in Serbia in the 1990s when scientists discovered the presence of the western corn rootworm, *Diabrotica vigifera*, in fields near the Belgrade airport. Probably introduced accidentally via military flights from the US, effective response, following what is believed to have been early detection, may have been able to prevent its subsequent spread. However, due to the military turmoil in the region, this did not occur and, by the end of the 1990s, it had spread into Italy and many of the eastern European countries, with potential spread throughout the rest of the corn growing regions in Europe and Asia.

Nowhere is early detection more important than when it involves the spread of human disease; but early detection of an outbreak is not enough. Currently, effective control of an outbreak can be jeopardized due to the fact that data and samples are not always promptly shared among countries and researchers, an unfortunate state of affairs resulting from various regulatory obstacles and issues involving intellectual property rights of individual researchers (Boyce 2007). Even with good intentions from a particular level of government, experience has shown that success in any prevention and early detection efforts of non-native invasive species is likely going to require a coordinated multi-level response. In a report by the Ecological Society of America regarding US policy and management of invasive species, Lodge *et al.* (2006) called for more coordinated efforts among different political units (e.g. local, state, federal) to reduce the transport and release of invasive species, more quantitative procedures for risk assessment of species proposed for import, more support for rapid responses when the early detection efforts identify a problem species, and more support for longer and larger scale mitigation efforts. The ESA report strongly recommended that the federal government take the leadership role in these efforts.

Efforts to identify invasive populations and susceptible environments

Another way to effectively utilize scarce management resources is to direct them more discriminately, rather than taking a broad-brush approach. Recognizing that not all populations of a species deemed invasive may actually pose a threat of spread (Bauer 2006), Smith *et al.* (2006) recommend that efforts be made to distinguish between populations that actually represent a threat from those that do not. This emphasis is consistent with growing interest in developing conservation and management plans for species that are population-based, and utilize now readily obtainable (in most cases) population genetic data (Palsbøll *et al.* 2007, Schwartz *et al.* 2007). If the genotypes of particularly invasive populations could be identified, then managers could use this information to help them distinguish between threatening and less-threatening populations of the invasive species being managed. This would enable the management efforts to be much more effective, both in time and money.

Likewise, not all environments will be equally susceptible to invasion. For example, Keller *et al.* (2008) developed an invasion model to identify which lakes in a county were at the greatest risk of being colonized by the non-native and invasive rusty crayfish (*Orconectes rusticus*). They also then embedded their results in an economic model to assess the economic benefits from a management program that targeted the susceptible sites. Of course, even among sites that become inhabited with an invasive species, not all the sites will be equally harmed, whether that harm be ecological, economic, or both. Thus, a similar approach would be to use an economic model like the one developed by Keller *et al.* in a triage approach, in which management efforts would be directed at those environments most likely to be colonized by non-native species and to experience significant harm.

Efforts to eradicate

While eradication of non-native invasive species may often be the most desired management option, accomplishing this objective once the species has become well-established is normally extremely difficult, if possible at all, and expensive (Wittenberg and Cock 2005). For example, Rejmánek and Pitcairn (2002) concluded that it is normally possible to eradicate an invasive plant population if its occupancy range is less than 1 ha, that eradication efforts involving areas between

1 and 100 ha sometimes succeed, and that eradication efforts involving areas greater than 100 ha seldom succeed. This would suggest, at least for plants, that eradication is unlikely to be possible unless the invasion is caught in its very earliest stage. Of course, one can always eradicate a species from a small site; however, if the species continues to inhabit surrounding environments, eradication from the site of interest will need to be a continual and never-ending project. These obstacles notwithstanding, eradication of well-established populations has been successful in some cases. Two of the most notable examples involving mammals are the eradication of the North American muskrat and the coypu from Great Britain (Gosling and Baker 1989, Gherardi and Angiolini 2004), and the recent eradication of pigs from Santa Cruz Island, California (Morrison *et al.* 2007). Each of these efforts involved meticulously organized and comprehensively implemented trapping and monitoring plans. Eradication efforts of certain pathogens have also been successful. Due to the immense incentives to eradicate deadly human and livestock pathogens, countries have often been willing to invest the very large resources required to accomplish eradication. In some cases, due to intense and coordinated international efforts, global eradication has been accomplished, as in the case of the virus causing smallpox, which is believed to no longer exist in nature (Bazin 2000). Another disease that may be on the verge of global extinction is rinderpest. The outbreak in the 1970s elicited another intensive vaccination program throughout Asia, the Middle East, and Africa, which, by 2008, had appeared to have eliminated the disease from most areas (Fig. 8.5), raising the prospect of possible complete eradication (Normile 2008).

Efforts to prevent subsequent persistence and spread

Should prevention efforts fail, as well as attempts to eradicate early detected individuals or established populations, and if eradication is determined not to be feasible, then management efforts need to be implemented to contain subsequent spread and/or the reduce the impacts of the problem species. Wittenberg and Cock (2005) identify five types of management efforts to limit the spread and impact of invasive species: mechanical control (firearms, traps, harvesters of various kinds), chemical control using pesticides, biological control, habitat management, and integrated approaches using a combination of these approaches. The literature of pest management using chemical and mechanical control measures is extensive and will not be reviewed here.

Biological control

Wittenberg and Cock's use of the term 'biological control' is consistent with its traditional usage in the pest control field, including the introduction and augmentation of enemies, practices that interrupt the reproduction of the pest (e.g. introduction of sterile males in the case of insect control), and induced resistance in the valued host species (e.g. a crop plant). The use of biological control to reduce the spread and impact of non-native invasive plant and arthropod species has been a common management approach that has met with some success (Hoddle 2004, Culliney 2005).

Similarly, mammalian predators have often been introduced to try to control other pest mammal species, such as rabbits and rats, an approach that often has resulted in undesirable impacts on native wildlife (Loope *et al.* 1988, King 1990). However, to date, few biological control efforts have targeted non-mammalian vertebrates or aquatic invertebrates (Hoddle 2004). Although the ctenophore *Beroe ovata* was not introduced into the Black Sea intentionally in 1997, its independent arrival and very heavy predation on *Mnemiopsis leidyi*, another non-native ctenophore that had been introduced in the late 1980s, resulted in a crash in the *M. leidyi* population, suggesting the potential effectiveness of an introduced aquatic invertebrate biological control agent on another introduced aquatic invertebrate species (Kideys 2002). Currently underway are studies to evaluate the effectiveness of introducing particular pathogens and parasites to control species such as the brown tree snake (*Boiga irregularis*), cane toads (*Bufo marinus*), zebra and quagga mussels (*Dreissena polymoprpha* and *D. bugensis*), and the Mediterranean snail (*Cernuella virgata*) (reported in Hoddle 2004).

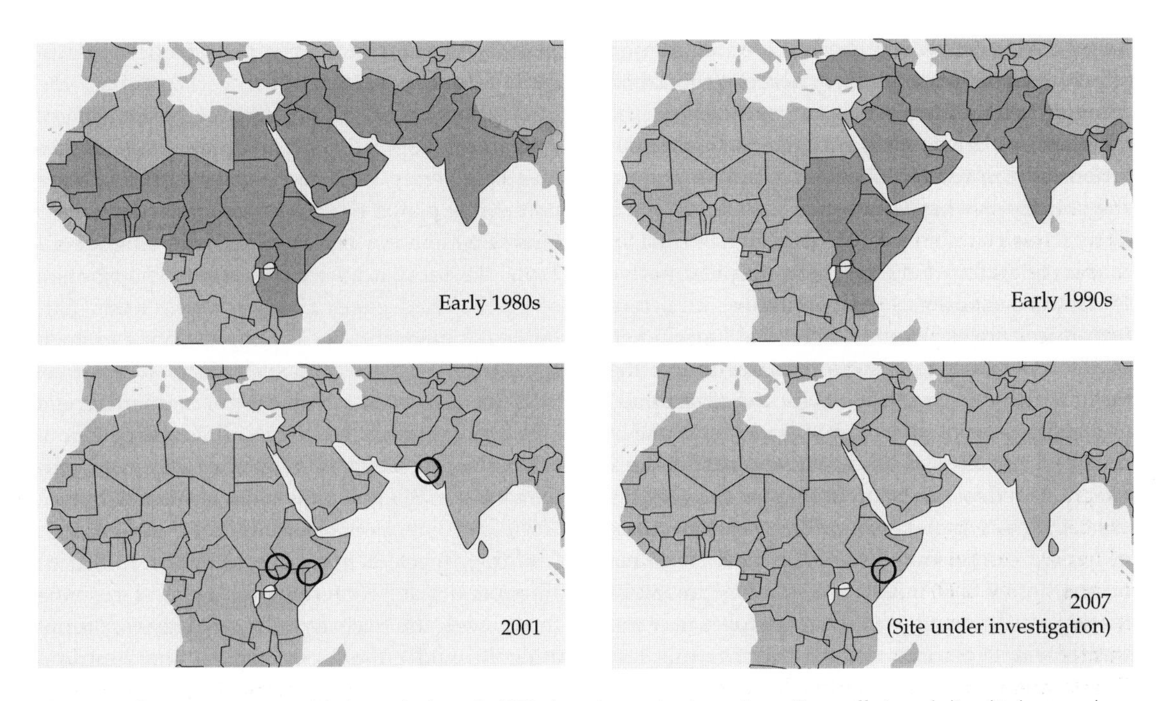

Fig. 8.5 Following a resurgence of rinderpest in the early 1980s, intensive vaccination and surveillance efforts are believed to have nearly eradicated the virus by 2007. Redrawn and printed, with permission from GREP (Global Rinderpest Eradication Programme) FAO (Food and Agriculture Organization, United Nations).

Managing the environment

Ewel (1986) argued that just focusing on the invasive species in a management program was like trying to cure the symptoms, believing that invasions are often primarily the result of changes in the environment. Under this view, then, invasive species might best be managed by managing the environment. As described in Chapter 6, one way to reduce the invasion pressure on a system is to shift it left in the *IP* map space by altering the traits of either the native or the invasive species. Another way to manage the system in the horizontal plane is to manipulate the environment to make it less invasible to a target species. Many studies have shown that certain environmental conditions tend to facilitate introductions of new species, particularly increases in resources (Davis *et al.* 2000), which are often associated with changes in the natural disturbance rate (Alpert *et al.* 2000). As a result, manipulating resource availability and disturbance regimes have been recommended as a way to manage invasive species (Green and Galatowitsch 2002, Blumenthal *et al.* 2003, Daehler 2003).

That effects of non-native invasive species often interact with the effects of land/seascape change means that in these circumstances, management efforts directed at non-native species will not be effective unless the land/seascape changes are also addressed (Didham *et al.* 2007). In these instances, it will be important to know whether the interactions are numerically or functionally mediated (Didham *et al.* 2007). Didham *et al.* used two examples to illustrate the different management approaches needed to address both types of interactions. In New Zealand, introduced and feral cats prey heavily on native skinks, *Oligosoma* spp., and this predation is higher when the natural skink habitats are surrounded by agriculturally-modified grasslands, which support increased densities of rabbits, *Oryctolagus cuniculus*. Since the rabbits serve as a prey subsidy for the cats, then the cat populations, and hence predation on the lizards, are maintained at a high level. In this case, reducing the rabbit population, e.g. by habitat management in the surrounding areas, would reduce the cat populations and in turn the level of cat predation on the skinks, in what Didham *et al.* refer to as an 'interactive chain effect' (Fig. 8.6a). In contrast, in many Australian *Eucalyptus* woodlands, native marsupials experience high predation rates from introduced foxes, *Vulpes vulpes*. Habitat change in these woodlands results in a decline in the number of structural refuges for the marsupials and an increase in the per capita predation rate of the foxes (Stokes *et al.* 2004), thereby producing a functional effect on the predation process. In this case, Didham *et al.* suggest that the restoration of habitat complexity may have a positive effect on the native marsupials by reducing predation levels, even though the numbers of foxes may not be affected. Didham *et al.* (2007) refer to this sort of synergistic interaction between habitat change and non-native species as an 'interaction modification effect' (Fig 8.6b).

Another example of how landscape change can influence the spread and persistence of non-native invasive species involves the dispersal of fruits of non-native plants by frugivores. In a fragmented landscape, the movement patterns of frugivorous birds are influenced by the spatial distribution of patches and edges (Buckley *et al.* 2006). This raises the interesting possibility of trying to manage perches and edges in such a way as to attract birds to these sites, where control measures could be focused to prevent successful establishment of the dispersed seeds, thereby creating sink environments for the non-native species (Buckley *et al.* 2006).

With respect to human health, the spread of zoonotic disease represents the greatest growing threat, with the majority of these diseases originating in wild animals in tropical and subtropical regions (Jones *et al.* 2008). The primary way

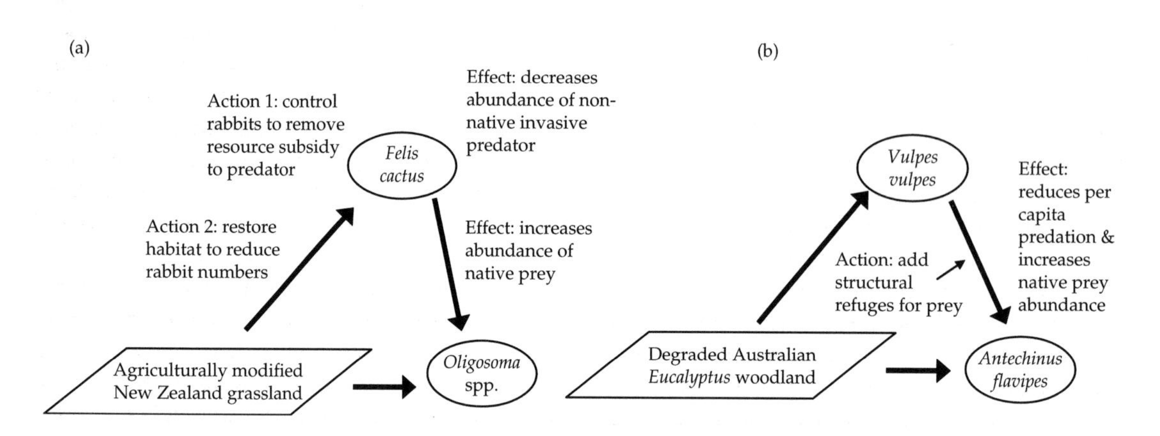

Fig. 8.6 Two ways to manage interactive effects of habitat modification and species invasion. (a) Predation from an introduced cat is reducing populations of native skinks in anthropogenically-disturbed New Zealand grasslands. High cat populations are maintained by high numbers of another prey species, an introduced rabbit, *Oryctolagus cuniculus*. Since it is normally difficult to reduce wide-ranging and elusive predators, an alternative approach is to try to reduce the cat populations by reducing their dominant prey species, the rabbits. This could be done through direct efforts to reduce rabbit abundance or by restoring the grassland so that it supports fewer rabbits. Either or both of these strategies would be expected to reduce the numbers of cats, and hence the predation on the skinks through a numerically mediated interaction chain effect. (b) Predation from introduced foxes in Australia is reducing populations of native marsupials in degraded *Eucalyptus* woodlands. Predation pressure on the marsupials might be reduced through management efforts in the woodlands that increase structural refuges for them. In this case, numbers of the predators would not necessarily be reduced, but the prey population would be expected to increase, or at least stabilize, through a functionally moderated interaction modification effect. Redrawn and printed, with permission, from Didham *et al.* (2007), copyright Elsevier Limited.

to prevent the emergence of these diseases is to reduce the frequency of contact between humans and wildlife hosts. This suggests that a useful management approach would be to conserve areas rich in wildlife and to minimize human activity in these reserves (Jones *et al.* 2008).

The *IP* model (Chapter 6) showed that propagule pressure and invasibility (an integrated function of the traits of the incoming species and the conditions of the environment) have approximately equal impacts on invasion pressure. Thus, theoretically, it would make equal sense to target management efforts at reducing propagule pressure as it would at reducing invasibility. However, while in some instances there may exist good opportunities to modify traits of organisms and/or to manipulate the environment, both done to reduce invasibility, effectively manipulating evolution and ecosystems is a very challenging endeavor. Thus, it is not surprising that, to date, most efforts to manage invasive species have focused on propagule pressure. In this case, the objective is to directly reduce the number of invasive individuals, whether this effort occurs prior to introduction (prevention efforts) or following introduction (control and eradication efforts).

As described in Chapter 3, the Allee effect has been shown to be able reduce establishment success and spread in non-native species. This suggests a possible management approach, in which specific efforts are undertaken to impose an Allee effect on founding or managed populations (Elam *et al.* 2007). This could be accomplished by keeping founding populations below the Allee threshold or reducing population sizes to below the threshold. For example, since pathogen spread is normally directly related to the abundance and density of its host, a primary approach to preventing the spread of an introduced pathogen is to restrict the pathogen to as few potential hosts as possible. This is usually accomplished in two ways. The infected organisms can be quarantined until the pathogen is eradicated or dies out due to lack of hosts, and/or uninfected but potential hosts can be immunized. Johnson *et al.* (2006) suggested that gypsy moths spread might be slowed by suppressing outbreaks occurring near the invasion front, which would reduce the number of dispersers beyond the front to below the Allee threshold.

Application of new technologies

While one should not look solely to technology as a panacea to invasive species problems, technology clearly has an important role to play in certain instances. For example, in 2007 the Great Ships' Initiative, a multi-national and multi-agency collaborative effort to minimize the import of aquatic invasive species via the Great Lakes St. Lawrence Seaway System (GLSLSS), solicited proposals to develop technologies that could be used to treat ballast water before it was released into the GLSLSS. According to its project manager at the time, it was expected that GSI would review technologies involving ultraviolet light, filtration, deoxygenation, heat and chemical treatments, as well as others (Passi 2006). Another approach to the ballast water problem is to get rid of ballast water entirely. Researchers are currently exploring the possibility of designing ships with large open tubes that run the length of the ship (Faden 2008). These tubes would fill with water while the ship is in motion and provide the stability normally provided by ballast water, but since the water is continually flowing through the tubes, only local water would be released. (When the ship is to be loaded with cargo, the tubes would be closed and pumped dry.)

New, and sometimes unique, technologies are also being developed to try to control the spread and impact of invasive species after they have been introduced. In some Hawaiian reefs, native corals are being killed by *Gracilaria salicornia* and *Eucheuma denticulatum*, two non-native invasive red algae, which develop thick mats capable of covering and smothering coral colonies. Until recently, control efforts involved removing algae by hand, a time-consuming and human-intensive enterprise. To make the removal process more efficient and effective, a joint venture involving the State of Hawaii, The Nature Conservancy, and the University of Hawaii developed a vacuum pump system (appropriately named 'Supersucker') to remove algal biomass without harm to other marine organisms. This machine, along with a smaller version, is deployed from a barge and a long vacuum hose (15 cm diameter) is managed underwater by divers who stuff algae into the hose opening. Although still requiring a five-person crew, the vacuum technology is able to remove approximately 750 pounds of algae per hour, equivalent to the efforts of 150 volunteers and 10 divers that had previously been used in half-day volunteer cleanups (Celia Smith, personal communication). Preliminary results indicate that once a substantial amount of the invasive algae is removed using the Supersucker, resident fish may effectively remove much of the remaining algae (Pala 2008).

Management by directed evolution

As discussed in Chapter 5, species introductions impact evolutionary processes, affecting both the long-term resident species and the newly arriving species. This raises the possibility of using evolution as a management tool. Ashley *et al.* (2003) and Stockwell *et al.* (2003) emphasized the value in taking a more evolutionary approach to management and Schlaepfer *et al.* (2005) proposed this idea with respect to the management of non-native invasive species, arguing that it may be a particularly effective approach when other efforts have not succeeded. Specifically, Schlaepfer *et al.* suggested two ways that adaptive responses by native populations to an introduced species might be facilitated through management, an approach that might be termed, management by directed evolution (MDE) (Matthews 2007). The first way involves creating or modifying conditions in the environment that will permit and encourage adaptive change on the part of the native species. For example, if a novel predator is introduced, and the native prey are experiencing high rates of predation due to predator naïvety (Cox and Lima 2006), then the provision of refuges may prevent rapid predatory extinction and give the naïve prey time to evolve more effective recognition and predator-avoidance behavior (Schlaepfer *et al.* 2005). Kilpatrick (2006) developed a model that showed how particular management techniques could be used to facilitate the evolution of resistance to avian malaria. Based on demographic data of native Hawaiian bird populations, Kilpatrick's model showed that the evolution of malaria resistance by the birds would be helped if the extent of rodent predation on both eggs and adult birds were reduced. His model showed that the higher population sizes of the birds, occurring following rodent control, would reduce the likelihood of rapid extinction due to avian malaria and would give the bird species more time to evolve resistance to *Plasmodium relictum*, the non-native parasitic avian malaria protozoa.

The second way Schlaepfer *et al.* proposed to manage the impacts of non-native invasive species through directed evolution was to actively manipulate the genetic composition of the native populations, e.g. 'inoculating' naïve populations with individuals from 'experienced' populations, in which natural selection had already elicited an adaptive response (Fig. 8.7). This ideas is consistent with Thompson's (2005b) geographic mosaic hypothesis, which predicts that gene flow between hot and coldspots will affect the coevolutionary dynamics of the respective populations. For example, moving individuals from hotspots into coldspots would be expected to increase the rates of evolutionary adaptation in the coldspots, a hypothesis which has been supported in a laboratory study of bacteria evolving resistance to viruses (Forde *et al.* 2004). Thus, land managers might use the genetic mosaic phenomenon to their advantage by actively moving native individuals from populations that had evolved resistance to an invasive species, to other yet vulnerable native populations, in an effort to speed up the evolutionary process of increased resistance.

Although Schlaepfer *et al.* did not address the possibility of manipulating the genetic composition of the invasive species, this could hold promise as well. If the invasive species was causing harm due to particular attributes characteristic of that population, introducing genes from non-invasive populations, e.g. via pollen for plants or via individuals for animals, could serve to dilute the invasive gene pool, thereby reducing the invasiveness of the problem population. There is good reason to believe that this dilution effect could work in certain situations. For example, it is known that a potential problem with captive breeding programs

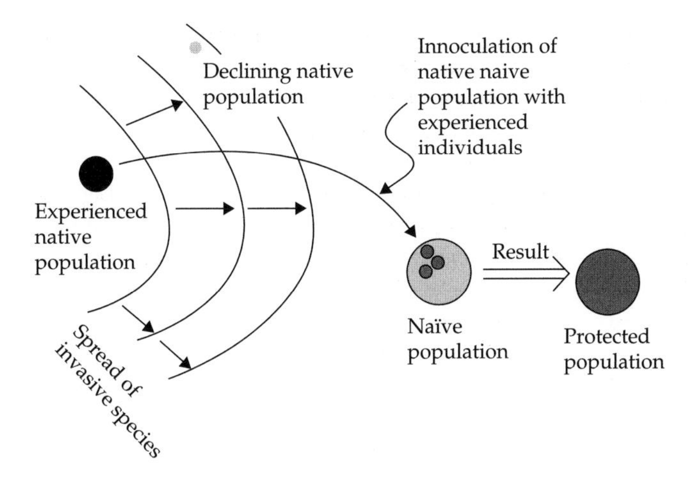

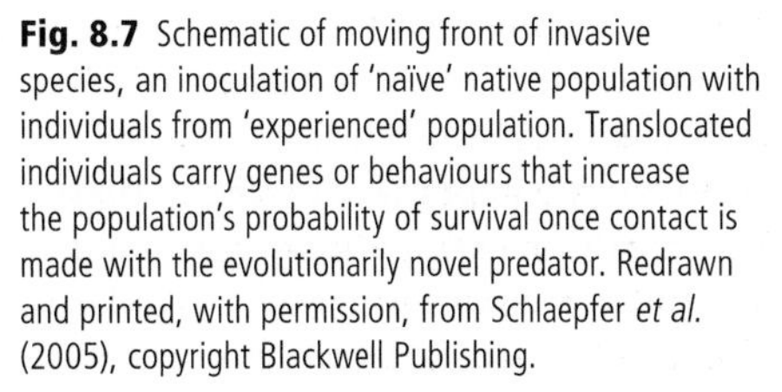

Fig. 8.7 Schematic of moving front of invasive species, an inoculation of 'naïve' native population with individuals from 'experienced' population. Translocated individuals carry genes or behaviours that increase the population's probability of survival once contact is made with the evolutionarily novel predator. Redrawn and printed, with permission, from Schlaepfer *et al.* (2005), copyright Blackwell Publishing.

is that captivity can select for traits that are deleterious in the wild (Goodman 2005). In a study of captive-bred steelhead following their reintroduction in the wild, Araki *et al.* (2007) documented a 40% decline in reproductive capabilities of the fish. This raises the possibility of intentionally rearing or producing individuals of the invasive species that possess maladaptive traits and then releasing them into the wild. There are at least two ways this could produce desirable effects. First, the captive/laboratory/greenhouse-reared individuals may reduce the numbers of invasive genotypes through competition. Second, they could reduce the invasive behavior of the population through hybridization with the invasive wild-type individuals.

Management by directed evolution (MDE) differs fundamentally from many traditional approaches to dealing with invasive species, the aim of which is often eradication. The objective of MDE is not eradication of the invasive non-native species but the adaptation of the native species so that the native and non-native species can coexist. Although one might wish for eradication of a particular species, experience has shown such efforts usually to be extremely difficult, if not impossible, very costly, and often protracted. This certainly does not mean that organizations or political entities should never decide on eradication as a goal. However, it does mean that advocates of eradication need to be realistic about the extent of societal investment required. Given constraints on time and money, management efforts with goals of coexistence are often going to be more feasible than ones with eradication in mind. On the other hand, if the management focus is on a small insular region, e.g. a small lake or small oceanic island, then eradication efforts may be more feasible, and they may make more sense than an approach trying to facilitate coexistence. However, a novel study by Russell *et al.* (2005), in which they released a single Norway rat, *Rattus norvegicus* onto a 9.3-hectare island, showed that eradication efforts, even in very small insular environments, may prove to be quite difficult. In this case, it took 18 weeks of concerted trapping efforts to capture this individual, even though the animal could be located via radio telemetry for the first 10 weeks, and nine different trapping and detection methods were used, including trained dogs, live traps, snap traps, buried traps, peanut butter bait, and poison bait. Not only did this study show how difficult it can be to capture the last individual, but the experience suggested that individuals may behave differently in the absence of conspecifics, and that bait may not be very effective in the absence of competition for food resources, meaning that eradication techniques may need to be modified once the population has been reduced to a few individuals (Russell *et al.* 2005).

Any management efforts involving the manipulation of traits, whether of the native or the non-native invasive species, would represent management taking place along the invasibility axis of the *IP* map (Fig 8.1, $A \rightarrow B$). These would involve efforts to shift the system leftward in the space defined by propagule pressure and invasibility. This can be accomplished by altering the traits of the native species to increase their resistance, thereby making establishment more difficult for the new species, as described by Schlaepfer, or by manipulating the traits of the invasive species to reduce its invasive ability. Biological-control efforts of pest insects, both native and non-native, have successfully utilized a similar approach for many years. Although not strictly an example of management by directed evolution, the intentional release of large numbers of captive-produced sterile insects (usually males), known as the sterile insect technique (SIT), causes reductions in the pest population due to the mating of large IUCN numbers of wild females with the sterile males (Dyck *et al.* 2005). The introduction of modified individuals of an invasive species as part of a control effort, is a practice that has been applied to vertebrates as well. In the Galapagos, sterilized, but hormone-injected and sexually motivated, female goats were released as a strategy of luring remaining goats into open areas where they could be shot (Guo 2006). In these instances, although the traits of the invasive species are manipulated, the intended purpose is to directly reduce the number of invasive individuals, and thus this management effort takes place along the axis representing propagule number in the *IP* map (Fig. 8.1, $A \rightarrow C$).

Managing biotic homogenization

One of the most pronounced impacts of introduced species is biotic homogenization (Lockwood and McKinney. 2001). Rahel (2002) described four ways to reduce this impact: reduce or prevent future introductions, eradicate species already naturalized, prevent extinctions of native species, and reduce habitat homogenization, which is a major driver of biotic homogenization, both in aquatic and terrestrial environments (McKinney and Lockwood 2001). Reducing habitat homogenization can be accomplished through preservation of existing diverse environments, and habitat restoration of environments already homogenized (Rahel 2002). For example, the biotic homogenization of many freshwater environments is believed to be due to the fact that so many of these aquatic systems receive the same anthropogenic impacts, e.g. pollutant runoff from cities and agricultural lands and diversion through dams and reservoirs (Marchetti *et al.* 2001, Paul and Meyer 2001, Poff *et al.* 2007). Thus, reducing pollutant inputs and restoring natural flow regimes would be expected to restore the abiotic heterogeneity among the aquatic environments, thereby promoting biotic heterogeneity as well (Rahel 2002, Poff *et al.* 2007).

Managing '*inter-situ*' reintroduction efforts

A particularly fascinating recent development in the conservation field is the idea, and practice, of what has been termed *'inter-situ'* reintroduction efforts or reintroduction biology. These are defined as 'the establishment of species by reintroduction to locations outside their current range but within the recent past range of the species' (Burney and Burney 2007, Armstrong and Seddon 2008). In cases of very rapid decline in a species, human memory and records may suffice to identify the extent of the recent historical range. In other instances, it may be possible to use paleo-ecological techniques to delineate range boundaries in the late pre-historical or early historical time periods, a technique that is being used in Hawaii to determine past ranges of species that declined precipitously once humans colonized the islands (Burney and Burney 2007).

In either case, in the context of invasion biology, it is ambiguous whether the reintroduced species should be regarded as native or non-native species in these new sites. In some instances, several millennia may have passed since the species occupied the sites of reintroduction, and no doubt community composition and ecosystem processes have changed during that time. Of course, there is no right answer to the question of native vs non-native. It all depends on one's perspective. Certainly, introducing species into ecosystems they have not inhabited for centuries or millennia would likely create novel or no-analog communities (Hobbs *et al.* 2006, Williams and Jackson 2007), in the context of our knowledge of current ecological environments. As a result, as is the case for traditional introductions of non-native species, *inter-situ* reintroduction efforts raise the possibility of 'ecological surprises' (Williams and Jackson 2007) owing to unanticipated impacts on ecosystem processes and other species (Armstrong and Seddon 2008). For example, reintroduction of the tule elk, *Cervus elaphus nannodes*, into northern California grasslands, where it had nearly gone extinct in the 1800s, has produced both desirable and undesirable results, decreasing the abundance of native perennial species (undesirable) and a highly invasive grass, *Holcus lanatus* (desirable), while increasing the cover of annuals, including both native (desirable) and non-native (undesirable) species (Johnson and Cushman 2007). In this case, the elk reintroduction has created a new type of grassland community—one dominated by native and non-native annuals.

During the nineteenth and twentieth centuries, populations of the wild turkey, *Meleagris gallopavo*, declined substantially throughout the US, even to the point of extinction in many areas. During the past several decades, turkeys have been reintroduced to many of these regions, and some introductions have taken place beyond its nineteenth and twentieth century range, making these non-native introductions. Many of these efforts have been extremely successful, so much so that in some areas, both within and outside their historic range, the proliferation of turkeys has become a problem. They eat crops, potentially compete with other native birds, such as the ruffed grouse, and represent a nuisance to residents in suburban areas due to their often aggressive behavior. Most departments of natural resources in states where turkeys have been reintroduced now have policies and tip sheets on how to deal with nuisance turkeys. Clearly, trying to manage the abundance of introduced species within a desired abundance range is a formidable challenge. Although invasive behavior on the part of the species reintroduced as a part of *inter-situ* reintroductions may not be likely in most instances, those involved in these reintroductions might consider borrowing some of the techniques and approaches used in the monitoring and management of invasive species to ensure that a valued reintroduced species does not end up causing undesirable impacts, and even becoming a pest.

Use of regulations and legal frameworks

Laws and various sorts of institutional regulatory frameworks and agreements have been implemented from the local to the international level to prevent introductions of new species, and to control the spread and impact of those already introduced. Shine *et al.* (2005) provide an excellent overview of many of the international treaties and conventions addressing non-native species that have been developed and implemented in recent years, including the 1992 Convention on Biological Diversity. While the international instruments typically deal with the transport of species across national borders, additional regulatory efforts are often imposed within countries to restrict movements of species already introduced. In some of these efforts, management of non-native invasive species has percolated down into the everyday lives of citizens, influencing where and when they can travel and what they can transport with them.

Due to the threat of spread of the emerald ash borer, *Agrilus planipennis*, and the Asian longhorned beetle, *Anoplophora glabripennis*, in the northern US and southern Canada, states and provinces are regulating the movement of firewood by restricting use to locally procured wood. With increasing concerns over the spread of the viral hemorrhagic septicemia virus throughout the Great Lakes region of North America, some of the US states have begun to try to reduce the transport of bait fish from one state to another through education programs and some legislation. In many regions, recreational boating is a primary vector type for the secondary dispersal of many aquatic organisms, both plant and animal (Muirhead and MacIsaac 2005). When quagga mussels, *Dreissena bugensis*, were discovered in Lake Mead in southern California, part of the initial response was to ground all National Park boats and restrict concessionaires from transporting rental boats (Stokstad 2007). The state of Minnesota, USA, has implemented a variety of boating regulations and educational programs in an effort to prevent activities or practices likely to introduce invasive aquatic species into Minnesota waters.

In 2007, approximately 50 watercraft inspectors examined 42,000 watercraft, a number that was supplemented by inspections on additional watercraft by citizen groups trained and equipped by the Minnesota Department of Natural Resources (MN DNR 2008). At some US golf courses, golfers are required to remove the spikes from their shoes and install new ones in order to reduce the spread of *Poa annua*, an introduced grass generally regarded as an undesired weed by course superintendents.

Use of citizen volunteers in management efforts

Experience during the past few decades in trying to manage invasive species has made it clear that successful management efforts often are very costly, not just in dollars but in people and time. One way that managers have begun to deal with this issue is to incorporate citizen volunteers in some of the management activities. For example, a challenge with the 'sentinel' approach for early detection, described earlier in this chapter, is finding the resources to monitor a wide network of sentinel sites on a regular basis. The Australian Commonwealth Scientific and Industrial Research Organization (CSIRO) has encouraged the use of citizen volunteers to help with the monitoring, a recommendation also made by Lodge *et al.* (2006) in their Ecological Society of America report. The Invasive Plant Atlas of New England (IPANE) has made citizen volunteers an integral part of their effort to track the distribution and spread of invasive plant species in New England. The volunteers are trained in day-long sessions and then assigned an area near where they live to survey on a regular basis, and report their findings back to IPANE.

In the Midway Atoll in the Pacific Ocean, the US Fish and Wildlife Service initiated a program in 2008 to utilize tourists in a campaign to remove debris and invasive species from the island, used by the US military until 1996 and which is now part of the largest marine reserve in the world. The program permits tourist visits in groups of 15, with no more than 40 tourists staying on the island overnight. The tourists pay more than $3500 for this opportunity. Of course, while well-intentioned, the tourists could serve as vectors for new introductions. Thus, the program includes inspection efforts to minimize the likelihood of such accidental introductions (Barry Christensen, Midway Atoll National Wildlife Refuge Manager, personal communication). Although perhaps not as dramatic as volunteering in a Pacific atoll, local volunteer efforts to remove invasive plant species have become commonplace in many parts of the world, e.g. 'buckthorn busters' and 'buckthorn roundups' in the upper Midwest, USA, in which citizen volunteers participate in group efforts to eradicate *Rhamnus cathartica* from parks and urban areas.

Invasive species control as a means to solve other conflicts

In some cases, efforts to control or eradicate an invasive species may help resolve other issues besides ones immediately at hand. In the global fishing industry, capture of unintended species, referred to as 'bycatch,' is a common event, one that has stirred considerable controversy over the years leading, in some instances, to litigation, social conflict, and fisheries' closures (Wilcox and Donlan 2007). Using a population-modeling approach, Wilcox and Donlan described an alternative approach to resolving bycatch of seabirds, referred to as compensatory mitigation. Their results showed that, if the primary conservation concern is the protection of regional seabird populations, efforts to increase reproductive success of the birds on their breeding islands would usually have a much greater positive effect than shutting down the fisheries, and thereby eliminating mortality caused due to bycatch. As an example, Wilcox and Donlan used the eastern Australian flesh-footed shearwater, *Puffinus carneipes*, population, which experiences bycatch mortality estimated to be between 1800 and 4500 birds per year (Priddel *et al.* 2006). The entire population of this species nests on Lord Howe Island, where, according to Wilcox and Donlan, the birds experience high predation rates from rats and feral cats. Through their modeling efforts, Wilcox and Donlan showed that removal of the non-native predators would be 23 times more cost-effective than closing the fisheries. In this model, the money for the predator-removal efforts came from bycatch levies imposed on the fisheries, which, while an additional cost, certainly impacts the fisheries less than closing them down. Not all support this approach, however. In a series of lively exchanges with Wilcox and Donlan, Pridell (2007, 2008) vigorously argued against the value of compensatory mitigation as a way to protect seabirds, claiming that seabirds killed by long-line fisheries are normally large, whereas the majority of seabirds killed by rats are small. Specifically, Pridell (2007) argued that he did not believe rats were a major source of mortality for *Puffinus carneipes*, the fundamental assumption of the approach proposed by Wilcox and Donlan. Irrespective of whether or not compensatory mitigation makes sense in this particular case, the general approach presented by Wilcox and Donlan seems worth considering for other situations involving non-native invasive species.

The need for extensive and accessible databases

Graham *et al.* (2008) described five components of an effective invasive species cyberinfrastructure, a term used to describe an internet environment that supports a variety of data management and research tasks, including data acquisition, storage, management, integration, mining, and visualization. According to Graham *et al.*, such an environment would:

(1) allow collection of data on the location and characteristics of invasive or potentially invasive species;
(2) provide watch lists of potential new invasive species by area;
(3) send alerts for early detection of new invasive species to appropriate managers and stakeholders;
(4) model the current range and predicted spread of invasive species; and
(5) provide information on best-management practices for rapid responses to new invasions and for control and restoration efforts.

Currently, such a cyberinfrastructure does not exist for invasive species. A major obstacle to developing effective approaches to prevent and mitigate the impacts of invasive non-native species has been that, until recently, much of the available data have been local in origin and not easily accessible (Mack *et al.* 2007). As a result of concentrated efforts by many individuals, agencies, organizations, and countries, networks of information are being created that are now facilitating the exchange of data and information. It is believed that these databases and communication networks in turn will facilitate more research on ways to integrate ecological, economic, and trade factors into effective management plans for invasive species (Meyerson and Mooney 2007). Isolated and small-scale management efforts directed at an invasive species may not require knowledge of the species' spatial distribution and temporal dynamics throughout the region or country or continent. However, larger scale and coordinated efforts to manage invasive species requires access to up-to-date information on the distributions and dynamics of the target species if they are to be successful, emphasizing the vital importance of extensive, current, and accessible databases.

Ideally, data-sets would be taxonomically, geographically, and temporally complete, meaning that the data-set would encompass many taxonomic groups over an area that had been extensively and systematically surveyed over time (Crall *et al.* 2006). Taxon-specific databases would still be effective, as long as they were complete with respect to the other two attributes. Crall *et al.* reviewed more than 300 data-sets for non-native species in the US. They found that the majority dealt with plants and that 43% of all data-sets were not available online (Fig. 8.8). Only 19% of the data-sets met their criterion for completeness with respect to the geographic criterion (how carefully and completely the area had been surveyed), while twice as many (38%) met their criterion for temporal completeness (data were collected continually for at least 10 years) (Fig. 8.8). In addition, nearly half (45%) of the data-sets did not have any type of quality control measures implemented. Not surprisingly, the authors emphasized the need for data-sets that are more complete, more accessible, and of higher quality. They also called for pooling of data and the use of a standardized protocol. To this end, the National Institute of Invasive Species Science, a consortium of US

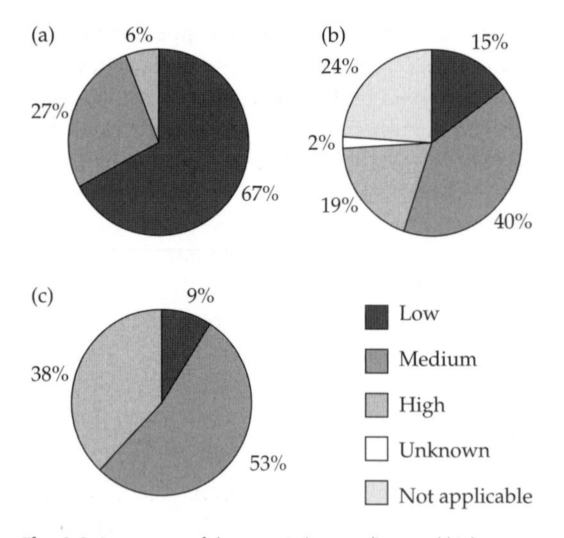

Fig. 8.8 Percentage of data-sets in low, medium, and high categories for (a) taxonomic, (b) geographic, and (c) temporal completeness. Redrawn and printed, with permission, from Crall *et al.* (2006), copyright 2006 Ecological Society of America.

government and non-government organizations, have developed and are promoting standardized approaches to collecting, entering and disseminating data (Crall *et al.* 2006).

Graham *et al.* (2008) came to a similar conclusion based on their assessment of existing data availability and highlighted a number of challenges that must be met if an effective cyberinfrastructure is to be created for invasive species. They emphasized that an effective invasive species cyberinfrastructure would require long-term and stable funding, which for the most part does not currently exist. Another challenge, they said, was to persuade researchers to make their data freely available, while still ensuring them credit for the original work that produced that data. They also noted that the quality of the data will inevitably vary and it will be important that users have information regarding the quality of the data.

A recent development has been the growing interest in developing a barcoding database of non-native invasive species. Following the first International Barcode Conference, held in London in 2005, the Consortium for the Barcode of Life (CBOL) formed the International Network for Barcoding Invasive and Pest Species (INBIPS). The goal of INBIPS is to provide existing groups and initiatives, already addressing invasive species, with information about the application of DNA barcoding to the problem of non-native and invasive species (more information is available at the INBIPS website: http://barcoding.si.edu/INBIPS.htm).

The 'LTL' approach

Presumably, few would question efforts by society to manage, or even eradicate, non-native species causing great health or economic harm, the latter including damage to vital ecological services. Although support would probably not be as widespread, many in the public and policy sphere would likely approve efforts to control or eradicate non-native species that are truly threatening resident species with imminent extinction. However, there may be some sense in questioning the control and management efforts involving non-native species producing other undesirable ecological impacts. These would include species that are altering the composition of historical communities, changing disturbance regimes, and/or affecting ecosystem processes, but that are not threatening our health, economies, ecological services, or causing other types of great ecological harm such as driving other species to extinction.

When confronted with something we do not like, we basically have two options. We can try to get rid of, manage, or otherwise exert our control over this thing. Or, we can try to change our attitude towards it. In many aspects of life, the latter option is often the most prudent. For species that we wish were not there, but are not actually causing any substantial health, economic, or ecological harm, the best management strategy may be the LTL approach, i.e. *Learn To Love 'Em*. For some readers, this may require a mind shift beyond what is possible at this point. If you can't *Learn To Love 'Em*, then perhaps you can *Learn To Like 'Em*, or at the very least you can try to *Learn To Live with 'Em*. By adhering tightly to a native preference, one essentially guarantees oneself a lifetime of frustration and disappointment, since one will likely increasingly be surrounded by more and more non-native species. Many in the general public do not have this problem. For example, in Minnesota, Queen Anne's Lace, *Daucus carota*, a species introduced from Europe, is the favorite summer wildflower for many people, and I suspect most Minnesotans enjoy the beauty of the colorful summer roadside flora, which consists almost entirely of non-native species such as *D. carota*, purple clover, *Trifolium pratense*, chicory, *Cichorium intybus*, and bird's-foot trefoil, *Lotus corniculatus*. There is value in trying to maintain a sense of perspective and pragmatism. The introduced species are not going back. Like it or not, they are our new residents, no matter what we choose to call them. Richardson *et al.* (2007) essentially argued this same point in their discussion of management options of urban riparian plant communities. They emphasized that the removal of non-native plant species in these landscapes, not only is almost certainly going to be futile but that it could be counter-productive with respect to maintaining ecosystem services.

In instances where non-native, and even non-native invasive, species are not causing significant harm but only change, altering one's perspective is certainly much less costly than any other sort of management program. That said, society may certainly go ahead and try to restore or maintain a portion of nature in some historical, e.g. 'pre-invasion,' condition. After all, societies commonly undertake other historical restoration and preservation efforts, and our natural environment normally plays a large part in a society's cultural heritage. At the societal level, the issue becomes one of priorities and ethics, since public resources used in such efforts could always be used to support other social and environmental projects.

Challenges, risks, and pitfalls of management efforts

Impact does not equal mechanism

It is often relatively easy to recognize, and even quantify, impact, particularly when it is strongly negative. However, in order to alleviate the damage, one normally needs to know something of the underlying causes. Unfortunately, it is usually much easier to assess impact than it is to determine the series of ecological causes for it. White *et al.* (2006) concluded that there are many more good studies of impacts of non-native species than there are detailed studies of the underlying ecological mechanisms. Ultimately, it is the latter type of study that is going to be most helpful to managers.

The playing field and the players keep changing

Just as a continually shifting environment inhabited by continually changing species makes it difficult for ecologists to understand and predict invasions, the same dynamism creates problems for managers (Peterson 2005). This is a central challenge for all types of conservation management (Pressey *et al.* 2007). The fact that the ecological systems often interact with social systems, makes predictability and management even more challenging (Westley *et al.* 2002). Strategies that work at one site may not be as effective at another site. An intervention that was effective one year may not be as successful at the same site the following year. If invasion-management efforts are to be successful over the long term, it is clear that managers will need to be flexible and responsive to changes in the systems, both natural and social, that they are managing.

Problems with eradication

In cases where eradication is the goal, managers are faced with a dilemma when it appears that eradication has been successful. When a survey has detected no individuals of the target species, should future surveys be terminated? Or, since there is usually some probability that some individuals may escape detection in a given survey, should surveys be continued to ensure that the eradication really has been successful and, if so, how long into the future should such surveys be conducted? If managers decide to terminate subsequent surveys once a survey comes up empty, they risk the possibility that some individuals were missed. On the other hand, if they continue surveys into the future, they may utilize resources unnecessarily.

One approach is to conduct the eradication efforts in a systematic way to minimize the probability that any individuals remain once eradication has been declared complete. This was the approach taken during the eradication of pigs from Santa Cruz Island, California (Morrison *et al.* 2007). During this eradication campaign, hunters progressed methodically through individual fenced zones, using a variety of way to locate the pigs (helicopters, bait, trapping) in order to reduce the likelihood that pigs were able to escape, using refuges. Efforts were also made not to take the easily captured pigs in the first pass, leaving pigs increasingly savvy in capture-avoidance. As described by Morrison *et al.*, if hunters felt that they would not be able to kill all individuals in a herd, they were instructed not to kill any of the pigs.

Regan *et al.* (2006) addressed this problem in a different way and suggested that, rather than making decisions regarding future surveys on the basis of arbitrary thresholds, e.g. 1% or 5% likelihood that a species is no longer present, decisions might be made based on an economic approach. In this approach, four factors need to be taken into account: the cost of conducting the surveys, the expected damage costs should individuals remain, the probability of detecting individuals if they are present, and the probability that the species is still present. The decision to stop further surveys would be made once the costs of continued looking exceed the expected benefits. Regan *et al.* showed that net expected costs (NEC) initially declines with increasing numbers of absent surveys, meaning that the risk of expected future damage costs of missed individuals exceeds the costs of conducting future surveys, but eventually NEC rises, meaning that the cost of additional surveys begins to exceed expected future damage costs. The optimal number of absent surveys, i.e. the point at which no further surveys should be conducted, occurs at the inflection point in Fig. 8.9.

Taxonomic uncertainty

In some cases, it is not clear whether a species should be considered an endangered native species, and hence protected, or a non-native and harmful species that should be controlled. For example, the Australian dingo was introduced into the continent about 5000 years ago, and recent DNA evidence indicates that the introduced individual(s) were likely domesticated dogs, probably from Indonesia (Savolainen *et al.* 2004). Although overhunting, habitat loss, and disease are believed to have driven the last remaining thylacine, *Thylacinus cynocephalus*, populations to extinction in Tasmania in the

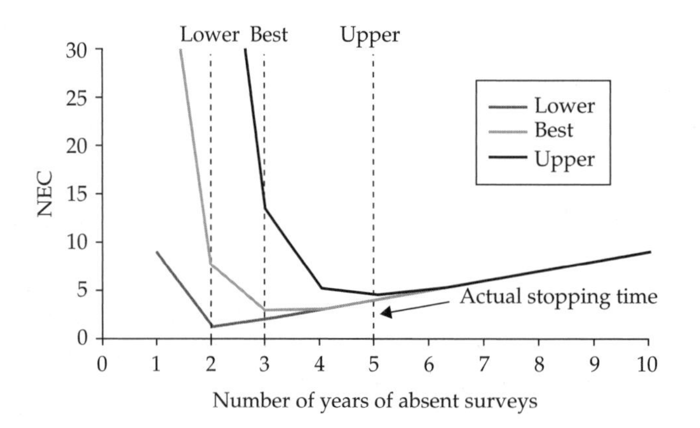

Fig. 8.9 Net expected cost (NEC) of conducting additional surveys as a function of the years of absent surveys. Shown is the predicted relationship involving *Helenium amarum*, a non-native and invasive annual perennial in Queensland, Australia, using the algorithm developed by Regan *et al.* (2006). Shown are results for various parameter range combinations. The optimal solution for each parameter combination is the lowest inflection point, shown via dotted lines. Redrawn and printed, with permission, from Regan *et al.* (2006), copyright Blackwell Publishing.

early 1900s (Bryant and Jackson 1999), it is thought that competition from the introduced dingoes contributed to their decline (Guiler 1985). Today, in many parts of Australia, the dingo is still considered a pest that preys on sheep and other livestock and should be eradicated (Miller 2007). Despite its introduced and domesticated origins, in 2007, the Australian state of Victoria, upon the recommendation of the scientific advisory committee for the state government's Flora and Fauna Guarantee Act, decided to consider listing the dingo as an endangered species, owing to the species decline, which is primarily due to hybridization with domesticated and wild dogs (Clarke 2007). Whether the species should be protected as an endangered native species (it has been an Australian resident for 5000 years after all), or a controlled introduced and feral species, which originated from, and still breeds with, domesticated dogs, is ultimately a matter of perspective. One argument to protect the dingoes is the fact that recent research has indicated that, since the dingo is a top predator, in areas where it is present, it reduces the number of other introduced predators, such as cats and foxes, thereby reducing predation pressure on many small native marsupials (Johnson *et al.* 2007).

The role that taxonomy can play in deciding whether to manage a species as endangered or non-native is also illustrated by the case of raccoons (*Procyon*) on Caribbean islands. In the Caribbean, raccoons are found on many islands and, in many cases, the island populations have been given species status, at least by the island human populations (Nichols 2007). However, recent molecular analyses indicate that all the island populations are likely recent reintroductions from the mainland, and are not deserving of any special taxonomic designation at all (Nichols 2007). According to Don Wilson (cited in Nichols 2007), taxonomist at the National Museum of Natural History in Washington DC, 'In the Bahamas, they were delighted. They instantly changed the raccoon's status from endangered to invasive species and set up a control program to eradicate them.'

Challenges of biological control

Biological control efforts involving the introduction of enemies can be an effective management strategy with low environmental impact (McFadyen 1998, Syrett *et al.* 2000). The benefits of biological control depend greatly on the relationship between abundance of the pest and the extent of harm produced by the pest (Thomas and Reid 2007). Biological control is going to be easier and positive effects will appear sooner when applied to high-threshold pests, i.e. pests that produce high levels of damage only at high densities (Fig. 8.10). In these instances, even a small decline in the population size of the pest species will substantially reduce damage. Biological control of low-threshold pests, pests that produce high levels of harm even at low densities, will be more of a challenge and more costly, since significant mitigation will only occur once the pest population has been reduced to very low numbers (Fig. 8.10).

The use of insects to control non-native invasive plant species has achieved some success in a number of cases (McFadyen 1998, Van Driesche *et al.* 2003). However, in most instances, little effort has been made to quantify positive impacts of the biological control agent beyond documenting a decline in the problem species (Thomas and Reid 2007). Presumably, of ultimate interest is not the decline in the problem species but improvements in ecological services, increased abundance of native species, and so on. Thomas and Reid expressed concern over the lack of uniform benchmarks with which to evaluate biocontrol programs for invasive species and called for more efforts to quantify community-wide and ecosystem impacts of these control efforts.

A major concern with biocontrol approaches to manage non-native species is that the introduced enemy, usually non-native itself, often does not confine its effects on the target species but also impacts desired native species. For example, as described in Chapter 7, the weevil *Rhinocyllus conicus*, introduced into North America to control several invasive non-native thistle species, *Carduus* spp. (Julien 1992), has been found also to feed on native thistles, substantially reducing their reproductive success (Louda *et al.* 1997). Significantly, the ability and tendency of many introduced control agents to switch to, or additionally impact, non-target organisms does not seem usually to require adaptation and evolution subsequent to introduction, at least in the case of macro-organism control

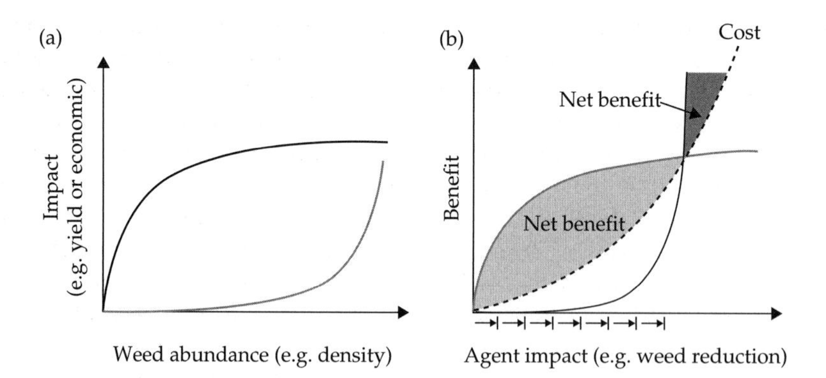

Fig. 8.10 The benefits of biocontrol for an agricultural weed. (a) The economic impact of individual weeds varies from those causing minimal impact until a certain threshold density is exceeded (gray line; e.g. Paterson's curse, *Echium plantagineum*), to those with high initial impact even at low densities (black line; e.g. wild radish *Raphanus raphanistrum*). (b) For the high-threshold weed (gray line), a single biocontrol agent that causes even a relatively small reduction in weed abundance from an initial high level (impact of individual agents indicated by arrows below the x-axis) can deliver a net benefit (light-shaded area). For the low-threshold weed (black line), there is only a net benefit (dark-shaded area) when weed abundance is reduced below a threshold, even if control is substantial (because weed impact is high, even at low abundance). The costs of biocontrol are represented by an increasing non-linear function to capture the probable escalating costs of reducing the abundance of a weed to ever-decreasing levels. An ineffective biocontrol agent could still deliver the net benefit, if it acts additively (as indicated by sequential arrows) or synergistically with other agents. Redrawn and printed, with permission, from Thomas and Reid (2007), copyright Elsevier Limited.

agents, like insects (Louda *et al.* 2003, Roderick and Navajas 2003). Instead, in these cases, the data indicate that it is more accurate to consider the impacted non-target species as part of the 'fundamental host range' of the introduced control agents (Roderick and Navajas 2003). Serious undesirable effects of biological control agents can occur indirectly as well. Pearson and Ragan (2006) found that two specialized insect herbivores (*Urophora spp.*), which had been introduced into western North America to control spotted knapweed (*Centaurea maculosa*), resulted in an increase in the resident deer mice (*Peromyscus maniculatus*) population by providing them with a food subsidy. More importantly, the increased density of deer mice resulted in an increased risk of infection by hantavirus for the mice, thereby raising the health risk to humans (Pearson and Callaway 2006).

Unintended consequences

Certainly, a first-order objective of any management effort should be to not make things worse. Zavaleta *et al.* (2001) urged that all efforts to remove invasive species be conducted in a whole-ecosystem context, their point being that the total impact of the species may include important and even desirable food web and ecosystem effects that may not be as immediately obvious as the undesirable impacts (Wittenberg and Cock 2005). Hobbs and Mooney (2005) and Peterson (2005) emphasized a similar point, stating that the interactions of a variety of drivers will make it difficult to predict impacts from particular management interventions. For example, reducing the rat population in a New Zealand forest by poisoning increased the predation of weasels on native prey (Murphy and Bradfield 1992).

Removal of a non-native top predator may result in a substantial increase in meso-predators, producing an undesired reduction in their native prey (Zavaleta *et al.* 2001). Removal of prominent herbivores also can have unintended consequences. In an effort to protect native plant species, pigs and goats were removed from Sarigan Island, an island in the Commonwealth of the Northern Mariana Islands (Kessler 2001). However, the eradication of the non-native herbivores resulted in the population explosion of a non-native vine, *Operculina ventricosa*, which had been preferred food for the goats. Within two years, coverage of this vine had spread over most of the small island (Courchamp and Caut

2005). Similarly, when cattle were removed from some California grasslands as part of an effort to protect native forbs, the cover of non-native grasses, which had been able to colonize the environment due to nitrogen deposition, increased dramatically, reducing abundance levels of the native forbs below that occurring under grazing (Weiss 1999). Initial efforts to eradicate South American fire ants in the southern US using insecticides ended up causing great harm to cattle and to many native ant species (Simberloff 1996).

In some instances, the non-native species may be providing desirable functions previously provided by native species, in which case the eradication of the non-native species would also eliminate these functions. For example, the endangered southwestern willow flycatcher, *Empidonax trailii extimus*, traditionally nested in stands of *Salix* and *Populus* that grew in riparian habitats in the southwestern US. Due to a variety of causes, including water diversion and changes in land use, the native woody species have substantially declined. However, the flycatcher has been able to persist in these environments because it has switched to nesting in saltcedar, *Tamarix* spp. (Zavaleta *et al.* 2001). Since the environment has been so altered due to human activity, it is not clear to what extent the native species would be able to re-establish in the absence of *Tamarix* and thus eradication of the *Tamarix* could threaten the persistence of this already endangered bird species. Because of these potential undesirable consequences of efforts to eradicate non-native invasive species, Zavaleta *et al.* recommended that, prior to implementing an eradication effort, managers should do their best to assess the nature and extent of trophic interactions between the non-native and native species, and also among the non-native species themselves, and to describe any key functional roles the non-native species may be playing in the ecosystem. Only when the managers can be satisfied that the eradication or substantial reduction of the target species will not have undesirable system-wide effects, should the eradication effort be undertaken.

Smith *et al.* (2006) argued that current approaches to control non-native plants growing in wildlands have largely adopted the agricultural approach to weed management, which involves extensive mechanical and chemical-based approaches. The problem with this approach, according to Smith *et al.*, is that the desired effects are often temporary, and the mechanical- and chemical-based approaches often produce other undesirable envirsonmental impacts, including the evolution of herbicide resistance, undesired impacts on non-target organisms, and the export of chemical herbicides beyond the managed ecosystem (Innes and Barker 1999, Matarczyk *et al.* 2002, Robertson and Swinton 2005, Wootton *et al.* 2005).

Another major weakness of the agricultural approach to controlling wildland invasive plants species is that the agricultural approach is primarily a species-focused strategy, in which the control method generally involves little more than a direct effort to reduce, ideally eradicate, the target weed. As Smith *et al.* (2006) pointed out, this ignores the real possibility that the ecosystem is altered in various ways due to the management practices or to changes driven by the invasive species. The success of these approaches is often short-lived due to the fact that the control efforts frequently end up functioning as major ecosystem disturbances, which just facilitate the re-establishment of the site by the same or other weed species (Smith *et al.* 2006). Moreover, it is now widely known that non-native plant species can alter the soil microbial community (Wolfe and Klironomos 2005) and the seed bank (Vilá and Gimeno 2007), and leave a legacy effect, even if the target species is removed. Smith *et al.* (2006) emphasized that management efforts should target the root of the problem, e.g. changes in disturbance regimes, resource availability, and propagule pressure, rather than simply targeting the species itself, which may be more a symptom than a cause of the problem.

The use of this high-impact agricultural approach has not been confined to wildlands but is being carried out in urban and suburban areas as well. In the metropolitan area of Saint Paul and Minneapolis, USA, extensive efforts by townships and neighborhoods have been undertaken to eradicate European buckthorn, *Rhamnus cathartica*, from areas into which it has spread, efforts that have included both mechanical and chemical approaches (Zamith 2007). The success of these approaches is being questioned by some local managers and researchers who argue that, not only does the buckthorn often return to pre-control levels after a few years but the control efforts frequently cause other problems, including soil erosion and damage to desired native plant species (Zamith 2007).

Describing and perceiving invasion harm

Conservation and land-management efforts of any kind are ultimately driven by the way a problem is perceived (Schwartz 2006). If the problem is characterized as mild and/or distant, it may be more likely that a low-impact approach would be taken. However, if the problem is presented as critical and imminent, a much more intensive and extensive approach likely will be used, one that may end up producing other undesirable impacts. In the case of non-native species, some have argued that many management practices have been influenced by the promulgation of a sort of simple-minded 'nativism' paradigm, in which native species are embraced and non-native species are vilified (Larson 2005, Smith *et al.* 2006). When a problem is framed using inflammatory language, e.g. militaristic metaphors and unjustified ecological hyperbole (e.g. non-native plants threaten native plants with extinction), it is more likely that managers and the public will support the use of more drastic measures, such as the application of chemical pesticides. An unfortunate example of this involves the concerted efforts begun during the late-1980s by departments of natural resources in the Great Lakes states of the US to eradicate purple loosestrife from many wetlands. Based on little actual scientific evidence that the species caused great harm to native wetland species, managers vilified the species and several states essentially declared war on the species, spending large amounts of human and financial resources in an effort to eliminate the species from the landscape, or at least drastically to reduce its coverage. Studies conducted in the late-1990s and early 2000s on this species found little in its ecological impacts to justify this output of resources (Hager and McCoy 1998, Houlahan and Findlay 2004). Obviously, studies of this type should precede extensive management efforts to avoid the misuse of limited conservation resources.

Diverse public perspectives

The extent to which efforts to manage non-native invasive species may negatively affect the resident human population is an important aspect that should always be assessed in any management plan. This should be done for humane reasons, as well as to increase the likelihood that the control effort will be successful, since the success or failure of conservation and management efforts usually hinges on the extent of support from local human residents. It is very important to recognize that harm is a social value, and hence what constitutes harm to one person, or group of people, may not be considered so by others, and, in fact, may be even considered desirable (McNeely 2005). Since, in many instances, the scientists and managers are not local residents, their own perceptions and values associated with the site to be managed may differ from those who live nearby. For example, it has been argued that scientists and land managers are often less place-dependent in their environmental attachments, meaning that they are attached more to a type of land/seascape, than is the general public, which often has very strong attachments to a particular place. This means that the public often wants to preserve the place the way they have known it, and may object to the removal of non-native species, which they may view as an integral part of the environment that they love (Ryan 2000, McNeely 2005).

In the Galapagos Islands, an intensive and extensive goat eradication program was undertaken, even though some of the poorer local human residents of the islands used the goats as an important source of food and income, and objected to the goal of complete eradication (Romero 2007). In this case, and from the perspective of the conservationists, the eradication program, funded by millions of dollars and carried out using helicopters, hunting dogs, and hired shooters, was largely successful. Similar instances have occurred elsewhere. In Australia, aboriginal peoples now hunt some of the introduced mammals, e.g. rabbits, water buffalo, and camels, and have objected to efforts by the government to control these invasive species (McNeely 2005).

When a restoration proposal was made in the 1990s to remove the non-native honeysuckle that had originally been planted on Montrose Point, a peninsula extending into Lake Michigan and part of Chicago's Lincoln Park, and to plant native prairie vegetation, birdwatchers vigorously objected (Ryan 2000). Due to the combination of the Point's location and the vegetation structure provided by the honeysuckle, Montrose Point had become a major stopover sanctuary for migrating song birds and the line of shrubs was being referred to as the Magic Hedge by birders. In the end, the decision was made not to convert the vegetation to native vegetation. Of course, given that Montrose Point is an entirely human-constructed land mass, the whole notion of native vegetation for the Point is somewhat questionable in the first place.

In some cases, scientists themselves have objected to measures proposed to control the spread of non-native invasive species. As described in Chapters 2 and 7, the chytrid fungus, *Batrachochytrium dendrobatidis*, causes the disease chytridiomycosis, a major source of mortality for frogs worldwide (Weldon *et al.* 2004, Marris 2008). The global commercial trade of the African clawed frog, *Xenopus laevis*, which can serve as a host for the fungus, is believed to be a primary dispersal vector for the fungus. In 2007, the UK Department for Environment, Food and Rural Affairs (DEFRA) included *X. laevis* on its list of potentially harmful non-native species and indicated that it wanted to ban all future sales and trade of this species into the country, citing not only its role as a vector for the fungus but the fact that escapees compete with, and sometimes eat, native amphibians (Vogel 2008). Because *X. laevis* is widely used as a research animal in laboratories throughout the world, a variety of scientific groups responded quickly to this announcement, raising their concerns over the proposed restrictions on the importation of this species. As of mid-2008, DEFRA and the research advocates were discussing the possibility of a revised plan to limit, but not ban, the import of the frogs (Vogel 2008).

Hattingh (2001) argued that humans draw the line between what is native and what is non-native, and thus different people may disagree over the status of particular organisms. Illustration of the extent to which the public can think of non-native species as part of their own identity is the fact that more than a dozen US states have adopted the non-native honeybee, *Apis mellifera*, as their state insect. Not only have Vermont residents named the honeybee as their state insect, they have named as their state flower another non-native species, purple clover, *Trifolium pratense*. (Their next door neighbor, New Hampshire, adopted the non-native common lilac, *Syringa vulgaris*, as their state flower.) For many human generations, whenever someone arrived in Vermont, whether by birth or immigration, they found the honeybee and purple clover already waiting for them in abundance. Thus, it is hardly surprising that the two species would be considered an integral part of the state's homeland.

Effective management projects will have gained the widespread and long-term support of the stakeholders. This support is most likely to be the end-product of a planning process in which all the stakeholders have had a voice. Understanding the social values and personal motivations behind the public's preferences for some non-native species, even for some species considered invasive by ecologists, should better prepare ecologists and managers for their discussions with other stakeholders (Mack 2001, Reaser 2001, McNeely 2005, Coates 2006). If this knowledge helps ecologists and managers understand that people can legitimately differ on what is considered beneficial or harmful, then this should improve the dialog between the public, and the ecologists and managers. However, problems are likely to occur if this knowledge is used by scientists and managers purely to devise more effective 'marketing campaigns.' If ecologists and managers proceed, convinced that contrary views are wrong, and their knowledge of the values and motivations behind citizens with different perspectives is used only to try to manipulate the public into changing their views, then ecologists risk being perceived, correctly, as arrogant, and will likely end up estranging the very people needed to support the management efforts.

The relevance of invasion research to invasion management

The last paragraph of most invasion research papers and proposals typically includes some boilerplate emphasizing the management value of the research findings or the proposed research. I have only talked informally with managers regarding the practical value of invasion research papers, and thus I am not able to comment conclusively on this issue. (I am sure that invasion researchers would be very interested to hear from managers, as well as from managers' professional societies, on this topic.) However, my impression is that most invasion research and theory is only generally relevant to the specific challenges faced by individual managers. For example, the fluctuating resource availability theory has proven to be among the most robust invasion theories to date, having been supported by research studies conducted in a wide variety of environments with a range of taxa (Daehler 2003). The theory has been promoted by management-oriented researchers as suggesting a promising management strategy:

> Reducing community vulnerability to invasion by manipulating resource availability appears to be a promising approach to invasive species management.'
>
> (Perry *et al.* 2004.)

> The approach I have described [using the fluctuating resource availability theory to help guide management efforts] requires that we take the additional step of evaluating the influence of climatic variation, natural disturbance, and management on resource availability.... Such an approach would be an important step in identifying sites with a high potential for weed invasion.
>
> (Svejcar 2005.)

While the fluctuating resource theory has been widely supported, there certainly have been exceptions (e.g. Lennon *et al.* 2003, described in Chapter 3). In a study of the invasibility of short tussock grassland communities in New Zealand, Walker *et al.* (2005) concluded that characteristics of the resident plant community may be more critical than resource fluctuations in determining the invasion success of *Hieracium pilosella*. Even in instances where data have generally supported the fluctuating resource availability theory, support has often been contingent on other factors as well (Hastings *et al.* 2005). These exceptions and contingencies indicate that, even for theories that have proven to be quite robust, it would be unwise to apply them indiscriminately to management efforts. Not only might the applications not be effective, they could potentially make matters worse.

Managers have been frustrated by the fact that research results are often not directly applicable to management priorities and needs (Hulme 2003, Richardson and Pyšek 2008). The managers with whom I have spoken have uniformly agreed that the most useful invasion research is that conducted on the same ecological system as the one they are managing, with the most useful of all being research undertaken on their actual management site with the specific species of concern. This makes sense given the important role that history and local and system-specific ecological contingencies play in the invasion process. But system- and site-specific research can have its drawbacks if it is unfocused. Ideally, theoretical and general invasion research could assist system- and site-specific research by helping the latter to focus on events and processes likely to play an important role in defining the invasion dynamics in the system being managed (Fig. 8.11).

Whether additional invasion theory still has a lot to offer management efforts is an interesting question. Of course, our ability to generate additional theory is endless. McIntosh (1987) characterized the field of ecology as a theory-laden discipline. Richardson and Pyšek (2008) observed that 'the invasion literature is accumulating a growing number of theories and generalizations, with much duplication, redundancy and reinventing the wheel.' This may be due partly to increased specialization, which, by focusing on current research, can leave the researcher disconnected from past contributions in the field (Graham and Dayton 2002). It may be that we are reaching the point of diminishing returns with respect to the contributions of new theory to management, meaning it is unlikely that additional developments in invasion theory will lead to any major management breakthroughs (Fig. 8.12). Given the importance of history and local idiosyncrasies in influencing the dynamics and outcomes of

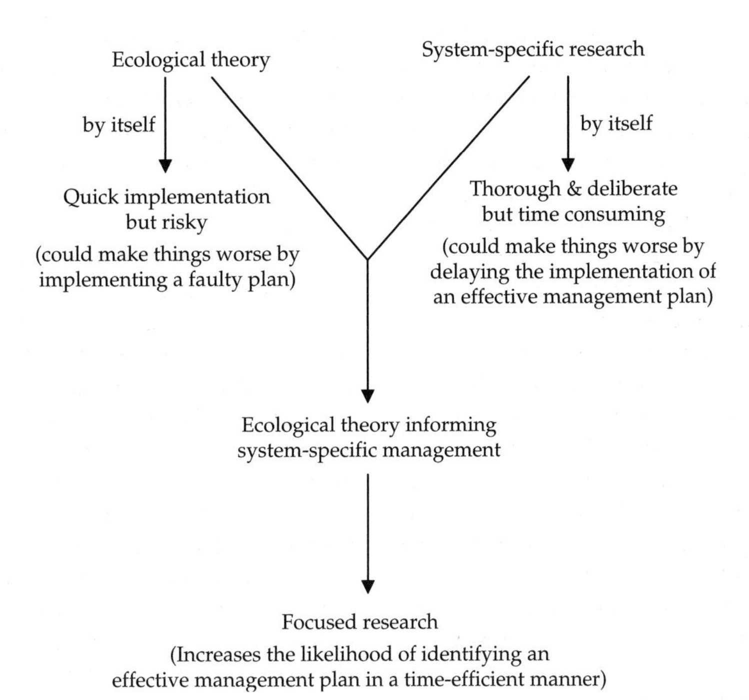

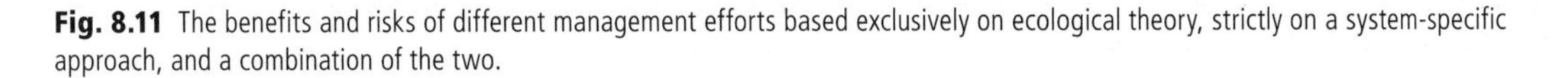

Fig. 8.11 The benefits and risks of different management efforts based exclusively on ecological theory, strictly on a system-specific approach, and a combination of the two.

invasions, it seems doubtful that any additional theory will be able to provide the guidance at the level of detail needed by managers on the ground. This is not to say that new theory has nothing to offer management. For example, new theory may be able to provide some additional guiding principles that might help to focus managers' priorities. While future management contributions from new theory may be modest, more well-informed system- and site-specific research should continue to be of great management value (Fig. 8.12). Readers can decide for themselves where they might place the current state of affairs regarding invasion theory and management in Fig. 8.12. If managers generally agree with my assessment, then it would seem that a productive approach would be one that promoted and supported invasion research conducted within managed environments and that focused on the actual species and processes of management concern.

Summary

It is normally not possible, or at least it is very expensive, to eradicate an undesired species that has spread throughout a large area. Thus, with respect to non-native invasive species, there is a strong incentive to intervene early in the invasion process. With international trade being the primary driver of species' introductions, both intentional and accidental, many countries, states, and provinces have established a variety of types of filters along their borders to prevent the initial introduction of harmful species. In many instances, filters have also been implemented at the transport sources. These filters include inspections of transported goods, laws preventing the export and import of certain organisms, regulations of the transport process to minimize the accidental introductions of undesired organisms, and various prophylactic treatments of the goods and transport vehicles, also to prevent

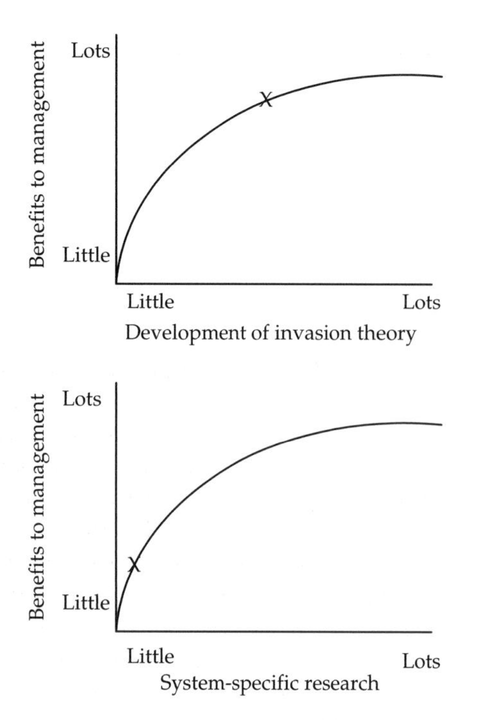

Fig. 8.12 Suggested benefits to management by the development of additional invasion theory (top) and by the implementation of additional system-specific research (bottom). The X on the respective lines indicates the proposed current location of the field with respect to invasion theory development and site-specific research. It is suggested that management has less to gain from additional invasion theory than it does from additional system-specific research.

accidental introductions. In some cases, management begins even prior to transport. In the case of infectious diseases, efforts are often undertaken in the infected region to eradicate, or substantially reduce, the prevalence of the disease and the number of infected individuals, in order to reduce the likelihood that the pathogens will even reach any of the border filters.

Since border filters will never be 100% effective, there is increasing interest in establishing monitoring systems that will provide managers with an early warning of a recently introduced invasive species, when it may still be possible to contain or even eradicate the species. These monitoring techniques range from hi-tech approaches, e.g. remote sensing using satellites and molecular techniques to identify fragments of DNA found in the environment, to vigilant citizens who have been trained to recognize the undesired species. If neither the border filters nor the early warning efforts have succeeded, and eradication from the entire region is no longer a viable option, then managers must focus on the control of the species and its impacts. Unless one is able to exterminate a harmful or otherwise undesirable species from the entire planet, managing it is a never-ending enterprise. Even if managers are able to eradicate a species from an area, they must remain vigilant if the species is to be prevented from recolonizing.

As many past efforts have shown, managers face a myriad of challenges, risks, and pitfalls. Despite good intentions, poorly conceived and implemented management efforts can cause more harm than good. Management efforts will be more likely to meet their objectives if the management strategy has been developed using reliable and comprehensive scientific knowledge of the specific system being managed, and if it acknowledges that the primary and inevitable constant of the natural world is change.

Since the final objectives of a management project will have arisen from the values of those involved in the planning process, the more informed the participants are of the nature and origins of their values, the greater are their chances of deciding on objectives that will enjoy widespread support and that will be achievable. Since ecologists are frequently involved in the planning process, often in leadership positions, it is important that we try to understand some of the potential implications of our own values and possible predispositions.

PART III

Reflections

Invasion biology has made great progress over the past several decades, but this progress has not come without scrutiny. At times, colleagues from the social sciences and humanities have called into question aspects of our work, particularly how we have chosen, or been inclined to, frame biological invasions. These critiques have created some tension in the field. Some have welcomed and agreed with these commentaries, while others have sharply criticized them. Individual invasion biologists undoubtedly will differ in how much value they attribute to analyses and assessments of science made by philosophers, historians, and sociologists. Some may side more with Nobel physicist Richard Feynman, who is reported to have said, 'Philosophy of science is about as useful to science as ornithology is to birds.' Others may believe that disciplines such as philosophy, history, and sociology can provide us with observations and perspectives not always apparent to us. And some may believe that science should never be separated from its history, and that scientists themselves must take the time to articulate 'what is implicit in the structure of scientific activity' (Simpson *et al.* 1961). Chemist and educator James B. Conant (1947) similarly stressed that there must be constant critical appraisal of the progress of science and in particular of scientific concepts and operations. William Cooper, the early twentieth century American botanist, emphasized the importance of self-reflection, arguing that 'a periodic inspection of foundations is most desirable for any edifice, and particularly so when the superstructure is being continually added to, as in the development of scientific knowledge' (Cooper 1926, cited in Pickett 2007). And Frank Egler (1951) called for ecologists 'to question... [their] own beliefs... in an attempt to realize the limitations of them.' Personally, I believe that our colleagues from other disciplines do us a service by offering us their observations and insights. I think that findings and observations of colleagues from other disciplines can inform our practice of science, sometimes in very important ways. I also agree with Simpson, Conant, Cooper, and Egler on the value of practicing science in a self-conscious and self-reflecting way.

In the first chapter in this part of the book, and using findings and perspectives from other disciplines, I suggest how certain inherent inclinations and tendencies we exhibit as scientists may not always be apparent to us, even though they may be contributing to the way we think about biological invasions. Some readers may view this inquiry as a distraction, a sort of unenlightening bout of navel-gazing. I feel differently, and I hope most readers will agree with me. In addition to discussing these issues, in the first chapter I also review some of the difficult decisions that must be confronted when invasion biologists communicate with the public and policy makers. All of these issues influence how we conceive, describe, and discuss biological invasions, both among ourselves and with the broader public. In the second and third chapters of this part, I reflect on what I believe have been some challenges that the research field of invasion biology has struggled with during the past 25 years, and suggest what I think are some promising paths the field might consider taking.

CHAPTER 9

Framing biological invasions

Although quantum physicists seemed to have recently questioned the veracity of realism (Gröblacher *et al.* 2007), and post-modernists argue that humans do not have any access to reality (Allen *et al.* 2005), most scientists believe, not only that an external reality exists but, that humans are able to access at least some of this reality through our senses and instruments (Ellis 2005). At the same time, most scientists would likely also agree that the extent of our access to this external reality is not unrestricted, and that our descriptions and understanding are approximations of the external world, often only rough approximations. It seems difficult to deny that our access and descriptions are limited and influenced by the filtering effects imposed on us by our senses and instruments, our language, and the physical nature of the human brain. However, if we can never achieve complete and uninhibited access to the ultimate nature of reality, we can at least do our best not to create additional impediments, filters, and distortions. Unfortunately, it seems humans may be inclined to do exactly this. Humans possess particular cognitive predispositions, prejudices, and values that lead us to experience and interpret aspects of reality in certain ways (Alpert 1995, Frith 2007). In some cases, these prejudices may have been selected for in the past, or they may simply be a byproduct of the structure and function of the human nervous system. In other instances, we are undoubtedly influenced by values, perspectives, and approaches associated with the time and place in which we live. And, some of our ways of thinking may be more a product of disciplinary inheritance, habits, and tendencies adopted from the scientists who preceded us (Simpson *et al.* 1961). But, at all times, every individual scientist brings a particular bias to his/her work (Pickett *et al.* 2007).

Issues of categories

Most invasion biologists have imposed a sharply dichotomous paradigm on the field, one in which there are two types of species: native vs non-native, native vs exotic, indigenous vs non-indigenous, invasive vs non-invasive. Elton (1958) certainly emphasized this dichotomy, and SCOPE sustained this perspective when it challenged the field with three questions, the first two of which illustrate a binary perspective:

What factors determine whether a species will be an invader or not?

What are the characteristics of the environment that make it either vulnerable to or resistant to invasions?

How can the knowledge gained from answering the first two questions be used to develop effective management strategies?

But does this dichotomous approach make good ecological sense? Although new arrivals and long-term residents may lack a common evolutionary history, this does not necessarily mean they should be considered categorically different. For example, the new species may have recently evolved with species that are phylogenetically and ecologically quite similar to the residents. In any case, by the second generation, the new arrival and any long-term residents with whom they interact will already have begun to respond to one another both ecologically and evolutionarily. Cox (2004) similarly noted the limitations of an approach that dichotomizes species into natives and non-natives, observing that due to global climate change, native species are expanding their ranges into new regions, expansions sometimes involving hundreds of kilometers. Cox concluded that determining which species are native and which are not is frequently a nebulous enterprise. Willis and Birks (2006) and Warren (2007) also observed that the distinction between what is native and what is non-native is often unclear.

Rather than considering new arrivals and longer term residents as discrete categories, they may be much better viewed as part of a gradient or continuum, both with respect to their time of residency and the extent of evolutionary and ecological interactions they have had with other longer term residents. Carlton (2002) emphasized this point, arguing that invasive species do not represent a separate category (evolutionarily, biogeographically, ecologically, or socio-economically) and thus that they are not amenable to a dichotomous approach. Speaking more generally, Gould (2003) believed that dichotomous paradigms represent nature very poorly and argued ardently against their adoption. Addressing the same general topic even more broadly, renowned anthropologist Mary Douglas (1966) came to the same conclusion as Gould, arguing that classifying is a universal human activity and that all cultures impose lines of demarcation to create a sense of order, e.g. within and without, with and against. However, she warned that while 'it is part of our human condition to long for hard lines and clear concepts... experience is not amenable and those who make the attempt find themselves led into contradiction.' Pickett (2007) expressed similar concerns, warning against rigid classification schemes, inappropriate reification of idealized concepts, and narrow conceptions of phenomena and their causes, all of which, he observed, are not uncommon behaviors in ecology.

In some respects, perhaps we should not at all be surprised that we have distinguished between native and non-native/alien/exotic species. Recent research in diverse fields suggests that humans may be predisposed to impose group boundaries, and to adversely characterize outsiders (McElreath *et al.* 2003, Bernhard *et al.* 2006, Choi and Bowles 2007). Whether this bias stems partly from our genes (Gould 2003), or whether it is largely culturally-based (Ingold 2000), the tendency to categorize the world in this way appears virtually universal in the human species. Humans do seem to seek every opportunity to identify with a homeland, a home tribe, a home religion, a home team, and to declare someone else the opposition or the enemy. This declaration of home identity extends to the natural world. Most countries, states, and provinces have designated particular species of animals and plants as signature species, declaring them the national bird or the state flower, thereby making them ours. Even whole landscapes and seascapes can assume national and state identities. Given what appears to be a powerful predisposition to identify oneself locally (whether this inclination is primarily genetically or socially rooted does not matter), it does not seem surprising that people, including ecologists, have shown a strong favoritism toward native organisms. Whether the creation by ecologists of the distinct categories of native and non-native species is entirely the product of sound scientific reasoning, or whether it may have been influenced by inherent human tendencies to organize the world in a particular way, is a question, I believe, that is worthy of reflection and discussion by invasion biologists.

It is clear that some in the field have already been pondering this question and its implications. In their characterization of the fauna in the Caspian Sea, Aladin *et al.* (2002) declared that 'all its residents can be considered as invaders with various times of introduction,' noting that 'the most ancient invaders are regarded as indigenous.' Reise *et al.* (2006) pointed out that since both biotic and abiotic factors are often being altered considerably by humans, the terminology of native and non-native species loses much of its intended meaning, since both the native and non-native species are encountering new environmental conditions and 'all are strangers in a strange environment.' Similar points were made by Reise *et al.* (2006), who emphasized that 'the status of being a non-native refers to a position in evolutionary history but does not qualify as an ecological category with distinct and consistent properties.' Recognizing that issues of xenophobia have plagued the research field of invasion biology, Reise *et al.* stressed the importance of trying to 'keep this research free of hidden assumptions.'

It is interesting to ponder to what extent invasion biology might have developed differently had ecologists not imposed a dichotomous perspective on the phenomenon of global species redistribution, an approach that also generally manifested distinct nativism tendencies. What if species were not labeled as native and non-native? What if non-native species were not typically referred to as exotic or alien? And, what if the introduction of non-native species were not generally framed as threatening and undesirable? Instead, what if ecologists had taken a more quantitative than qualitative perspective, emphasizing a continuum rather than categories, and referring to the new species as 'new species,' or 'recently arrived species,' or 'new residents,' to distinguish them from 'longer term residents'? Might researchers have asked different questions? Instead of the three questions posed by SCOPE, perhaps others might have been proposed, such as:

What are the impacts on communities and ecosystems of the addition of new species?
Why do some populations remain relatively static for a period of time and then abruptly begin a period of spread?
Why do some populations sometimes grow and spread rapidly for a period of time and then diminish in size and extent?
In what ways do the recently arrived species affect the longer term residents, ecologically and evolutionarily?
How do the longer term residents impact the ecology and evolution of the newer residents?
What combinations of traits and environmental factors result in the establishment and rapid spread of some species?
How can we take advantage of the introductions of new species to learn more about fundamental ecological and evolutionary processes, such as population establishment and spread, community assembly, genetic and phenotypic adaptation, and the ongoing feedback between organisms and physical processes in ecosystems?
In what ways are the new species exacerbating health, economic, and/or ecological problems already caused by the longer term residents?
In what ways are the new species ameliorating some of the problems caused by the longer term residents?
In what ways are the new species creating new problems?
In what ways are the new species providing new benefits?

These questions acknowledge the dynamic nature of communities and ecosystems but do not signal or assume ecological upheaval and calamity. They do not cast the new introductions as a unique ecological phenomenon, but instead present them within the context of traditionally defined ecological processes. In this way, the new introductions are viewed as an opportunity to illuminate fundamental ecological, evolutionary, and biogeographical processes, a perspective that has been widely emphasized only recently by invasion biologists (e.g. Sax *et al.* 2005a, Cadotte *et al.* 2006). Some readers might argue that had species introductions not been characterized as a unique phenomenon, and had the negative impacts and threats not been the primary theme of invasion biology during the 1980s and 1990s, society would not have been motivated to mobilize resources to combat the damage caused by some of the new species, and to prevent their spread and introductions. But, countries, states, and municipalities have always been motivated to combat and prevent damage caused by harmful species and to work to check the spread of these species. As long as the harm is real, it should not be necessary to promote a native vs 'alien' dichotomy to get society to respond.

Issues of change

Cultural historian Svetlana Boym (2001) observed that nostalgia tends to emerge most strongly when people and societies have experienced, or are experiencing, great change in their lives. More specifically, Boym asserted that 'globalization has deepened nostalgic longings.' She observed that 'the twentieth century began with a futuristic utopia and ended with nostalgia.' Boym was referring to social and cultural globalization; however, the analogy to biological globalization is too obvious to ignore. Goldstein (2009) argued that much of the rhetoric used in invasion biology literature is very similar to that of the anti-globalist movement, which laments the replacement of authentic, local cultures with a synthetic and homogenized culture.

Boym argued that nostalgia manifests itself in two ways. One, which Boym described as reflective nostalgia, occurs when 'you don't deny your longing but you reflect on it.' According to Boym, this type of nostalgia 'is a positive force that helps us explore our experience and can offer an alternative to an uncritical acceptance of the present.' The other type of nostalgia Boym called restorative nostalgia. Boym described restorative nostalgia as 'not about memory or history but about heritage and tradition. It's often an invented tradition—a dogmatic stable myth that gives a coherent version of the past.' 'Restorative nostalgia,' Boym explained, 'does not think of itself of nostalgia, but rather as truth and tradition.' She added, 'Restorative nostalgia knows two main plots—the return to origins and the conspiracy.... Home is forever under siege, requiring defense against the plotting enemy.' Thus, in the face of rapid change and globalization, a climate of restorative nostalgia can be seen as giving rise to passionate efforts to restore an idealized vision of the past and a largely dichotomous view of the world, in which the local descendants of the past are sanctified, while any and all newcomers, often deemed responsible for defiling the historical paradise, are vilified.

Legendary US environmentalist Stewart Brand has criticized many in the current environmental movement from a perspective similar to Boym's, using the concept of romanticism instead of restorative nostalgia. In a recent interview with the New York Times (27 February 2007), he expressed his concerns regarding certain environmental perspectives, including tendencies toward apocalyptic thinking and romanticism. 'I keep seeing the... terrible conservatism of romanticism, the ingrained pessimism of romanticism. It builds in a certain immunity to the scientific frame of mind.'

The reflections of Boym, Goldstein, and Brand also hearken back to some of the ideas presented by Douglas (1966), especially Douglas's contention that humans are preoccupied with 'separating, purifying, demarcating, and punishing transgressions' in order to impose structure on an inherently 'untidy' world. In particular, Douglas argued, having constructed what usually are stark and unambiguous categories, cultures tend to be obsessed with 'matter out of place.' When 'pollution,' i.e. matter out of place, does occur, she explained that people consistently respond with anxiety, discomfort, and even anger. Douglas stressed that cultures respond predictably to pollution, vigorously condemning 'any object or idea that is likely to confuse or contradict cherished classifications.' In sum, Douglas argued, 'purity [all matter in its rightful place] is the enemy of change, of ambiguity, and of compromise.'

It should not be difficult for the reader to think of examples from the invasion literature, both public and scientific, that seem to illustrate the contentions of Boym, Goldstein, Brand, and Douglas. The dichotomous characterization of species based on their geography of origin is obvious, as is the normative language often used by invasion biologists with respect to non-native species, including 'purity' references such as 'biological pollution' (McKnight 1993, IUCN 2000). Although some have declared as unconvincing criticisms that the field of invasion biology has roots in a nativism paradigm (Simberloff 2003), some of our colleagues from other disciplines would argue that the language in much of the scientific literature during the last 20 years belies this defense (Gobster 2005, O'Brien 2006, Warren 2007). No doubt, readers' opinions on this issue will vary widely. However, the titles of some invasion books seem almost like billboards of a restorative nostalgic mindset, of apocalyptic thinking and romanticism, and of purity obsessions, e.g. *Killer algae: the true tale of a biological invasion* (Meinesz 1999), *A plague of rats and rubbervines* (Baskin 2002), *Nature out of place* (Van Driesche and Van Driesche 2000) *Tinkering with Eden: a natural history of exotic species in America* (Todd 2002), *Strangers in paradise* (Simberloff *et al* 1997).

In recent years, criticisms and misgivings regarding language and nativism perspectives have been coming from within the field of invasion biology as well (Brown and Sax 2004, Gurevitch and Padilla 2004, Gurevitch 2006). Summarizing their findings, regarding the impacts of non-native species on coastal marine systems, Reise *et al.* (2006) stated that there has often been an expectation of negative impacts by introduced species, arguing that 'current evaluations often rest on prejudice and not on science.' This is quite a blunt indictment coming from colleagues within the field of invasion biology.

To be clear, I am not trying to suggest that the attention that ecology has given to introduced species is nothing more than a universal human obsession with order. However, I do think there is value in reflecting on the extent to which largely universal psychological/emotional reactions to phenomena such as rapid change and boundary crossings may have influenced how ecologists and science writers have thought about and characterized the global redistribution of species.

Issues of communication

Given the nature of our expertise, ecologists are uniquely positioned to warn and advocate on matters of the environment. In many ways, scientists have joined religious figures as prophets of the modern world. However, while scientists have a responsibility to report their findings to the public, there is disagreement over the extent to which ecologists should try to remain objective in their communications on topics such as invasive species (Slobodkin 2001, Brown and Sax 2004, Larson 2007). Mixed feelings regarding advocacy are nothing new in ecology. During the first half of the twentieth century, the Ecological Society of America (ESA) was actively engaged in advocacy, e.g. through its Committee for the Preservation of Natural Conditions. However, unease over this advocacy by some in ESA resulted in a curtailment of ESA advocacy and the founding in 1946 of a separate advocacy organization called the Ecologist's Union, which became The Nature Conservancy four years later (summarized by Foreman 2004).

Accompanying this debate over advocacy is the question of whether the field of ecology really is just a natural science (Bradshaw and Bekoff 2001, De Laplante 2004, Larson 2007). Rather than considered as a natural science, some have suggested that ecology is better viewed as a 'bridge between science and society' (Odum 1997, cited in Larson 2007). Whatever one's opinions on these subjects, it is certain that many ecologists, and their organizations, will continue to make, not just their findings but their concerns and recommendations known on the environmental issues of the day, including non-native invasive species. The question, then, is what should be the responsibilities of invasion biologists who choose to issue warnings or advocate particular policies?

In an essay on science communication in the journal *Science*, communications researcher Matthew Nisbet and author and reporter Chris Mooney (Nisbet and Mooney 2007a) argued that scientists need to more consciously and actively frame their messages to society at large. The authors described several possible functions of framing, including organizing central ideas, simplifying complex material to give certain aspects greater emphasis, and defining a controversy to resonate with core values and assumptions. Nisbet and Mooney (2007a) closed their essay with a statement sure to stir the pot, 'in many cases, scientists should strategically avoid emphasizing the technical details of science when trying to defend it.' Nisbet and Mooney (2007b) also recommended that science organizations work with communication experts to determine the most effective way to reach particular audiences. These efforts could involve activities such as surveys and working with focus groups (Nisbet and Mooney 2007b). Not surprisingly, Nisbet and Mooney's remarks drew some lively response. Holland (2007) accused the authors of encouraging dishonesty by scientists. Pleasant (2007) expressed concern that Nisbet and Mooney were helping to perpetuate the myth that complexity cannot be successfully communicated and also argued that winning the daily mass media wars may not best serve the long-term relationship between science and society. In an earlier article in *Science*, Chew and Laubichler (2003) also expressed concerns about how scientists have framed their ideas in their communications to the public, as well as to one another, using metaphors. They acknowledge the value of metaphors, e.g. providing accessibility to new and/or complex ideas, but also emphasize that they can impede understanding as well as facilitating it.

It is easy to for this discussion on framing to assume a dichotomous nature, but it may be helpful to frame the process of framing as occurring along a continuum. Since one cannot communicate in absence of any context, the question is not 'to frame or not to frame', but how to frame. Proceeding on either end of the spectrum is problematic. If invasion biologists simply report naked findings, leaving it completely up to the audience to derive meaning from them, the size of the informed audience will almost certainly be very small, and not likely to include many outside the field of invasion biology. Thus, the scientist who reports the bare facts risks generating little interest and response from a broader audience. On the other hand, the scientist who constructs a very prescriptive social message, perhaps embellishing it with vivid metaphor and language, may be quite successful in attracting attention to the message, and even in securing short-term support and responses. However, the problem with operating on this end of the framing spectrum is that the framer assumes too much responsibility and forces a very particular interpretation onto the audience. In the former case, the scientist puts too much faith in the audience's ability to discern or manufacture a message, while in the latter the scientist disrespects the audience, believing that it can and should be manipulated from above, with scientists pulling the strings.

When I have discussed this issue with colleagues, some have argued that 'message enhancement' of invasion issues is a necessary strategy when dealing with the public and policymakers. It is necessary, they contend, in order to get the public's attention, and to balance out the hyperbole and misrepresentations coming from the other side. However, I am concerned there are serious risks with this approach. If framing occurs to the point of substantial distortion and misrepresentation, the message can end up eliciting well-intentioned but misguided responses on the part of the audience. One also risks sacrificing ones future credibility and of perpetuating misconceptions (Scheiner 2008). In addition, many in the general public recognize, even if the scientists do not, that what is often being presented to them is not just science, but a particular social or environmental agenda. In an essay on the role of subjectivity and implicit values in biology, Alpert (1995) observed that when science involves politically sensitive issues, the interpretations, and even sometimes the results, of scientists tend to correlate with their own political views. Despite good intentions, value-laden science communication can precipitate a strong public backlash, particularly if the public feels they have not been full partners in the discussion (Helford 2000). Finally, even if such approaches manage to secure rapid responses from the target audience, these responses may not promote effective long-term solutions (Owens 2005). While scientists certainly can influence public policy, the process is normally very time-consuming and a long-term perspective is much more realistic (Lawton 2007).

Although we may differ on what we believe constitutes appropriate framing, I am sure that we are unanimous in agreeing with Sykes (2007) that it is crucial for scientists to maintain a high quality in our dialogue with the public and policy makers, since otherwise time is wasted, we lose the public's trust, and poor policy decisions will be made. If neither end of the framing spectrum seems like a good way to proceed, then what would framing look like from somewhere near the center? The challenge faced by a scientist planning to communicate to colleagues, policy makers, or the general public, is how to frame the material in such a way as to effectively communicate one's message, i.e. make it accessible, interesting, and meaningful, without compromising the science behind the message. No formula exists to guide one on how to accomplish this balance. I suggest the following guidelines as possibly providing some worthy direction.

Responsible framing simplifies, but does not conceal. Most framing will involve some simplification of the topic of interest, but this should not be carried to the point where complexity, variability, and uncertainty are largely excluded from the discussion or presentation. While framing normally involves leaving out some of the details of the whole story, it seems reasonable to expect responsible framing to communicate the basic features of the entire narrative, not just one side of the story.

Responsible framing emphasizes but does not exaggerate. The essence of framing is calling attention to a particular issue or aspect of a problem. A common and simple way to attract attention to an issue is to exaggerate its importance or impact. There may even be incentives that might prompt individuals to embellish their message in this way, and for journals to publish them. Gitzen (2007) argued that the personal rewards from publishing inflated or overly liberal conclusions far outweigh any personal risks involved. However, in addition to being disingenuous, the use of hyperbole can backfire, since intelligent readers will typically view hyperbole as a tactic to gain support for problems that are not sufficiently compelling in themselves. After all, if a serious concern truly exists, there should not be a reason to exaggerate it. It is clear that an increasing number of invasion ecologists are conscious of this issue. For example, after describing some of the undesirable impacts of some of the species recently introduced into the Caspian Sea, Aladin *et al* (2002) emphasized that 'there is no reason to dramatize the impact of them.' Exaggeration of certain invasion impacts, as well as over-emphasis of particular ecological processes, can also slow the field's progress. If readers, and particularly students, do not receive balanced perspectives, then they may proceed with their own research, management, and careers, guided by invalid assumptions and mistaken understandings of the relative importance of the different threats and ecological processes involved.

Responsible framing counsels but does not frighten. Certainly, a time-honored way to get people to respond is to scare them. The problem is that when people are frightened, they often do not make the best decisions. In particular, their ability to consider complexity and longer-term consequences of responses is often diminished. As a result, prompt and urgent responses may not only be ineffective in dealing with the problem at hand, but such responses may end up producing new problems. Some have argued that the environmental movement has relied so long on fear and threats of impending disaster as public motivators that citizens have become desensitized to such claims, responding with increasingly indifference as they continue to be bombarded with seemingly endless dire forecasts (Gobster 2005).

Responsible framing is explicit, not implicit. Responsible framing recognizes that decisions regarding environmental management are based on values. While scientists may have greater expertise in describing the processes of the natural world, responsible framing recognizes that non-scientists are equally qualified to discuss and decide values (Hull and Robertson 2000, Warren 2007). The answer is not that invasion biologists should not promote particular social and political paths; rather it is that we should be clear in our intentions. Usage of terms and metaphors such as 'biological pollution' (McKnight 1993, IUCN 2000), 'invasional meltdown' (Simberloff and Von Holle 1999), crimes against ecosystems (Clergeau and Nuñez 2006), and even the term 'invasion biology' are examples where we have embedded particular social and/or personal values into the language we use to communicate with one another, as well as managers, policy-makers, and the general public. In many ways, such exhortations have moralistic overtones. While it is possible that humans are genetically predisposed to interpret human actions from a moral framework (Hauser 2006), this does not mean we are obliged to do so.

It does not really matter whether the use of value-laden language has been done intentionally or not by invasion biologists. In either case, the result has been the creation of a sort of hybrid language that has wedded values with scientific concepts. Picket *et al.* (2007) made the same point, stressing the importance of recognizing that complex social values are implied in terms such as ecosystem health. Personally, I do not believe blending values and science in this way is good practice. It confounds our efforts to understand and describe the world as it exists separate from our own perspectives. I am not arguing that we try to divorce values from science. Not only would this be impossible, it may very well lead to lower quality and ineffective science (Alpert 1995). Rather, I think we owe it to the general public, policy makers, and to one another to be explicit, not implicit, in the statement of our values and preferences regarding the natural world. Values will always accompany our science, but I believe it is important that we make a conscious effort not to cloak them, but instead to make sure we wear them on our sleeves. Readers may differ with me on this issue, but again, I think this is another topic that merits reflection and discussion within the invasion field.

Summary

Science is about discovery and solving problems, not maintaining a tradition or promoting an ideology, other than its emphasis on free and independent thinking and empirically verifiable claims.

However, while we may strive to ensure that our theories and perspectives are rooted in empirical findings, there is considerable evidence that our thinking is profoundly shaped by both genetic and cultural predispositions, the latter rooted in both the scientific cultures within our disciplines and the larger social cultures in which we live. More often than not, we are probably not aware of these inclinations. To the extent that we are bringing a particular mindset to the study of biological invasions we are inclined to view the global redistribution of species in certain ways. This particular vantage point also influences the questions we ask as invasion biologists and the way in which we communicate our findings and perspectives to the public.

It may very well be that future advances in human genetic engineering and bio-technology, as well as possible social and cultural developments, may alter some of our cognitive limits and tendencies. However, we will always be separate from what we are observing, and thus our knowledge will inevitably be incomplete and filtered in some ways. We cannot change this fact. Nevertheless, having inclinations does not mean we must act on them, nor must we unwittingly embed personal values into our science. My intent in this chapter was to emphasize the importance of taking an explicit self-conscious approach to practicing science, and invasion biology in particular. The more we are cognizant of the nature and origins of our predispositions, perspectives, and biases, the better we will be able to correct for them. These are not idle reflections. Most scientists would agree that the goal of science is to try to describe the external world as it is, not to describe it in our own image.

CHAPTER 10

Researching invasion biology

The development of specialty areas seems to be an inevitable consequence of scientific activity. Certainly, ecology has developed many of them (Graham and Dayton 2002, Pickett *et al.* 2007). Without question, there are benefits to be gained from such focused attention. Researchers with similar interests may be able to work more efficiently by collaborating on research with one another, and findings may be disseminated within the group very quickly. As long as specialists in different groups continue to regularly communicate with one another, it remains possible to pursue a narrow path while still remaining connected to the larger scientific landscape. However, the drawbacks can begin to exceed the benefits if specialization leads to intellectual isolation (Graham and Dayton 2002). Pickett *et al.* (2007) described three negative, but perhaps inevitable, outcomes of specialization and development of sub-disciplines: gaps in understanding develop at the interfaces between sub-disciplines; specialized sub-disciplines often tend to focus on specific scales or levels of organization; and, specialty areas often develop their own distinctive definitions, viewpoints, and lexicons. In turn these developments can lead to missed opportunities for synthesis and integration.

Issues of specialization and dissociation

Elton's 1958 classic is a curious mixture of disciplinary integration and dissociation. On the one hand, as pointed out by Richardson and Pyšek (2007), Elton's book embodied a rich multi-disciplinary outlook, addressing invasion biology in the context of biogeography, conservation biology, epidemiology, human history, as well as community and population ecology. On the other hand, Elton clearly and decisively described invasions as a unique ecological phenomenon. In 2001, Phil Grime, Ken Thompson, and I suggested that the field of plant invasion biology had become substantially dissociated from mainstream ecology and related sub-disciplines Davis *et al.* 2001). We argued that the driver for this separation was the approach that regarded biological invasions as a unique ecological and evolutionary phenomenon. If one regards invasions as ecologically distinct, then it is not a large step to similarly assume that unique perspectives, and even a unique sub-discipline, are required to understand it. Following a 2003 plant invasion workshop in České Budějovice, Czech Republic that focused on this topic of dissociation in invasion biology, participants conducted a study of the plant invasion literature to assess the validity of our 2001 claim. The study examined both the key words and the bibliographies of a sample of the plant invasion literature, as well as samples from three other ecology sub-disciplines that studied vegetation change—succession ecology, gap-dynamics, and climate change ecology. We analyzed the extent to which the key words and the bibliographies referred to the other sub-disciplines (Davis *et al.* 2005c).

We chose to look at both key words and bibliographies because we felt that the respective results would tell us different things. Selecting key words is a conscious decision on the part of the author to characterize the article in a particular way. Thus, key words should reflect an author's assessment of the paper's scope and relevance. An analysis of key words in articles within a single specialty area should reveal whether the researchers in that field tend to take a narrow or broad view with respect to the significance and impact of their studies. If the selection of key words represents an explicit decision by authors, a bibliography represents more an empirical documentation of the author's use of findings and ideas from the literature while writing the article. Thus, analyses of key words and bibliographies both should be informative, but in different ways. For example, authors may characterize their papers in a quite specialized way even though they use a much broader conceptual context to research and write their papers. Or, the reverse could be true; authors might tend to characterize their papers as being quite broad in their scope and significance, while they actually rely on quite a narrow body of literature to conceive and write them.

The analyses supported our 2001 contention that plant invasion biology had become intellectually quite isolated from other plant ecological research. The analysis of key words showed that invasion ecologists rarely cross-referenced their articles, with only 6% of the 500 invasion articles surveyed being cross-referenced to any of the other sub-disciplines via key words (Davis *et al.* 2005c; Fig. 10.1). It should be noted that, although invasion biology exhibited the lowest rate of cross-referencing via key words among the four sub-disciplines, rates were low in the other three research areas as well, with none of the cross-referencing rates (per cent of articles with cross-referenced key words) exceeding15% (Fig. 10.1). The story was similar with the bibliographies, although not quite as extreme. The majority of articles listed in bibliographies were still determined to be in the same research area as that indicated by the paper's key words, with invasion biology and succession ecology exhibiting the highest rate of intra-field citations, 80 and 81%, respectively (Fig. 10.2). It is interesting that the results suggest that use of the research literature used by the authors to write the articles tended to be broader than the way they actually chose to characterize their papers. Addressing this same issue of dissociation, others have similarly concluded that invasion biologists have not made good use of findings and advances in conservation biology, paleo-ecology, restoration ecology, and weed science (Richardson and Pyšek 2007, van Kleunen and Richardson 2007).

In some respects, this behavior is quite understandable. Clearly, there can be very real personal motivations, and often institutional pressures, to

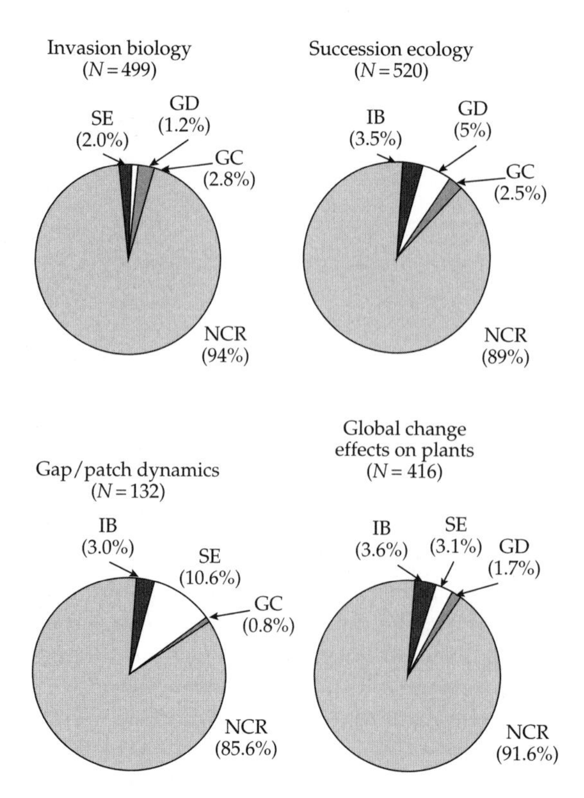

Fig. 10.1 Patterns of cross-referencing in articles from four research areas that study vegetation change; based on an analysis of key words. Cross-referencing is defined as using key words typically associated with one of the other research specialty areas. Analyses were conducted using the basic BIOSIS electronic database for articles published from January 1999 to December 2003. Percentages indicate the per cent of the sample of articles that used a keyword typically associated with one of the other specialty area. Sample sizes (number of articles for which key words were analyzed) are listed for each research specialty area. (IB, Invasion Biology; SE, Succession Ecology; GD, Gap/Patch Dynamics; GC, Global Change Effects on Vegetation; NCR, No cross-referenced key words used.) Redrawn and printed, with permission, from Davis *et al.* (2005c), copyright Elsevier.

work within a narrow framework. For example, if one is trying to establish or maintain a particular research identity, one may feel publishing more narrowly makes the most strategic sense. Due to publishing pressures, some investigators may choose to formulate narrow questions or hypotheses, which may permit more rapid data collection and result in more narrowly defined papers. Others may respond to the same publishing pressures by spreading the results of a single large study to

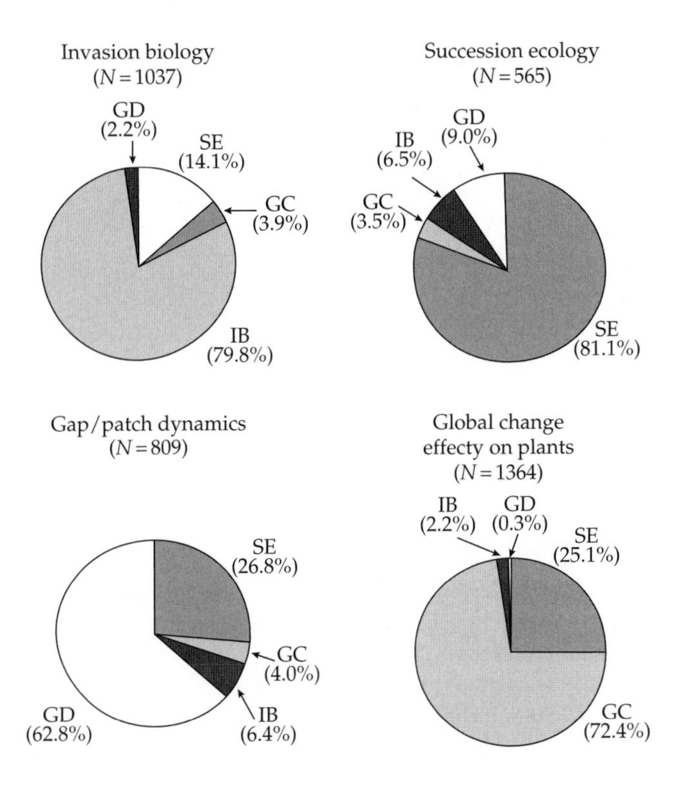

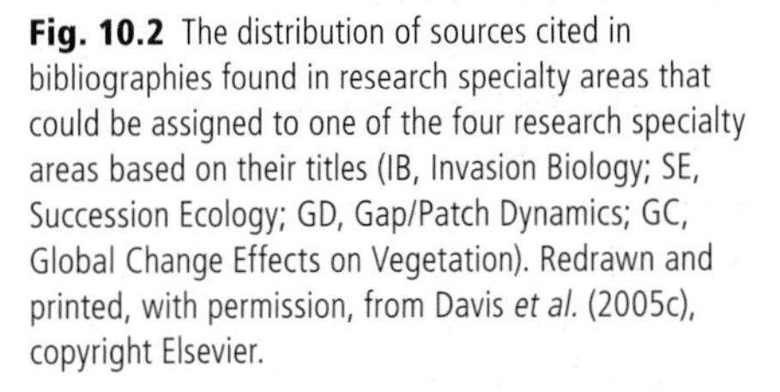

Fig. 10.2 The distribution of sources cited in bibliographies found in research specialty areas that could be assigned to one of the four research specialty areas based on their titles (IB, Invasion Biology; SE, Succession Ecology; GD, Gap/Patch Dynamics; GC, Global Change Effects on Vegetation). Redrawn and printed, with permission, from Davis *et al.* (2005c), copyright Elsevier.

several papers, each focusing on a different aspect of the study and submitted to a specialty journal in that area. Another possible explanation for narrowly defined articles is that it is simply usually more difficult and time-consuming to write a paper that draws substantively from different research areas. By writing more broadly, one also exposes oneself to criticism from a much broader group of colleagues, including experts from diverse areas, than the more focused, and smaller, community of readers one would encounter within a narrow research area. In addition, funding opportunities often aggregate around 'hot' topics of the day. Acid rain, climate change, species extinctions, biological invasions are all examples of topics that have attracted, or are still attracting, the particular interests of funding agencies. If one wants to increase the chance of being funded, one is likely to portray one's research within the specific conceptual framework of whatever topic is receiving the most attention and support.

During the 1980s and 1990s, the numerous invasion-specific conferences, symposia, task forces, international working groups, and publication initiatives, likely helped to promote and maintain the notion that biological invasions are an ecologically distinct phenomenon requiring special attention. However, as illustrated by our analysis of plant invasion biology, special and separate treatment may come with a cost. The four specialty areas are essentially studying the same fundamental processes of colonization, establishment, turnover, persistence, and spread, all of which influence and are influenced by both biotic and abiotic factors (Fig. 10.3). Furthermore, phenomena under particular scrutiny in the different areas often interact, e.g. gaps and climate change may facilitate invasions. Given this overlap, one might expect, or hope, there to be considerable information exchange among these research areas. However, on the basis of our analysis, this has not seemed to have been the case in recent years (Davis *et al.* 2005c).

Should we be surprised over this apparent dissociation of invasion biology from the rest of ecology? One could make a case either way. It certainly is not as if the dangers of over-specialization in

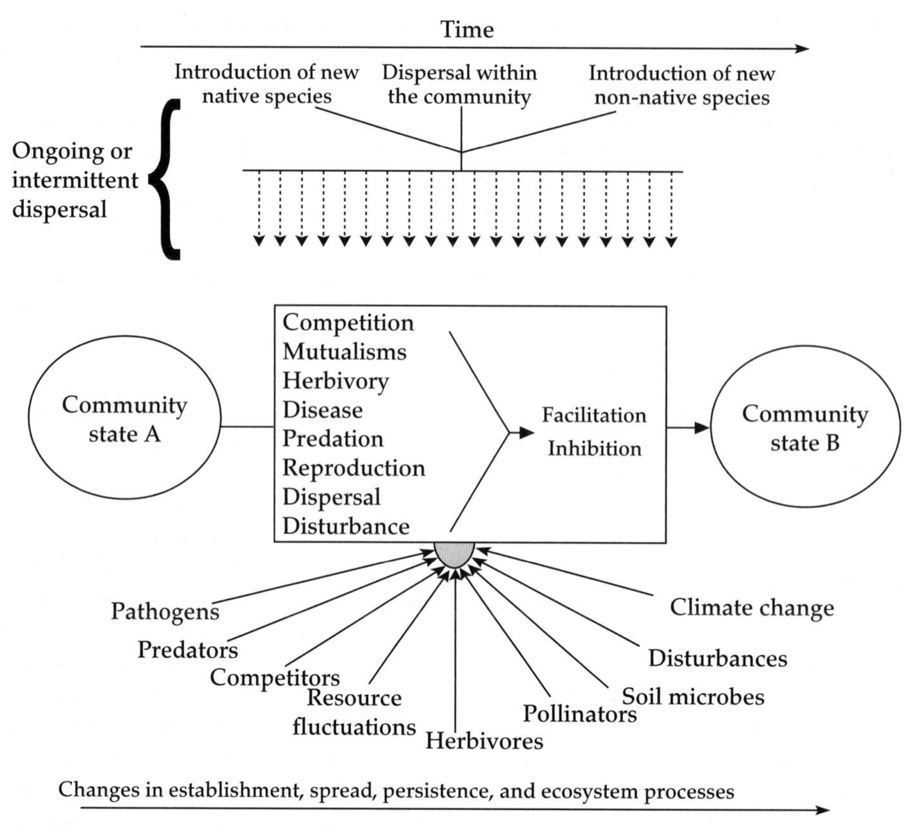

Fig. 10.3 The same factors change plant communities regardless of the specialty area in which the research is conducted. Redrawn and printed, with permission, from Davis *et al.* (2005a), copyright Elsevier.

ecology had never been raised before. Concerns over parochial tendencies among ecologists have been raised a number of times during the past quarter century (Egler 1951, Bartholomew 1986, McIntosh 1987, Pickett *et al.* 1994, Graham and Dayton 2002, Pickett *et al.* 2007). There have also been a number of prominent researchers who did not take the narrow path but instead maintained a broader and more integrative perspective. In his many studies of New Zealand flora, Williams (e.g. 1981, 1992) often studied succession and invasion together in an integrated fashion. Johnstone (1986) offered that 'the ideas and concepts from successional theory provide some insight into the process of invasion.' Huston (1994) argued similarly, claiming that that the same processes that 'ultimately regulate community structure and species diversity in 'natural' communities, also regulate the success and failure of invading species.' In his 1977 book, Harper stated as clearly as possible his belief that plant invasions should not be considered as a special or unique ecological phenomenon:

> Superficially there appear to be two distinct contexts in which one can consider dispersal: (i) that of the expanding range and increasing population size of an invading species into a new area, island, or continent and (ii) as part of the process by which an established and stabilized population maintains itself within the ever-shifting 'islands' that constitute the pattern within established vegetation. In reality these two situations are not contrasting, but parts of the same system.

Speaking more broadly, Vermeij (1991, 2005) has consistently maintained that ecologists have tended to exaggerate the fundamental differences between biotic interchanges of the past and human-assisted invasions of today.

In the face of these admonitions and arguments against the ecological uniqueness of invasions, one could maintain that it is surprising that this dissociation developed. On the other hand, one could argue that imposing some clear boundaries around the field was beneficial as the young discipline was developing in the 1980s and 1990s, perhaps even inevitable. Whatever the initiative, be it political, social, or scientific, a narrow and clearly defined agenda is often most successful in promoting a sense of purpose and direction. In addition, when such initiatives are presented as novel or distinct, and/or as part of a particular ideological or value-based perspective, they may be more successful in forging a collective atmosphere of enthusiasm and commitment.

Efforts to re-associate

Although a narrowly focused vision may have fueled an initiative's early progress, at some point, it is likely to begin to impede the program's development. Experience and pragmatic realities may reveal inadequacies in the initial perspective. Also, sustained development may require the support of more than the initial group of enthusiastic founders, including individuals who may not fully endorse some of the narrow founding principles, beliefs, and perspectives. Initiatives and movements vary in how they respond at this critical juncture. Some reinforce the wall around them, hunker down, and try to sustain themselves as defined by their founders. Others are more open to change. They adapt and continue to evolve, usually in an increasingly integrated way with the larger world. Invasion biology has vigorously chosen the second option. Callaway and Maron (2006) describe the field as now helping to 'catalyze a healthy fusion between fields and sub-disciplines that have historically operated in isolation.' Increasingly, invasion biologists are studying invasions in concert with other changes, such as climate change, disturbances, and land use. In a recent review of the field, Richardson and Pyšek (2007) came to a similar conclusion regarding the current status of the field, emphasizing that the invasion biology is now actively looking outward, and employing knowledge and adopting techniques developed in the fields of GIS, molecular biology, resource economics and risk analysis.

There are a number of things that can be done by individual researchers to ensure that this vigorous, outward-looking perspective, which has begun to characterize invasion biology, is sustained. For example, researchers can design and conduct studies that cross typical, and usually arbitrary, research boundaries. It is becoming more difficult to find environments that are completely free of recently introduced species, or ones that are not being affected by human land/water use practices, or atmospheric nitrogen deposition, or climate change. It makes more sense to undertake more integrated studies of community change, rather than trying to set a particular research project within the narrow framework of a particular sub-discipline.

When writing their proposals and papers, researchers can also try to think of their research in a broader conceptual context. Rather than primarily looking to other invasion research for guidance and citation, investigators should seek insights from research conducted in other specialty disciplines that are studying similar ecological phenomena, just not in an invasion context. Investigators should also think carefully about their choice of key words for their articles. If researchers characterize their articles quite narrowly, as at least plant ecologists have done in recent years (Davis *et al.* 2005c), then, somewhat ironically, the use of electronic search engines may actually limit our exposure to relevant articles. Journals might be able to help by providing guidelines and/or some standardization of key words. This effort could use as a guide ecological metadata language (EML), developed to provide consistent specifications to ecologists to document their data, in order to facilitate the searching and retrieval of data.

It is inevitable that invasion researchers, managers, writers, and policy makers will participate in the revolution, which, at the time I am writing this, is being referred to as Web 2.0. Web 2.0 refers to the social networking aspect of the internet, characterized by sharing and collaboration. At this time, it is not clear exactly what form this 'cyberscholarship' or 'e-science' (Murray-Rust 2008) might take. However, various exploratory initiatives and experiments are already underway. For example, some scientific organizations are beginning to try to take advantage of the social networking internet phenomenon. In late 2007, the American Institute of Biological Sciences (AIBS) established social networking groups, associated with Facebook and LinkedIn internet sites, intended to increase connectivity opportunities for its members (more information provided at the AIBS website: http://www.aibs.org/online-social-networking/).

By the end of the first decade of the new century, we are all benefiting enormously from being able to distribute our papers to colleagues and to access others in original published format. However, at this point, we are simply moving around static facsimiles (Butler 2007). In the world of Web 2.0, material posted on the internet would be, not so much a final product as an invitation to begin new conversations. For invasion biologists, and scientists in general, this would mean that, instead of our articles representing end products, and rather inert ones at that, forever defined by the title and key words that we assigned them, they would represent a more vital and ongoing act of engagement with our colleagues, both within and outside our specialized fields.

By the time this book is published, most readers likely will be familiar with tagging, a method developed to provide an efficient way to organize and access large amounts of data, which explains why librarians have been at the forefront of tagging research and development. The unique aspect of tagging is that the tags, the designated labels that can be used to identify particular items (e.g. articles, photographs, videos, data-sets), are assigned by the users, not the original creators, of the material. In other words the relevance and meaning of different items, and their connections to other items, is an ongoing and organic process, such that the network created by this process actually increases in its connectivity over time. This process holds great promise for organizing and accessing invasion ecology literature (as well as the literature of any discipline). For example, using key words, an invasion researcher may have described his/her article only in the narrow context of invasion ecology, even though the paper also addressed some issues associated with climate change. Although the original author characterized the paper quite narrowly, readers of the paper could 'tag' the article with a climate change label, as well as any other labels that might describe additional associations. Thus, future searches conducted by users using 'climate change' as an identifier would now also identify this article, even though the author described it only as an invasion paper. In addition to tagging items, readers can also leave comments, to which, then, other readers, including the original author, can respond. In a real sense, under this system, all invasion articles would become part of a collective knowledge base to which all members of the community would be able to provide input.

By 2008, PLoS One (an interactive open-access science journal published by Public Library of Science) already had instituted an option for readers of PLoS papers to attach comments to specific parts of the papers, as well as to rate the paper as a whole (Giles 2008). More radically, they have been exploring an alternative approach to identifying important papers. Rather than screening papers with a peer-review process prior to acceptance, PLoS publishes papers as long as they meet basic requirements. However, it then keeps track of how many times papers are accessed and how they are rated by readers, and then highlights and promotes those papers that have been highly rated. Quality control is still maintained by peer review, but it occurs after, not before publication. It is possible that the approach used by PLoS may be more egalitarian than the traditional peer-review process, potentially making it easier for young and/or lesser known invasion biologists to be heard.

While there are some very innovative things the discipline of invasion biology can do to encourage broader perspectives and more inclusive discussions, there are also some very conventional steps it could take toward this end. At national meetings, a conscious effort could be made to organize more mixed sessions. These would not be random collections of papers, but they would provide attendees, and the speakers, the opportunity to hear different perspectives and learn about alternative approaches relating to a similar topic. For example, plant and animal ecologists studying invasions, typically hold separate symposia, or even separate conferences, the result being that there is typically very little communication between the two groups. Symposia and conferences incorporating multiple-organism types, but still organized around a unifying theme, would be one such example of a mixed session. Another would be to use a particular organism type as the organizing component and incorporate papers describing research that, although similar at a fundamental level, has been conceived and characterized using different conceptual frameworks. For example, a symposia or conference on vegetation change could include papers on plant invasion ecology, range-shifts due to climate change, and plant succession. A good illustration of the latter was an international workshop for plant ecologists held in Ascona, Switzerland, in February 2007, sponsored by the University of Zurich and the Swiss Federal Institute of Technology. The stated aim of the workshop was 'to work at the interface of native-species range expansion and non-indigenous species invasion,...by bring[ing] together scientists from these different, but closely linked ecological fields.'

Issues of taxonomic and geographic specialization

A final issue in this discussion of specialization involves the study organisms selected by invasion biologists during the past several decades. This is an issue that involves taxonomy, habitat type, and geography. Although biological invasions involve all major taxa, terrestrial organisms have received more attention than those from freshwater and marine systems (Fig. 10.4). More specifically, plants have been most studied by far in terrestrial systems (Bruno *et al.* 2005, Pyšek *et al.* 2006, 2008; Fig. 10.5), while near-coastal marine animals have been most studied in aquatic systems (Pyšek *et al.* 2006). Given that many marine animals are either sessile (e.g. hydroids, mussels, bryozoans, and sponges) or very sedentary (e.g. spionid polychaetes, and wood-boring and tube-dwelling crustaceans) (Fofonoff *et al.* 2003), this means that much of invasion theory has been developed on the basis of sessile or sedentary organisms.

In our paper on invasibility and fluctuating resource availability (Davis *et al.* 2000), we concluded that it was not clear at that time how well the theory of fluctuating resource availability, developed for plants, could be applied to animals. Subsequent study has shown that the theory has applied well to plant-like animals, e.g. benthic organisms, as well as to freshwater zooplankton and fish (Thompson *et al.* 2001, Jiang and Morin 2004, Havel *et al.* 2005b, James *et al.* 2006, Stachowicz and Byrnes 2006). While the theory has proven quite robust in these systems for these organisms, it still remains to be seen whether the theory is reliably applicable to others, such as mobile, terrestrial animal species. As we stated in 2000, further study will be required to establish the extent to which aspects of behavior (e.g. aggression and dominance) reduce the dependence of introduced animals upon a supply of unused resources. Other topics of invasion biology are also much more associated with a particular group of organisms. For example, issues involving biodiversity, invasibility, ecosystem processes, population genetics, and the role of disturbance characterize the plant invasion literature much more than do animal studies (Pyšek *et al.* 2006). Conversely, issues related to impact, dispersal pathways, biogeography, and, to a lesser extent, dispersal and competition, have been addressed proportionately more frequently in animal invasion studies (Pyšek *et al.* 2006).

Pyšek *et al.* (2008) raised an interesting question regarding the selection of additional species to study. They suggested that selecting species similar to those already studied might be similar to increasing the area of a habitat to census species. For a while there may be a substantial payoff, i.e. important new information would continue to be discovered. However, in time the payoff would drop as the amount of novel information still to be learned from this group of similar species would decline. Pyšek *et al.* (2008) acknowledged that species-specific information would always be needed to inform management of particular species, but they suggested that beyond a certain point, the value of additional information from similar species would not likely contribute substantially to robust generalizations or theory.

Pyšek *et al.* (2008) concluded that, although terrestrial plants were the most studied taxon by far, most major taxa of organisms have been thoroughly researched, at least sufficiently to enable researchers to formulate general principles and

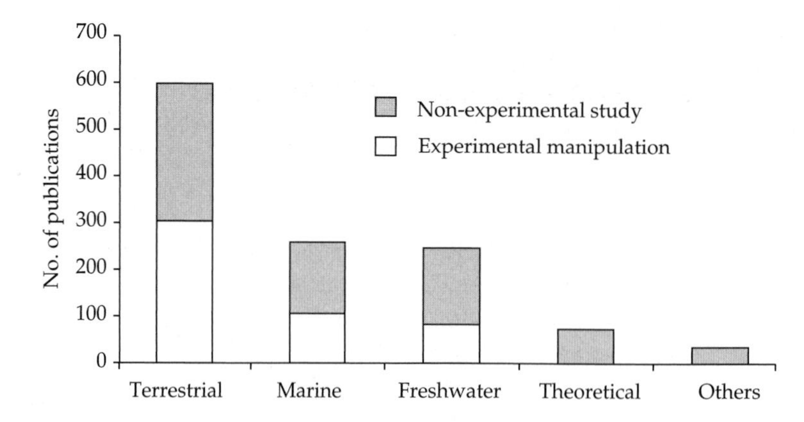

Fig. 10.4 Bioinvasion papers from the ecology literature 1995–2005 by system and study type. Published papers from terrestrial systems outnumbered those from marine 2-to-1. Of these, approximately half of the terrestrial studies involved an experimental manipulation, while only 42% of the marine and 35% of the freshwater studies did. The white portion of each bar represents the number of studies that included an experimental manipulation; the black portion represents all other types of studies. The system category 'other' includes studies that either cut across all systems or did not specify a system. Information for this figure came from an analysis of 14 journals: *Ecology, Science, Nature, Oecologia, OIKOS, Ecological Applications, Biological Invasions, Ecology Letters, Journal of Ecology, Journal of Animal Ecology, Proceedings of the National Academy of Sciences USA, Marine Ecology Progress Series, Journal of Experimental Marine Ecology and Biology, and Hydrobiologia.* The Web of Science was used to search these journals using the following search terms: invas*, invad*, exotic, alien, nonnative, non-native, nonindigenous, non-indigenous. Redrawn and printed, with permission, from Olyarnik *et al.* (2008), copyright Springer Science and Business Media.

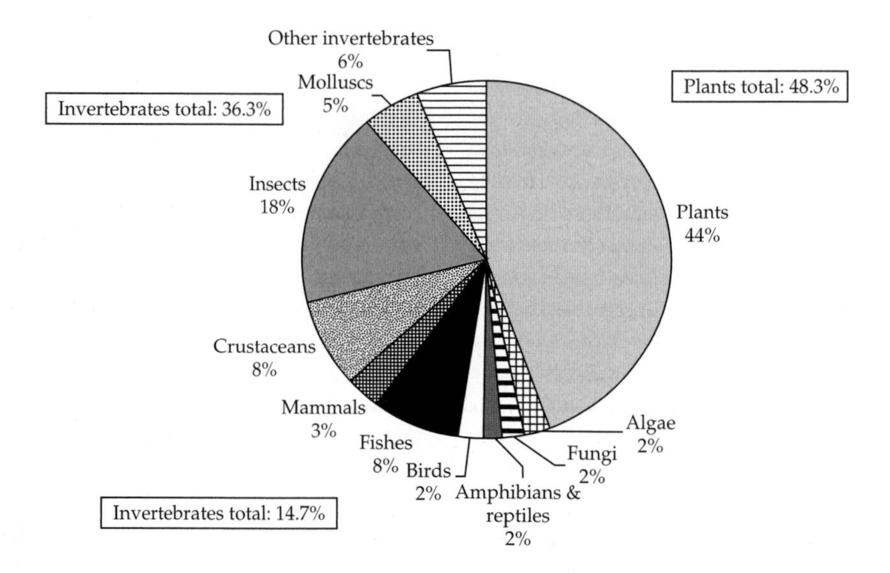

Fig. 10.5 Taxonomic structure of species case studies on biological invasions (892 species from all over the world). Derived from the Web of Science (http://portal.isiknowledge.com, accessed 4 September 2006), searched for the relevant key words and their derivatives (invasive, exotic, alien, naturalized) in combination with main taxonomic groups (plant, animal, mammal, bird, insect). The search yielded 4475 papers on various aspects of biological invasions, from which were selected studies that focused on individual species and were investigated in detail various aspects of their biology and ecology. This screening yielded 2670 case studies that were classified according to the taxonomic group and geographical region of invasion. Plants and insects together account for two thirds of the species studied. Redrawn and printed, with permission, from Pyšek *et al.* (2008), copyright Elsevier Limited.

theories. However, the fact that all major taxa are reasonably represented in invasion studies does not mean that there are not important ecological or functional types of organisms that are under-represented. Given that plants and sessile/sedentary animals have dominated much of the research to date, as described above, we should be careful not to over-generalize our theories or conclusions. While certain very general statements and frameworks may be mostly universal in their applicability, e.g. 'invasion success is influenced by the traits of the organism, resource availability, and propagule pressure,' it may very well be the case that more specific theories and conclusions will often need to be presented and considered in a contingent manner, e.g. associated with particular types of organisms and/or environments.

The geographic bias in studying invasions is another type of specialization. Based on their analysis of the literature, Pyšek *et al.* (2008) found that nearly half of all the species studied by invasion biologists were related to North America, and more than half of all the studies conducted worldwide were related to North America. Conversely, Africa (excluding South Africa), oceanic islands, and Australasia were found to be under-studied, relative to the number of naturalized species in the respective regions. The reasons behind this geographic disparity in invasion research are not difficult to imagine, and certainly involve differences in financial and educational resources, as well as national priorities (Pyšek *et al.* 2008). Pyšek *et al.* believed that the geographic disparity in invasion research constitutes a more serious impediment to our understanding of invasions than any taxonomic bias.

Summary

Ultimately, invasion biology faces the same fundamental challenges of any scientific discipline. Urges to specialize may compromise a field's development by isolating it from other, but related, developments. In turn, this intellectual isolation can reduce opportunities for synthesis and new ideas. Since there are numerous incentives to specialize, conscious efforts by the discipline and individual scientists are necessary to prevent a field from experiencing the potential problems associated with specialization. There is evidence that during the 1980s and 1990s, the field of invasion biology experienced a high degree of specialization and became somewhat dissociated from related research areas. Moreover, recent analyses of the invasion literature have revealed research specialization with respect to taxonomic groups and geographical areas, which also has the potential to impede the field's progress. Fortunately, recent developments suggest the field is currently engaging in active efforts to reconnect with other fields of study. Additional efforts by individual scientists, as well as by the discipline as a whole, can increase the extent and quality of intellectual exchange, both among ourselves and with others outside of invasion biology. These efforts could include such activities as intentionally pursuing research initiatives that cut across traditional disciplinary boundaries, implementing a more effective key word selection process, taking better advantage of web-based social networking opportunities, and organizing more meetings, conferences, and symposia that are multi-disciplinary, multi-taxonomic, and/or multi-geographic in nature.

CHAPTER 11

Disciplinary challenges

In terms of the number of research papers published, the field of invasion biology has been experiencing exponential growth during the past several decades (see Fig. 1.1). It is difficult for any human enterprise, whether an organization, a discipline, a country, or a business, to grow very rapidly in a short time without experiencing some growing pains. Given the speed with which invasion biology has grown, it should not be surprising that it has experienced some of these pains as well. The intellectual isolation described in the previous chapter is one example of such growing pains. Owing to the benefits of detachment, usually provided by the passage of time, historians of science are normally able to evaluate past scientific developments with relative dispassion. However, such a perspective is much more difficult to achieve when the science is ongoing and dynamic. These obstacles notwithstanding, I suggest that invasion biology has encountered some challenges during its rapid growth into a prominent research discipline. Specifically, I suggest that problems have emerged involving pluralism, authority, and paradigms.

Issues of pluralism

While one cannot deny the competitive aspects of science, more than anything, science is a collaborative enterprise. It proceeds in the context of knowledge and theory discovered and developed by a community of scientists. When new discoveries are made or new theories conceived, they are communicated to colleagues. Thus, it is imperative that one be familiar with the work others are doing, so that one can build on their findings and provide an effective context and meaning for one's own findings. It has been argued that cognitive progress in the sciences is not so much an individual process as it is a collective one, in which cognitive activity is distributed among many individuals, in much the same way that computer processors networked together work both independently and collaboratively to solve problems (Knorr-Cetina 1999, Giere 2006). This approach, sometimes referred to as distributed cognition, or a distributed cognition system (Giere 2006), is believed to have provided the natural sciences with much of its power. Thus, by collaborating and sharing individual models, the scientific community is able to build better models than could be developed by any single individual (Frith 2007).

There is general agreement that in order to maximize opportunities for progress and breakthroughs in disciplines based on collaboration, it is vital to accommodate diverse perspectives. As emphasized by Longino (1990), a primary benefit of participating in a diverse community is that the community is able to recognize and cancel out the biases of individuals brought, either intentionally or unintentionally, to the table. The importance of embracing plurality in ecological thought has been emphasized many times over the years (McIntosh 1987; Pickett *et al.* 1994, 2007, Graham and Dayton 2002, Cuddington and Beisner 2005). Thus, although science is ultimately a group effort, individual creativity and imagination remain a crucial part of the system. The value of collaboration in solving problems ultimately stems from the synergy resulting from independent and diverse perspectives (Page 2007). The impact of reduced independence among individuals on the whole group's performance is illustrated by some of the examples provided by Surowiecki (2004). As described earlier, the mean of a group of independent estimates is generally much more accurate than any single estimate. However, if the group operates as a committee, it usually yields a much less accurate estimate than the mean value based on independent estimates of each individual of the group. Surowiecki (2004) explained that the poorer performance by the group when operating as a committee is because the small-group dynamics reduce the impact of independent thinking in the group. For example, the discussion may be controlled by a few dominating individuals, effectively reducing the degrees of freedom in which the group is able to operate.

There have been times when invasion biology has seemed a bit prescriptive to me, when it has not been as welcoming as it might have been of diverse perspectives. Like all sciences, if invasion biology is to maximize its progress, it needs to encourage diverse perspectives, to be open to criticism, both from inside and outside the discipline, and to effectively network thousands of independently-minded researchers and managers. In recent years, more and more diverse perspectives are populating the invasion literature, many from young investigators. If permitted, better yet, encouraged, these diverse views will invigorate the field, facilitating its progress, not impeding it.

Issues of authority

In any discipline, it is important that preliminary ideas or tentative conclusions made on the basis of one or a few studies do not acquire a life of their own, eventually assuming a level of validity and generality that is unjustified on the basis of the actual data. Unfortunately, with common citation practices, it is very easy for this to happen. When citing a particular finding or conclusion for the first time, authors often take the time to describe the particular context in which the specific finding or conclusion was made. The same author may then cite this same finding in another manuscript, or other researchers, without having actually read the original source, may utilize the information provided by the author's first citation to cite the original work. In either case, it is common for these subsequent references to leave out the details needed to assess the reliability and generality of the finding or conclusion. As time goes on, the finding/conclusion is often simply stated as fact, with a perfunctory citation of the original author. At this point, the original findings or conclusions may often be included as boilerplate in introductions and conclusions of articles and proposals. After enough of these iterations, the original finding can become such an integral part of accepted ecological wisdom that many authors feel comfortable in reporting it without citing any source at all. The general problem is that the more often that preliminary ideas and tentative conclusions are presented as boilerplate, and as an axiomatic starting point for further discussion and research, the more likely it is that practitioners, particularly young practitioners, begin to regard the statements as factual, i.e. having been thoroughly and comprehensively empirically confirmed.

A particular striking example of this phenomenon in invasion biology is the conclusion by Wilcove *et al.* (1998) that non-native species are the second greatest threat to the survival of species in peril. This statement has been cited in hundreds of scientific articles since its publication (more than 700 in the decade following its publication) and in countless research proposals, management documents, and college classes. By the early 2000s, it had become common boilerplate for invasion literature, the conclusion often presented as fact. Given limitations and some biases in the information used by Wilcove *et al.* to come to their conclusion, it is difficult to believe that all those who have cited this article actually have read it. First, little of the information used to declare non-native species the second greatest threat to species survival involved actual data at all, as the authors were careful to make very clear:

> We emphasize at the outset that there are some important limitations to the data we used. The attribution of a specific threat to a species is usually based on the judgment of an expert source, such as a USFWS employee who prepares a listing notice or a state Fish and Game employee who monitors endangered species in a given region. Their evaluation of the threats facing that species may not be based on experimental evidence or even on quantitative data. Indeed, such data often do not exist. With respect to species listed under the ESA [Element Stewardship Abstract, prepared by The Nature Conservancy], Easter-Pilcher (1996) has shown that many listing notices lack important biological information, including data on

past and possible future impacts of habitat destruction, pesticides, and alien species. Depending on the species in question, the absence of information may reflect a lack of data, an oversight, or a determination by USFWS that a particular threat is not harming the species. The extent to which such limitations on the data influence our results is unknown.

Second, the article deals only with species in the United States. Third, the findings are dramatically affected by the inclusion of Hawaii, which, while of course part of the United States, clearly has a dramatically different invasion history than does the continental, and substantially majority, portion of the country.

A similar review of extinction threats in Canada found introduced species to be the *least* important of the six categories analyzed (habitat loss, over-exploitation, pollution, native species interactions, introduced species, and natural causes, the latter including stochastic events such as storms and factors inherent to the species, e.g. limited dispersal ability) (Venter *et al.* 2006; Fig. 11.1). When the Hawaiian species were excluded from the Wilcove *et al.* data, the US and Canada did not differ with respect to the threats posed by introduced species (Venter *et al.* 2006), meaning that non-native species would have ranked very low on the list of threats to American species survival. Other studies that have examined species threats over a much larger global area have come to similar conclusions. For example, an analysis of the causes of species depletions and extinctions in estuaries and coastal marine waters concluded the threat of non-native species to be negligible compared to exploitation and habitat destruction (Lotze *et al.* 2006).

Wilcove *et al.* are not the ones primarily responsible for their preliminary and region-specific conclusion ascending to the status of general ecological canon. After all, as shown above, they explicitly described the limitations of their data. And, the title of the article makes it clear that their focus was just regional (the US) and not global. However, the authors should accept a significant portion of the responsibility. For example, they concluded that 57% of imperiled US plants were threatened by predation or competition from alien species. Since predation is unlikely to be a common threat to plants, one must assume the authors meant to imply that most of the threat to native plants came from non-native plant species. However, it is widely known that the impacts of non-native plants on biodiversity are much less than those of non-native

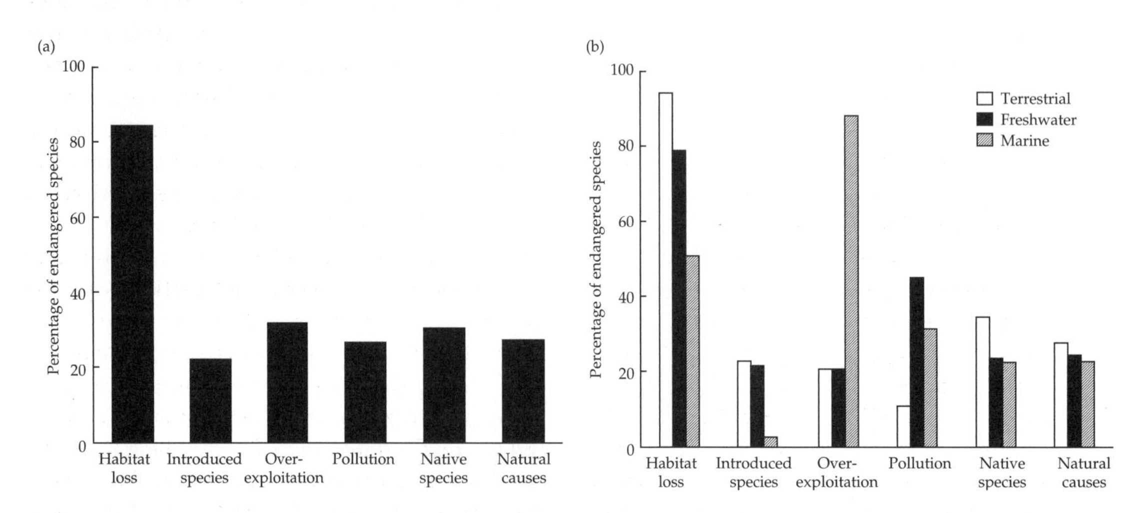

Fig. 11.1 (a) The percentage of endangered species in Canada ($N = 488$) identified by the Committee on the Status of Endangered Wildlife (CSEW) in Canada in June 2005 as threatened by habitat loss, introduced species, over-exploitation, pollution, native species interactions, or natural causes. (b) The percentage of Canadian terrestrial ($N = 231$), freshwater ($N = 154$), and marine ($N = 43$) endangered species that are listed by the CSEW as threatened by the factors listed above. Redrawn and printed, with permission, from Venter *et al.* (2006), copyright American Institute of Biological Sciences.

pathogens, herbivores, and predators (Rejmánek *et al.* 2005b). Moreover, given that when the paper was written there was not any evidence that a single native North American plant species had been driven to extinction, or even extirpated within a single US state, due to competition from an introduced plant species (John T. Kartesz, Biota of North America Program, University of North Carolina, personal communication), concluding, or implying, that non-native plant species threaten a large portion of the US flora with extinction seems quite unjustified, even reckless. (As of November, 2007, there was still no evidence of plant extinction in a US state being caused by competition from a non-native plant species, Kartesz, personal communication). In addition, the authors certainly would have known that their conclusion regarding non-native species would have been quite different without the inclusion of Hawaii.

While Wilcove *et al.* should be held accountable for framing their data as they did, the primary responsibility for the conclusion's ascendancy to boilerplate must lie with those who have continued to cite the article's region-specific conclusion as a generally accepted global fact, even in the face of considerable and increasing evidence presented in recent years showing that non-native species do not represent a major extinction threat to most species in most environments, islands and other insular environments being the primary exception. Even in some recent and prominent publications, the article continues to be cited as justification for making global statements regarding non-native species being one of the top two extinction threats (e.g. Perrings *et al.* 2005, Carrete and Tella 2008). In some cases, the statement that non-native species are the second most important cause of species extinctions is made without any citation at all, but simply stated as fact (e.g. Shine *et al.* 2005). Addressing the general phenomenon illustrated by this example, Gitzen (2007) lamented the vitality and longevity of many inflated scientific claims, observing that 'once bold claims about... a weak result are published, their sins are forgiven and they can be worked into future introductions and discussions at will.'

One way to prevent the ascendancy of preliminary claims to unquestioned lore is quite obvious. When citing the claims, continue to provide the context of original work, along with any limitations of the original study. In reality, practical considerations such as space constraints will prevent authors from doing this for every bit of information they cite. Nevertheless, it seems reasonable that we should make a conscious effort not to misrepresent the generality or definitiveness of conclusions or findings that we cite. There are some fairly simple things invasion biologists, managers, and science writers can do to prevent preliminary or contingent ideas from evolving into conclusive and general ones. One is to emphasize that most ideas in the field do not have the standing of fact or truth, but instead are better characterized as hypotheses or perspectives or interpretations of data described and proposed by particular researchers. For example, 'Davis *et al.* (2000) argued that resource availability is a key factor controlling invasibility in plants' communicates something quite different than, 'Resource availability controls invasibility (Davis *et al.* 2000).'

In the past decade, thousands of individuals researched, managed, and wrote about biological invasions. Several studies have shown that certain categories of ecology and invasion papers are more cited than others, including the geographic location of the authors (US authors tend to be cited more frequently) (Leimu and Koricheva 2005, Pyšek *et al.* 2006). As pointed out by Leimu and Koricheva, the fact that US-authored publications are more likely to be cited than papers from researchers in other countries could be the result of higher quality research conducted by US ecologists, or it could be due to the parochial citation practices among US researchers; or, it could be due to a combination of both factors. Whether or not particular papers are of higher quality, once they begin to be cited, the probability of subsequent citation increases as well. In fact, there is a bit of a run-away selection dynamics associated with the citation process, in which small initial events in the citation process can lead to major differences in the impacts of different articles, even though the articles may be of comparable quality and relevance to the field. If true, this represents a problem, since it would mean that many high-quality articles may not be widely recognized. Conversely, it may also mean that some articles may be recognized due to factors other than, or in addition to, the quality of the particular paper, including the country of origin of the author(s) and the prominence of the author.

Issues of paradigms

One might imagine a gradient of system types. At one end are highly deterministic systems. These are potentially quite predictable and replicable systems, and they can often be operated or understood using universals. Newtonian physical systems are a good example of this type, which is why rocket science is actually a comparatively straightforward and manageable enterprise, at least as compared to systems at the other end of the continuum. At the other end are systems in which history and local contingencies play a very large role in the systems' operation. These systems have a low degree of predictability and replicability, and history and local idiosyncrasies normally trump universals. Where biology falls on this gradient remains a point of disagreement (Enquist and Stark 2007, Keller 2007). Not all readers will agree with me, but I would put nature, and hence ecology, on the latter side of the gradient (which would explain why ecology is not rocket science; it is inestimably more difficult).

Influences from community ecology

The development of invasion theory and its paradigms over the past 25 years has often mirrored that of community ecology. In many ways, invasion biology is a disciplinary offspring of community ecology, so it should not be surprising that developments in community theory have greatly influenced those in invasion biology. Moreover, many of the debates and controversies within the invasion field ultimately originated within the field of community ecology.

The past 50 years of community ecology have involved a tug of war between camps on either side of the continuum described above, with the deterministic camp dominating most contests until recently. The impact of ecologists like Hutchinson and MacArthur on the field in the 1950s and 1960s resulted in the domination, for several decades, of local deterministic paradigms in community ecology, particularly in North America, where processes such as keystone predation and competition in a niche-limited environment were viewed as the primary drivers of community structure (Cooper 2001, Holt 2005). In fact, if community ecology has a signature concept, most would probably agree that it would be the niche. First proposed by Grinnell in 1917, further developed by Elton (1927) and Gause (1934), who used it as the basis for his competitive exclusion principle, refined by Hutchinson and MacArthur in the 1950s and 1960s, used extensively throughout the remainder of the century (Tilman 1982, Chesson 2000), and reconceived by Chase and Leibold in 2003, the niche concept has been guiding ecological thought for nearly a century.

When invasion biology emerged as a distinct area of research and focus in the early 1980s, the field of community ecology was still largely dominated by the niche-based theories of MacArthur and Hutchinson. It is not surprising, then, that the early years of invasion biology were shaped by a perspective that emphasized determinism and local processes. For example, the three questions originally articulated by the 1983 SCOPE scientific advisory committee on biological invasions, which were intended to focus subsequent research, focused on species traits and local processes. The most prominent application to invasion biology of the niche-based paradigm is probably the diversity-invasibility hypothesis, which holds that species-diverse environments should be more resistant to invasion than species-poor environments. The essence of this argument has changed little in the 50 years since Elton articulated this line of reasoning. Those currently advocating the role of diversity in conferring invasion resistance to a community, whether it is species diversity or functional diversity, typically have invoked a niche-limitation argument (Fargione and Tilman 2005).

Increasing discontent

Recent years have seen increasing expressions of discontent in both community and invasion biology regarding progress and paradigms. Frustrated by a perceived lack of progress in community ecology by the end of the twentieth century, Lawton (1999) referred to the state of community ecology at that time as 'a mess' and questioned whether community ecology even had a future. Lawton (2000) particularly criticized the localized approach to understanding communities, noting that 'the details and many of the key drivers appear to be different from system to system in virtually every published study... and we have no means of predicting which processes will be important in which types of system.' Others have also continued to express their concerns over a perceived lack of progress in the field (Cuddington and Beisner 2005, Ricklefs 2006). Castle (2005) acknowledged the abundant discussions in community ecology over theory during the past 50 years, but questioned how much new knowledge actually has been acquired.

In community ecology in general, and in invasion ecology in particular, it often seems difficult for discovery to resolve debates regarding competing hypotheses and theories. In an unpublished presentation to the British Biological Society in 2004, Peter Grubb expressed dissatisfaction with progress in ecology, claiming the failure of ecologists to reject wrong ideas and faulty interpretations (cited in Grime 2007). Craine (2005) raised similar concerns, noting that some theories in community ecology have persisted in the face of empirical data that have contradicted them. In many cases the formulation of theories and hypotheses does not even make it clear what sort of data is required to reject them (Craine 2005). This can occur because of methodological limitations (Agrawal *et al.* 2007) or poorly constructed theories (Craine 2007). A similar point was made a few years earlier by Graham and Dayton (2002), who argued that without the clarity provided by empirical discovery, some ideas, theories, and approaches may dominate a field despite being supported by little empirical data. Graham and Dayton (2002) also suggested that the support from influential ecologists can instill a theory with considerable inertia. For example, despite an overwhelming lack of supporting evidence for the diversity-invasibility hypothesis, beyond that provided from very small constructed communities, the hypothesis still exhibits considerable vitality and little sign that it is about to be retired any time soon. Why has the field of invasion biology been so hesitant, unwilling, or unable to reject this hypothesis? This is a question I believe the field needs to squarely address.

In the larger field of community ecology, although the niche-based approach, with its emphasis on local processes, particularly competition, has guided much of the research and discussion of community assembly for nearly half a century, not all ecologists have been equally enthusiastic with this paradigm. For critics of the niche-based deterministic approach to community assembly, part of the problem is that ecologists have been trying to apply this approach to a system, nature, in which local determinism is often, some would say usually, overwhelmed by regional processes and the contingent forces of history (Hubbell 2001, Webb *et al.* 2002, Davis 2003, Ricklefs 2004, Vermeij 2005, Cardinale *et al.* 2006, Nekola and Brown 2007, Pierce *et al.* 2007, Stohlgren *et al.* 2008a, b). While criticisms of the niche-based paradigm are expressed more commonly now, they are not new. Reservations regarding this approach were raised several decades ago by Connor and Simberloff (1979, 1983). Thirty years ago, Connell (1978) expressed similar reservations when describing rainforests and coral reefs, emphasizing that 'equilibrium is seldom maintained, [and] disruptions are so common that species assemblages seldom reach an ordered state.' Continuing, Connell stated, 'Communities of competing species are not highly organized by coevolution into systems in which optimal strategies produce highly efficient associations whose species composition is stabilized.' Janzen (1985) agreed, arguing that a more accurate characterization of ecosystems and communities may be that they are the cumulative outcome of largely accidental colonization events. Recent studies have shown that a historically fern-dominated ecosystem on a mountain peak on Ascension Island has been replaced by a forest consisting almost entirely of introduced tree species, which supports complex food webs and the usual myriad of ecosystem processes (Wilkinson 2004). The events on Ascension Island support Vermeij's (2005) view that ecosystems 'are quite robust, in which the rules of engagement and the identity of players are flexible.'

The same souring on the niche-based approach that has been occurring in community ecology in recent years has been taking place in invasion biology. For example, Williamson (1996) did not believe a niche-based approach to studying invasions held much promise, bluntly concluding, 'it looks as if models of invasion based on niches will be as disappointing as other community studies of niches.' More recently, the use of niche theory in invasion biology has been criticized by Bruno *et al.* (2005), who charged the field with uncritically accepting the niche-based competition paradigm for several decades. Certainly, if the impacts of biotic interactions on community assembly are very weak, whether because species respond similarly to one another and to the environment, and/or because regional and stochastic processes normally overwhelm the effects of biotic interactions, then it would seem a niche-based model would be a poor choice to represent community assembly involving recently introduced species. Nevertheless, despite increasing reservations by many regarding the utility of a niche-based and competition approach to understanding invasions, niche-based invasion models have continued to play a major role in invasion theory (e.g. Shea and Chesson 2002, Fargione *et al.* 2003, Tilman 2004, Melbourne *et al.* 2007).

Calls for change

In 1987, Ricklefs called for community ecologists to reject the paradigm that characterized community structure as the outcome of biotic interactions taking place within the local community. He advocated a new perspective in which regional processes played a central role in determining community composition and diversity. As more and more ecologists have recognized the limitations of the local and niche-based approach to understanding community assembly processes, many have joined Ricklefs's call for the incorporation of regional processes into community theory and the recognition of the important role played by history in community assembly (Ricklefs and Schluter 1993, Cornell and Karlson 1997, Hubbell 2001, Bond and Chase 2002, Cuddington and Beisner 2005, Cardinale *et al.* 2006, Nekola and Brown 2007, Pierce *et al.* 2007). Some have promoted the notion of meta-communities (Holyoak *et al.* 2005), and 19 years following his initial plea for change, Ricklefs urged that we abandon completely the traditional notion of local communities (Ricklefs 2006).

Levins (1966) observed that good theory rests on the three pillars of generality, precision, and realism. With respect to biological invasions, I think niche theory scores high on the first pillar, but low on the other two. To some, questioning niche theory may be a bit like questioning one's national flag. Thus, a question sure to generate lively debate is whether the localized and niche-based approach to invasions has hindered more than it has helped progress in the field of invasion biology. It is possible that this approach, which, in the early 1980s, likely led the emerging field of invasion biology to focus attention on species traits and local processes, may have delayed the development and emergence of the contemporary view of invasions, which emphasizes history and regional factors, as well as local processes in the invasion process (Lockwood *et al.* 2005 Rejmánek *et al.* 2005a, Colautti *et al.* 2006). In addition, the niche-based approach is at the base of the diversity-invasibility theory, which dominated the field for more than 20 years until recent studies and analyses have shown this theory to be quite inadequate under most conditions and spatial scales. One might argue that, with its emphasis on local and deterministic processes, the diversity-invasibility theory contributed to a perspective that impeded the recognition and acceptance of the importance of regional processes and history in the invasion process.

Summary

To a great extent, humans are creatures of habit, inclined to proceed as we have done so in the past. It is often takes an unexpected jolt from the periphery to wake us up and prompt some self-reflection and consideration of alternative, and perhaps more successful, paths. As scientists, we should look forward to these disruptions. Trainers put blinders on horses to restrict their vision, but scientists need to be open to distraction. It is precisely the unexpected findings and the new perspectives that stimulate creativity and lead to progress in the field. Disciplines will be more successful in discovering solutions if they encourage diverse and independent perspectives. Like all disciplines, it important that invasion biology encourages diverse perspectives within the discipline, and welcomes observation and commentary regarding our behavior and practices, even critical commentary, from colleagues, both within and outside the field.

In disciplines in which knowledge is easily verifiable, preliminary findings or tentative conclusions are seldom regarded as facts or accepted as part of the discipline's canon. However, in invasion biology, and ecology more generally, much of what is regarded as knowledge is not easily confirmed. Because of the large number of invasion articles and books published every year, it is not feasible for individuals to read them all, making all of us dependent on one another to demonstrate sound citation practices. If the context and limitations of conclusions and statements made in the original source are lost during the citation process, then it is all too easy for preliminary results and provisional remarks to gradually assume a level of generality and factualness that is unjustified given the actual empirical support for them.

In many ways, the extensive and furious development of invasion biology in the past several decades has revealed, or at least highlighted, cracks in the aging foundations of some ecology's historically central paradigms and perspectives. Specifically, invasion biology has illuminated some of the shortcomings of community ecology. In this chapter, I said that the future of invasion biology will partly be dependent on the future of community ecology. I think one could argue the opposite as well. It is entirely possible that changes and initiatives in invasion biology may influence developments in community ecology at large. Depending on developments during the next decade, it is possible that future historians of ecology will view invasion biology as a pivotal agent of change within the larger discipline.

CHAPTER 12

Conclusion

Nature is continually changing, but it does not have a goal. Nor, is there only one valid nature that can or should exist at a particular place and point in time (Diamond 1987, Hull and Robertson 2000). As scientists who both study and value nature, we need to operate somewhere between passively accepting the inevitability of change and obsessively trying to preserve the world as we have known it. As individuals, and as a discipline, we need to find a place on this operational continuum where we are able to work toward remedying harm where it truly exists, without becoming compulsive and parochial in our perspectives and behavior, without mistaking change for harm. In urging ecologists not to descend into a siege mentality, Vermeij (2005) made a similar point, emphasizing that ecological change is a reality that needs to be dispassionately studied, not wished away or denied.

Invasion research priorities

What should be the focus of future research in the field? Ultimately it is best that individuals decide for themselves the topics that are of greatest importance and interest. Thus, my suggestions below should not be considered anything like an invasion manifesto, but simply a conversation starter.

In order to develop more successful strategies to prevent introductions of unwanted new species, researchers will need to discover the importance of the different vector elements, as described by Carlton and Ruiz (2005), in determining the likelihood of a successful invasion by the species in question.

In order to better understand the extent to which native and non-native species may be able to persist together for long periods of time, theories and models with relaxed coexistence requirements need to be developed, ones that focus on long-term, but not necessarily permanent, coexistence.

Additional studies are needed to discover more about the mutualistic and facilitative interactions occurring between already established species, including both native and non-native species, and those that are just arriving to a site.

Empirical studies are needed to discover the generality of the 'invasion cliff,' as described in Chapter 6.

Since the impacts of individual factors (e.g. resources, propagule pressure, enemies, mutualists, traits of the species involved) are likely mediated by the other variables, experiments involving the simultaneous manipulation of multiple factors are needed to discover how these factors interact to affect the invasion process.

Due to the fact that non-native invasive species often exhibit considerable geographic variation in morphological, physiological, chemical (plants), and behavioral (animal) traits, comparison studies involving native and non-native populations will need to sample a large number of populations in both regions.

Since species introductions are not occurring in isolation, but in concert with other types of global change (e.g. climate change, increases in atmospheric CO_2 and atmospheric nitrogen deposition, changes in land use and disturbance regimes, and changes in hydrologic and geo-biochemical processes in aquatic systems), more invasion studies should be conducted in the context of some of these other factors.

Long-term experimental and monitoring studies are needed to discover how phenotypic and/or genetic adaptations alter the nature of the interactions between new and long-term resident species, as well as how the impact of the new species on the community and ecosystem changes over time.

Since a common effect of introduced species is to increase species-richness, more studies on the impacts on ecosystem processes of increased species richness are needed.

Since it is important to discover if changes in invasiveness may sometimes involve epigenetic changes (changes in gene expression without changes in genotype), invasive and non-invasive populations should be compared on the basis of gene expression, as well as genotype.

Research is needed that applies the emerging discipline of 'eco-devo' (which focuses on the molecular and cellular mechanisms involved during development) to our understanding of phenotypic plasticity and its role in the invasion process.

Since so much of invasion theory has been developed from studies of plants and sessile/sedentary animals, it is not yet clear how well these ideas apply to more mobile animals, and what role behavior plays in the invasion process of these animals. Thus, more invasion studies of these animals are needed to discover the generality of much of the current invasion theory.

To facilitate the development of effective management approaches, studies that document and describe impacts need to also discover the ecological mechanisms responsible for producing the impact.

Management efforts would be helped enormously if research could identify genetic markers that could be used to distinguish between invasive and non-invasive populations of the same species; i.e. management resources could be utilized much more efficiently by focusing on just the invasive populations. The use of genetic barcoding may hold some promise in this area.

Management of non-native invasive species will especially benefit from invasion research conducted on-site and and within the actual ecosystem(s) of management interest.

Invasion biology might want to follow the lead of climatologists and participate in more ensemble forecasting endeavors in order to try to predict future changes involving non-native species.

Invasion management considerations

One can trivialize anything by increasing the spatial or temporal scale under consideration. It is true that the impacts of invasive species today, and in ensuing decades, matter little in the context of geological time; but this is irrelevant. In the spatio-temporal scale of our lives, and those of our children and grandchildren, what happens on this planet during the next century, and in our own countries, and backyards, matters much. This means that as long as we have deemed that some non-native species are causing problems, we will be motivated to impose some control over their spread and impact.

From an ecological perspective, it is clear that the simplest, and usually least expensive, management approach to prevent undesirable invasion impacts is to prevent the invasions from occurring in the first place. Unless eradication of the incoming species is accomplished almost immediately after arrival, things usually become much more complicated and difficult quite quickly. Once on site, the new species begin to interact with other species already present and to influence a myriad of ecosystem processes, thereby becoming part of the site's ecological system. Once this occurs, it will be impossible to impose any management strategy on the undesired species without also affecting other species and ecosystem processes. Predicting those effects normally will be very difficult. The impacts of management, just like the invasions themselves, are very much context dependent. We know that the ecological consequences of invasions of the same species differ from site to site owing to biotic and abiotic differences among the sites. The same will be true for any particular management intervention. In fact, the same holds true for management interventions at the same site at different times. Due to possible ecological changes at a site, and/or to changes in other factors, such as climate or weather patterns, one cannot assume that a management intervention that produced particular results in the past will have the same effects if imposed again, even if at the same site. These uncertainties present a major challenge to managers. While managers may certainly acquire insight from management outcomes at other sites, as well as from their own experience, possibly even including experience at the site of interest, every management initiative is unique. Thus, managers can never be completely confident ahead of time of the outcome of any management efforts.

How should managers proceed under these circumstances? Cautiously! Like researchers, managers need to do their work self-consiously, with their eyes wide open. The adaptive management approach (Holling 1978, Nyberg 1999, Murray and Marmorek 2003) is a particularly effective way to proceed when it comes to managing invasive species. Active management (AM) is viewed as a problem-solving approach. As described by Murray and Marmorek, 'it involves synthesizing existing knowledge, exploring alternative actions, making explicit predictions of their outcomes, selecting one or more actions to implement, monitoring to determine whether outcomes match those predicted, and using these results to adjust future plans.' The iterative circular nature of the AM approach, involving an ongoing process of evaluation and adjustment, is one of the defining characteristics of AM, which ultimately is also a continual learning experience (Murray and Marmorek 2003; Fig. 12.1). Due to the extent and complexity of ecological systems, unintended consequences will always be a possibility for any management initiative. However, the AM approach affords managers of invasive species the best chance to meet their objectives while minimizing the extent of unintended and undesired consequences. In a discussion of the management implications of 'ecological surprises', Doak *et al.* (2008) came to a similar conclusion, arguing that ecological management should be guided by flexibility and caution.

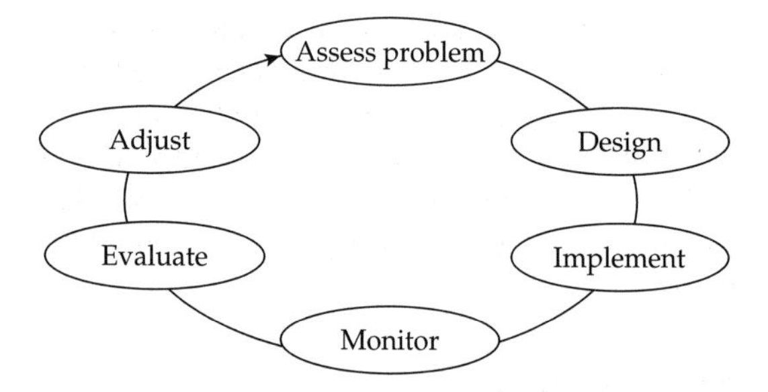

Fig. 12.1 The circular iterative process of adaptive management. Redrawn and printed, with permission, from Nyberg 1999, copyright British Columbia Ministry of Forests and Range.

The first step in the AM approach is assessment (Fig. 12.1), in which the nature and extent of the problem is defined. Practicing self-conscience management at this point is probably more important than at any other time during the AM process. No one would disagree that invasive species' problems are of our own making. We have laid the foundation for these problems by transporting non-native species around the world. However, like anything else in the world, non-native species become problematic only when we deem their impacts to be undesirable. In reality, most management initiatives begin after much of the assessment has already been completed, in that the issue has already been declared a problem. While all harm is in the eye of the beholder, there would seem little to dispute regarding the problematic aspects of invasive species causing great economic harm or threatening human health, as described earlier. However, whether or not ecological impacts constitute harm, or simply change, is more a matter of opinion, one that is likely often to vary considerably among different stakeholders. Thus, whether or not invasive species causing ecological changes should be considered harmful, and hence a problem, is not as obvious as for those species causing economic harm or human health concerns. Since pressure is put on society to allocate resources to address environmental changes once they have been declared as problems, managers can play an important role during the assessment phase of AM by actively reviewing and questioning any already existing assumptions and perspectives regarding the problematic nature and extent of the non-native species involved. Ultimately, society needs to know whether it is in its best interest to declare as harmful particular ecological impacts of non-native species, and managers should actively participate in helping to make these decisions.

Looking ahead

This may be the first time that an author has concluded a book, the title of which is the same as the discipline being reviewed, by recommending that participants consider abolishing their discipline. This certainly would seem to be an effective way to guarantee a short shelf-life for one's book. I am certainly not suggesting that biologists cease their research on invasions. Rather, I am recommending that it should be practiced more as part of mainstream ecology, and integrated with the goals and perspectives of related sub-disciplines. Thus, what I am really proposing is a merger. In fact, the field is already moving in this direction. Two recently edited volumes on invasion ecology were both organized around the goal of highlighting the connections between invasion biology and the rest of ecology (Sax *et al.* 2005a, Cadotte *et al.* 2006). Richardson and Pyšek (2007) also noted the field's increased collaboration and integration with related disciplines. As already described, more and more ecologists have come to recognize that the study of invasibility is really the study of diversity and coexistence (Huston and DeAngelis 1994, Shea and Chesson 2002, Levine *et al.* 2004, Davis *et al.* 2005a, 2005, Ejrnæs *et al.* 2006, Renne *et al.* 2006, Melbourne *et al.* 2007). By now, it seems clear that, from an ecological perspective, there is little about biological invasions that make them so unique that a specialized sub-discipline need be sustained to study them. This recommendation is consistent with Pickett *et al.* (2007), who stressed the importance of integration among sub-disciplines as a way to advance ecology.

Any call for the retirement of invasion biology as a distinct sub-discipline will not likely be heeded, at least in the near future. After all, we now have journals, institutes, graduate programs, research consortia, publication series, and local and national agencies, councils, and initiatives specially created around the notion of biological invasions. The institutionalization of invasion biology through these efforts has created an inertia that will ensure that the field will continue for some time. However, perhaps it is not out of the question to hope for the end of invasion ecology as we have known it. If post-invasion biology seems a bit extreme, impractical, and unlikely, perhaps we could consider moving forward with some sort of a neo-invasion biology.

How would neo-invasion biology (or invasion biology 2.0) differ from the invasion biology of the past 25 years? Neo-invasion biology would use more simply descriptive language. In this way, we would be following Cooper's (1926) recommendation more than 80 years ago: 'We must, accordingly, rigorously exclude all stock ecological terms and phrases, and treat the phenomena in a purely descriptive manner.' The new invasion biology would avoid usage of hybrid language that mixes values with scientific concepts. It would be more pragmatically inclined, and values and ideology would be explicitly presented. It would be better connected with related specialty disciplines and fully integrated within mainstream ecology. More than this, it would acknowledge and emphasize the importance of contributions from colleagues outside of biology, including economists, anthropologists, sociologists, philosophers, and ethicists. In this way, it could become a leader in the development of integrative science, a multi- and transdisciplinary approach to addressing complex societal problems (Wake 2008).

Neo-invasion biology would recognize that invasibility and invasiveness are two sides of the same coin, and that the study of invasibility and the study of diversity are really much the same. Recognizing the importance of history and contingencies on the invasion process, neo-invasion biology would be less preoccupied with searching for generalizations with a capital G, and more content with accepting the merits of lower-case generalizations. The good news is that in recent years the field of invasion biology has already begun to exhibit all of these characteristics and behaviors. In other words, the field is currently in the midst of an active redefinition of itself. Fifty years since the publication of Elton's book, the field is in transition between the invasion biology of the 1980s and 1990s, which operated mostly within the conservation-oriented Eltonian paradigm, and, as proposed and described above, full-fledged neo-invasion biology, the invasion biology of the twenty-first century.

SPRED ecology

Of course, no legitimate neo-invasion biology discipline would include the term invasion in its name. Are there any other options? Gorman (2005) suggested fusion ecology. I do not believe fusion is an apt characterization from an ecological perspective, although I suppose it would describe those instances in which hybridization occurs between native and non-native species. Invasion biology is fundamentally about the redistribution of species throughout the world. Thus, we might refer to neo-invasion biology as SPRED ecology, the ecology of SPecies REDistribution. There are a number of appealing aspects associated with this new name. First, the acronym is descriptive of the defining phenomenon of the invasion process, species spread. Second, the new characterization would help the ongoing reintegration of invasion ecology with the rest of ecology. For example, SPRED ecology would incorporate, not only invasion research but also, research of range shifts due to climate change, as well as any ecology research in which dispersal (whether natural or human-assisted) and the redistribution of species plays a prominent role, e.g. succession ecology and restoration ecology. There would be great value in the reunification of these often disparate research areas. Third, a major benefit would come from discarding the military metaphors and value-laden language that has plagued the field from its inception, and replacing it with more simply descriptive language. Finally, the name SPRED ecology would emphasize the biogeographical underpinnings of the field. Invasion biology's close association with community ecology during the past 25 years strongly influenced its development. I believe that the field will benefit from a stronger connection with the discipline of biogeography. Biogeography generally studies species' distributions occurring over larger areas and longer time periods than those characterizing most community ecology studies. While shorter term and smaller scale events and processes will always be important in understanding invasion processes occurring at a specific location, it is clear that invasions cannot be fully understood without taking a regional, even global, perspective. An approach that promotes and maintains strong connections to both community ecology and biogeography should be well equipped to study biological invasions in the twenty-first century.

In closing

The field of invasion biology has exhibited remarkable growth and change since the creation of the first SCOPE scientific advisory committee in 1983. As I have emphasized, the field of invasion biology is in a period of rapid transition. This is partly due to the influx of many young investigators, who have been attracted to the intellectually rich and socially relevant field developed during the 1980s and 1990s. This bodes very well for the field's future. Disciplines begin to stagnate in the absence of new participants and perspectives. The influx of new minds and perspectives into the field is exactly what will ensure the field's vitality in upcoming decades. The new investigators and ideas represent opportunities, perhaps not unlike the way that introductions of new species can provide new ecological and evolutionary opportunities.

In writing this book, I have tried to identify and discuss many of the ideas and theories that have defined the field in recent years. In addition to attempting to document and describe the current state of invasion research, I have tried to identify some issues that I think still need some attention, as well as to describe some paths that might prove fruitful as the discipline strives to develop the most effective approaches possible to understanding and managing biological invasions. It is likely the reader will agree with some of my assessments and suggestions and disagree with others.

If there is one outcome that I hope for this book, it is that it provokes focused discussions on topics in the field of invasion ecology, and in ecology in general, that I believe should be addressed for the field to remain vital and forward-moving. Many of these issues are not peripheral ones, or even relevant only to ecology. Some go to the heart of what it means to be a scientist and of what science has to offer society. While we can never pretend that personal priorities do not enter into some of our decisions and behavior, it is important that they do not unduly compromise our science. Actually, it is not really *our* science; it is only ours to tend. Others have tended it before us, and now it is our responsibility to tend it well, so that it will be in good condition when we pass it off to those who will follow us.

I hope this book is read by science writers and those who write ecology textbooks. It is important that the public and incoming students to the field receive a balanced characterization of invasions. By this I mean that positive as well as negative impacts of introduced species should be reported, as well as the variety of factors that have been discovered to facilitate or inhibit invasions. In addition, authors should communicate the current diversity of perspectives and theories regarding the invasion process. While certain theories can be used to create a clear and tidy story of invasions, e.g. that increased diversity resists invasion, or, invasions occur due to the escape of species from their enemies, recent discoveries in invasion biology have shown that simple generalizations of this sort often do not widely apply. Certainly, one of the usual objectives of a textbook or public science writer is to help the reader reach some sense of understanding with respect to the subject being communicated. However, I think most would agree that if biological invasions are greatly influenced by contingency and history, and if invasion biology is best characterized right now by its diverse perspectives, then writers should be careful not to oversimplify the invasion process and foster a false sense of clarity on issues currently distinguished more by debate and controversy than consensus.

I particularly hope this book finds its way into the hands, and its ideas into the minds, of young ecologists. Invasion biology's future lies with the younger generation of scientists and managers. All scientific disciplines confront epistemological challenges, but ecology may experience some more intensely than other disciplines, due to the idiosyncrasies and historical vagaries of its subject—nature. The better that all ecologists, old as well as young, appreciate that the truly important challenges of our discipline do not involve the vicissitudes of working outdoors in inclement conditions, or the arcaneness of some of our statistical manipulations, but instead have to do with the way we think and communicate, the more self-consciously we will be able to pursue our work as scientists, and the better the science we will produce. There is nothing new in this exhortation. More than 80 years ago, Cooper urged the same in his call for ecologists to inspect the foundations of their discipline, emphasizing that 'the mode of thought is more fundamental that the result thereof' (Cooper 1926).

My advice to young ecologists is to be open to new ideas and not to follow accepted dogma with blind allegiance. There is nothing new in this counsel. More than 50 years ago, using almost identical words, Egler (1951) warned colleagues of the dangers of 'placidly accept[ing]... traditional dogma.' It is important to learn about current and dominant paradigms, but avoid marinating yourself in them. In order to be able to look at nature in different ways, it is necessary to be able to step outside traditional perspectives. To the extent that we can, we should always try to be in control of our paradigms, and not the other way around. Do not uncritically believe everything your adviser says, or everything said by leaders in the field. Do not absorb ideas indiscriminately. Ponder them; ruminate over them; but above all, be skeptical of them, including the ideas presented in this book. Skepticism is probably the most important attribute of a scientist. If an American state were to be the home of science, it would have to be the 'Show Me' state, Missouri. The most common belief as to the origin of this phrase is that it was made in speech by a Missouri US Congressman, Willard Duncan Vandiver in 1899. Addressing a naval banquet in Philadelphia, he enthusiastically exclaimed, 'I come from a state that raises corn and cotton and cockleburs and Democrats, and frothy eloquence neither convinces nor satisfies me. I am from Missouri. You have got to show me.' Better advice one could not give a young scientist.

In the end, I hope that readers will take the time to reflect on the questions, issues, and ideas I have presented in this book. As I stated earlier, what gives science its power is that thousands of independent investigators are cognitively networked with one another. Utilizing continually incoming data, this distinctive system is particularly effective in ultimately making the right decisions regarding alternative and competing ideas. With respect to the ideas presented in this book, I have faith in my colleagues. Ultimately, I have no doubt that the field of invasion biology will correctly separate the wheat from the chaff in what I have offered, and, like all good sciences, it will move forward.

References

Abbott RJ, Ireland HE, Joseph L, Davies MS, and Rogers HJ (2005) Recent plant speciation in Britain and Ireland: origins, establishment and evolution of four new hybrid species. *Proceedings of the Royal Irish Academy B*, **105**, 173–183.

Ackerly D and Sultan S (2006) Mind the gap: the emerging synthesis of plant 'eco-devo'. *New phytologist*, **170**, 648–653.

Agami M and Waisel Y (1998) The role of fish in distribution and germination of seeds of the submerged macrophytes *Najas marina* and *Ruppia maritima* L. *Oecologia*, **76**, 83–88.

Agrawal AA, Ackerly DD, Adler F, Arnold AE, Cáceres C, Doak DF, Post E, Hudson PJ, Maron J, Mooney KA, Power M, Schemske D, Stachowicz J, Strauss S, Turner MG, and Werner E (2007) Filling key gaps in population and community ecology. *Frontiers in Ecology and the Environment*, **5**, 145–152.

Agrawal AA and Kotanen PM (2003) Herbivores and the success of exotic plants: a phylogenetically controlled experiment. *Ecology Letters*, **6**, 1–4.

Ainouche ML, Baumel A, and Salmon A (2004) *Spartina anglica* C. E. Hubbard: a natural model system for analyzing early evolutionary changes that affect allopolyploid genomes. *Biological Journal of the Linnean Society*, **82**, 475–484.

Aladin NV, Plotnikov IS, and Filippov AA (2002) Invaders in the Caspian Sea. In E Leppäkoski, S Gollasch, and S Olenin, ed. *Invasive aquatic species in Europe. Distribution, impacts and management*, pp. 315–359. Kluwer Academic Publishers, Dordrecht.

Allan HH (1936) Indigen versus alien in the New Zealand plant world. *Ecology*, **17**, 187–193.

Allen TFH, Zellmer AJ, and Wuennenberg CJ (2005) The loss of narrative. In K Cuddington and B Beisner, ed. *Ecological Paradigms Lost: Routes of Theory Change*, pp. 333–370. Elsevier Academic Press, London.

Allendorf FW and Lundquist LL (2003) Introduction: population biology, evolution, and control of invasive species. *Conservation Biology*, **17**, 24–30.

Alpert P (1995) The boulder and the sphere: subjectivity and implicit values in biology. *Environmental Values*, **4**, 3–15.

Alpert P (2006) The advantages and disadvantages of being introduced. *Biological Invasions*, **8**, 1523–1534.

Alpert P, Bone E, and Holsapfel C (2000) Invasiveness, invasibility, and the role of environmental stress in preventing the spread of non-native plants. *Perspectives in Plant Ecology, Evolution and Systematics*, **3**, 52–66.

Alsos IG, Eidesen PB, Ehrich D, Skrede I, Westergaard K, Jacobsen GH, Landvik JY, Taberlet P, and Brochmann C (2007) Frequent long-distance plant colonization in the changing arctic. *Science*, **316**, 1606–1609.

Andow DA (2005) Characterizing ecological risks of introductions and invasions. In HA Mooney, RN Mack, JA McNeely, LE Neville, PJ Schei, and JK Waage, ed. *Invasive alien species: a new synthesis*, pp 84–103. Island Press, Washington DC.

Antsulevich A and Välipakka P (2000) *Cercopagis pengoi*: new important food object of the Baltic herring in the Gulf of Finland. *International Review of Hydrobiology*, **85**, 609–619.

Aplet GH, Anderson SJ, and Stone CP (1991) Association between feral pig disturbances and the composition of some alien plant assemblages in Hawaii Volcanoes National Park. *Plant Ecology*, **95**, 55–62.

Araki H, Cooper B, and Blouin MS (2007) Genetic effects of captive breeding cause a rapid, cumulative fitness decline in the wild. *Science*, **318**, 100–103.

Araújo MB and New M (2007) Ensemble forecasting of species distributions. *Trends in Ecology and Evolution*, **22**, 42–47.

Arenas F, Sánchez, Hawkins SJ, and Jenkins SR (2006) The invasibility of marine algal assemblages: role of functional diversity and identity. *Ecology*, **87**, 2851–2861.

Armstrong DP and Seddon PJ (2008) Directions in reintroduction biology. *Trends in Ecology and Evolution*, **23**, 20–25.

Aronson RB, Thatje S, Clarke A, Peck LS, Blake DB, Wilga CD, and Seibel, BA (2007) Climate change and invasibility of the Antarctic benthos. *Annual Review of Ecology and Systematics*, **38**, 129–154.

Arrington D, Toth L, and Koebel Jr, J (1999) Effects of rooting by feral hogs, *Sus scrofa* L. on the structure of a flood plain vegetation assemblage. *Wetlands*, **19**, 535–544.

Arroyo MTK, Cavieres LA, Peñaloza, A, and Arroyo-Kalin MA (2003) Positive interactions between the cushion plat *Azorella monantha* (Apiaceae) and alpine plant species in the Chilean Patagonian Andes. *Plant Ecology*, **169**, 121–129.

Arthington AH and Mitchell DS (1986) Aquatic invading species. In RH Groves and JJ Burdon, ed. *Ecology of biological invasions*, pp 34–56. Cambridge University Press, London.

Ashley MV, Willson MJ, Pergams ORW, O'Dowd DJ, Gende SM, and Brown JS (2003) *Biological Conservation*, **111**, 115–123.

Ashton IW and Lerdau MT (2008) Tolerance to herbivory, and not resistance, may explain differential success of invasive, naturalized, and native North American temperate vines. *Diversity and Distributions*, **14**, 169–178.

Austerlitz F, Jung-Muller B, Godelle B, and Gouyon P (1997). Evolution of coalescence times, genetic diversity and structure during colonization. *Theoretical Population Biology*, **51**, 148–164.

Ayres DR, Zaremba K, Sloop C, and Strong DR (2008) Sexual reproduction of cordgrass hybrids (*Spartina foliosa* x *alterniflora*) invading tidal marshes in San Francisco Bay. *Diversity and Distributions*, **14**, 187–195.

Badano EI, Villarroel E, Bustamante RO, Marquet PA, and Cavieres L (2007) Ecosystem engineering facilitates invasions by exotic plants in high-Andean ecosystems. *Journal of Ecology*, **95**, 682–688.

Baker HG (1948) Stages in invasion and replacement demonstrated by species of *Melandrium*. *Journal of Ecology* **36**, 96–119.

Baker HG (1965) Characteristics and modes of origin of weeds. In HG Baker and GL Stebbins, ed. *The genetics of colonizing species*, pp 147–172. Academic Press, New York.

Baker HG (1974) The evolution of weeds. *Annual Review of Ecology and Systematics*, **5**, 1–24.

Baker HG and Stebbins GL, ed. (1965) *The genetics of colonizing species*. Academic Press, London.

Balanyá J, Oller JM, Huey RB, Gilchrist GW, and Serra L (2006) Global genetic change tracks global climate warming in *Drosophila subobscura*, *Science*, **313**, 1773–1775.

Balter M (2007) In search of the world's most ancient mariners. *Science*, **19**, 388–389.

Banks PB, Newsome AE, and Dickman CR (2000) Predation by red foxes limits recruitment in populations of eastern grey kangaroos. *Austral Ecology*, **25**, 283–291.

Barney JN and DiTomaso JM (2008) Nonnative species and bioenergy: are we cultivating the next invader? *Bioscience*, **58**, 64–70.

Barney JN and Whitlow TH (2008) A unifying framework for biological invasions: the state factor model. *Biological Invasions*, **10**, 259–272.

Barry E (2007) A taste of baboon and monkey meat, and maybe of prison too. *New York Times*, 17 November 2007.

Bartholomew GA (1986) The role of natural history in contemporary biology. *Bioscience*, **36**, 324–329.

Baruch Z and Jackson RB (2005) Responses of tropical native and invader C_4 grasses to clipping, fire, and increased CO_2 concentration. *Oecologia*, **145**, 522–532.

Baskin Y (2002) *A plague of rats and rubbervines: the growing threat of species invasions*. Island Press, Washington, DC.

Bates JM and Granger CWJ (1969) The combination of forecasts. *Operational Research Quarterly*, **20**, 451–468.

Bates M (1956) Man as an agent in the spread of organisms. In WL Thomas, ed. *Man's role in changing the face of the earth*, pp. 788–804. University of Chicago Press, Chicago.

Bauer BD (2006) The population dynamics of tansy ragwort (*Senecio jacobaea*) in northwestern Montana. MS thesis, Montana State University, Bozeman, MT.

Bazin H (2000) *The eradication of smallpox: Edward Jenner and the first and only eradication of a human infectious disease*. Academic Press, London.

Becker T, Dietz H, Billeter R, Buschmann H, and Edwards P (2005) Altitudenal distribution of alien plant species in the Swiss Alps. *Perspectives in Plant Ecology, Evolution, and Systematics*, **7**, 173–183.

Beever EA, Tausch RJ, and Brussard PF (2003) Characteristic grazing disturbance in semiarid ecosystems across broad spatial scales using multiple indices. *Ecological Applications*, **13**, 119–136.

Beisner B, Haydon DT, and Cuddington K (2003) Alternative conceptions of alternative stable states in ecology. *Frontiers in Ecology and the Environment*, **1**, 376–382.

Beisner BE and Cuddington K (2005) Why a history of ecology? An introduction. In K Cuddington and BE Beisner, ed. *Ecological paradigms lost: routes of theory change*, pp. 1–6. Elsevier Academic Press, Amsterdam.

Belote RT, Jones RH, Hood SM, and Wender BW (2008) Diversity-invasibility across an experimental disturbance gradient in Appalachian forests. *Ecology*, **89**, 183–192.

Benning TL, LaPointe D, Atkinson CT, and Vitousek PM (2002) Interactions of climate change with biological invasions and land use in the Hawaiian Islands: modeling the fate of endemic birds using a geographi information system. *Proceedings of the National Academy of Sciences USA*, **99**, 14246–14249.

Berenbaum MR and Zangerl AR (2006) Parsnip webworms and host plants at home and abroad: trophic complexity in a geographic mosaic. *Ecology*, **87**, 3070–3081.

Berger J, Swenson JE, and Persson I-L (2001) Recolonizing carnivores and naive prey: conservation lessons from Pleistocene extinctions. *Science*, **291**, 1036–1039.

Bermejo M, Rodriguez-Teijeiro JD, Illera G, Barroso A, Vilà, and Walsh PD (2006) Ebola outbreak killed 5000 gorillas. *Science*, **314**, 1564.

Bernhard H, Fischbacher U, and Fehr E (2006) Parochial altruism in humans. *Nature*, **442**, 912–915.

Bertness MD and Callaway RM (1994). Positive interctions in communities. *Trends in Ecology and Evolution*, **9**, 191–193.

Bertness MD, Ewanchuk PJ, and Silliman BR (2002) Anthropogenic modification of New England salt marsh landscape. *Proceedings of the National Academy of Sciences USA*, **99**,1395–1398.

Bezemer TM and van der Putten WH (2007) Ecology: diversity and stability in plant communities. *Nature*, **446**, E6–7.

Bially A and MacIsaac HJ (2000) Fouling mussels (*Dreissena*) colonize soft sediments in Lake Erie and facilitate benthic invertebrates. *Freshwater Biology*, **43**, 85–98.

Blackburn TM, Cassey P, Duncan RP, Evans KL, and Gaston KJ (2004) Avian extinction and mammalian introductions on oceanic islands. *Science*, **305**, 1955–1958.

Blaney CS and Kotanen PM (2001). Effects of fungal pathogens on seeds of native and exotic plants: a test using congeneric pairs. *Journal of Applied Ecology*, **38**, 1104–1113.

Blossey B and Nötzold R (1995) Evolution of increased competitive ability in invasive nonindigenous plants: a hypothesis. *Journal of Ecology*, **83**, 887–889.

Blumenthal DM (2005) Interrelated causes of plant invasion: resources increase enemy release. *Science*, **310**, 243–244.

Blumenthal DM and Hufbauer RA (2007) Evolution of increased size but not competitive ability among fourteen invasive plant species. *Ecology*, **88**, 2758–2765.

Blumenthal DM, Jordan NR, and Russelle MR (2003) Soil carbon addition controls weeds and facilitates prairie restoration. *Ecological Applications*, **13**, 605–615.

Blumstein DT (2002) Moving to suburbia: ontogenetic and evolutionary consequences of life on predator-free islands. *Journal of Biogeography*, **29**, 685–692.

Bohlen PJ, Scheu S, Hale CM, McLean MA, Migge S, Groffman PM, and Parkinson D (2004) Non-native invasive earthworms as agents of change in northern temperate forests. *Frontiers in Ecology and the Environment*, **2**, 427–435.

Bond EM and Chase JM (2002) Biodiversity and ecosystem functioning at local and regional spatial scales. *Ecology Letters*, **5**, 467–470.

Borer ET, Hosseini PR, Seabloom EW, and Dobson AP (2007) Pathogen-induced reversal of native dominance in a grassland community. *Proceedings of the National Academy of Sciences USA*, **104**, 5473–5478.

Bossard CC (1991) The role of habitat disturbance, seed predation and ant dispersal on establishment of the exotic shrub *Cytisus scoparius* in California. *American Midland Naturalist*, **126**, 1–13.

Bossdorf O, Auge H, Lafuma L, Rogers WE, Siemann E, and Prati D (2005) Phenotypic and genetic differentiation between native and introduced plant populations. *Oecologia*, **144**, 1–11.

Bossdorf O, Prati D, Auge H, and Schmid B (2004) Reduced competitive ability in an invasive plant. *Ecology Letters*, **7**, 346–353.

Bossdorf O, Richards CL, and Pigliucci M (2008) Epigenetics for ecologists. *Ecology Letters*, **11**, 106–115.

Bossenbroek JM, Kraft CE, and Nekola JC (2001) Prediction of long-distance dispersal using gravity models: zebra mussel invasion of inland lakes. *Ecological Applications*, **11**, 1778–1788.

Boyce W (2007) Earth monitoring: vigilance is not enough. *Nature*, **450**, 791–792.

Boym S (2001) *The future of nostalgia*. Basic Books, New York.

Bradshaw GA and Bekoff M (2001) Ecology and social responsibility: the re-embodiment of science. *Trends in Ecology and Evolution*, **16**, 460–465.

Bradshaw RHW and Lindbladh M (2005) Regional spread and stand-scale establishment of *Fagus sylvatica* and *Picea abies* in Scandinavia. *Ecology*, **86**, 1679–1686.

Brasier CM (2000) The role of *Phytophthora* pathogens in forests and semi-natural communities in Europe and Africa. In EM Hansen and W Sutton, ed. *Phytophthora diseases of forest trees*, pp. 6–13. Oregon State University Press, Corvallis, Oregon.

Bridle JR and Vines T (2007) Limits to evolution at range margins: when and why does adaptation fail? *Trends in Ecology and Evolution*, **22**, 140–147.

Bright C (1999) Invasive species: pathogens of globalization. *Foreign Policy*, **116**, 51–63.

Britton-Simmons KH (2006). Functional group diversity, resource preemption and the genesis of invasion resistance in a community of marine algae. *Oikos*, **113**, 395–401.

Brock MT, Weinig C, and Galen C. (2005) A comparison of phenotypic plasticity in the native dandelion *Taraxacum ceratophorum* and its invasive congener *T. officinale*. *New Phytologist*, **166**, 173–183.

Broennimann O, Treier UA, Müller-Schärer H, Thuiller W, Peterson AT, and Guisan A (2007) Evidence of climatic niche shift during biological invasion. *Ecology Letters*, **10**, 701–709.

Brooks ML (1999) Alien annual grasses and fire in the Mojave Desert. *Madroño*, **46**, 13–19.

Brooks ML, D'Antonio CM, Richardson DM, Grace JB, Keeley JE, DiTomaso JM, Hobbs RJ, Pellant M, and Pyke D (2004) Effects of invasive alien plants on fire regimes. *Bioscience*, **54**, 677–688.

Brooks ML and Pyke D (2001) Invasive plants and fire in the deserts of North America. In KEM Galley and T Wilson, ed. *Proceedings of the invasive species workshop: the role of fire in the control and spread of invasive species. Fire conference 2000: the first national congress on fire, ecology, prevention and management*, 1–14. Miscellaneous Publications No. 11. Tall Timbers Research Station, Tallahassee, FL.

Brown AHD and Marshall DR (1981) Evolutionary changes accompanying colonization in plants. In GGE Scudder and JL Reveal, ed. *Evolution today*, pp. 351–363. Carnegie-Mellon University Press, Pittsburgh, PA.

Brown BJ and Ewel JJ (1987) Herbivory in complex and simple tropical successional ecosystems. *Ecology*, **68**, 108–116.

Brown BJ and Mitchell RJ (2001) Competition for pollination by invasive plants: Effects of pollen from *Lythrum salicaria* (Lythraceae) on seed set in native *Lythrum alatum*. *Oecologia*, **129**, 43–49.

Brown BJ, Mitchell RJ, and Graham SA (2002) Competition for pollination between an invasive species (purple loosestrife) and a native congener. *Ecology*, **83**, 2328–2336.

Brown CM (2008) Reaping the whirlwind? Human disease from exotic pets. *Bioscience*, **58**, 6–7.

Brown JH and Sax DF (2004) An essay on some topics concerning invasive species. *Austral Ecology*, **29**, 530–536.

Brown JH and Sax DF (2005) Biological invasions and scientific objectivity: reply to Cassey *et al.* (2005). *Austral Ecology*, **30**, 481–483.

Brown JKM and Hovmøller MS (2002) Aerial dispersal of pathogens on the global and continental scales and its impacts on plant disease. *Science*, **297**, 537–541.

Brown PD and Morra MJ (1997) Control of soil-borne plant pests using glucosinolate-containing plants. *Advances in Agronomy*, **61**, 167–231.

Brown RL and Peet RK (2003) Diversity and invasibility of southern Appalachian plant communities. *Ecology*, **84**, 32–39.

Bruno JF, Fridley JD, Bromberg K, and Bertness, MD (2005) Insights into biotic interactions from studies of species invasions. In DF Sax, JJ Stachowicz, and SD Gaines, ed. *Species invasions. Insights into ecology, evolution, and biogeography*, pp. 13–40. Sinauer Associates, Sunderland, MA.

Bruno JF, Kennedy CW, Rand TA, and Grant MB (2004) Landscape-scale patterns of biological invasions in shoreline plant communities. *Oikos*, **107**, 531–540.

Bruno JF, Stachowicz JJ, and Bertness MD (2003) Inclusion of facilitation into ecological theory. *Trends in Ecology and Evolution*, **18**, 119–125.

Bryant SJ and Jackson J (1999) *Tasmania's threatened fauna handbook: what, where and how to protect tasmania's threatened animals.* Threatened Species Unit, Parks and Wildlife Service, Hobart (Tasmania).

Buckley YM, Anderson S, Catterall CP, Corlett RT, Engel, T, Gosper CR, Nathan R, Richardson DM, Setter M, Spiegel O, Vivian-Smith G, Voigt FA, Weir JES, and Westcott DA (2006) Management of plant invasions mediated by frugivore interactions. *Journal of Applied Ecology*, **43**, 848–857.

Bukovinszky T, van Veen F, Jongema Y, and Dicke M (2008) Direct and indirect effects of resource quality on food web structure. *Science*, **319**, 804–807.

Burdon JJ and Chilvers GA (1977) Preliminary studies on a native Australian eucalypt forest invaded by exotic pines. *Oecologia*, **31**, 1–12.

Burney DA and Burney LP (2007) Paleoecology and 'inter-situ' restoration on Kaua'I, Hawai'i. *Frontiers in Ecology and the Environment*. **5**, 483–490.

Burns JH (2004) A comparison of invasive and non-invasive dayflowers (Commelinaceae) across experimental nutrient and water gradients. *Diversity and Distributions*, **10**, 387–397.

Burns JH (2006) Relatedness and environment affect traits associated with invasive and noninvasive introduced Commelinaceae. *Ecological Applications*, **16**, 1367–1376.

Butler D (2007) Data sharing: the next generation. *Nature*, **446**, 10–11.

Byers JE, Cuddington K, Jones CG, Talley TS, Hasting A, Lambrinos JG, Crooks JA, and Wilson WG (2006) Using ecosystem engineers to restore ecological systems. *Trends in Ecology and Evolution*, **21**, 493–500.

Byrnes JE, Reynolds PL, and Stachowicz JJ (2007) Invasions and extinctions reshape coastal marine food webs. PLoS ONE 2(3): e295. doi:10.1371/journal.pone.0000295.

Cadotte MW (2006) Darwin to Elton: early ecology and the problem of invasive species. In MW Cadotte, SM McMahon, and T Fukami, ed. *Conceptual ecology and invasions biology: reciprocal approaches to nature*, pp. 15–33. Springer, Dordrecht, The Netherlands.

Cadotte MW, McMahon SM, and Fukami T, ed. (2006) *Conceptual ecology and invasions biology: reciprocal approaches to nature*. Springer, Dordrecht, The Netherlands.

Caldow RWG, Stillman RA, Durell SEA le V dit, West AD, McGrorty S, Goss-Custard JD, Wood PJ, and Humphreys J (2007) Benefit to shorebirds from invasion of a non-native shellfish. *Proceedings of the Royal Society B*, **274**, 1449–1455.

Callaway E (2008) The green menace. *Nature*, **452**, 148–150.

Callaway RM, Cipollini D, Barto K, Thelen GC, Hallett SG, Prati D, Stinson K, and Klironomos JK (2008) Novel weapons: invasive plant suppresses fungal mutualists in America but not in its native Europe. *Ecology*, **89**, 1043–1055.

Callaway RM and Maron JL (2006) What have exotic plant invasions taught us over the past 20 years? *Trends in Ecology and Evolution*, **21**, 369–374.

Callaway RM and Ridenour WM (2004) Novel weapons: invasive success and the evolution of increased competitive ability. *Frontiers in Ecology and the Environment*, **2**, 436–443.

Callaway RM, Ridenour WM, Laboski T, Weir T, and Vivanco JM (2005) Natural selection for resistance to the allelopathic effects of invasive plants. *Journal of Ecology*, **93**, 576–583.

Callaway RM, Thelen GC, Rodriguez A, and Holben WE (2004) Soil biota and exotic plant invasions. *Nature*, **427**, 731–733.

Campbell DH (1926) *An outline of plant geography*. Macmillan, New York.

Campbell LG, Snow AA, and Ridley CE (2006) Weed evolution after crop gene introgression: greater survival and fecundity of hybrids in a new environment. *Ecology Letters*, **9**, 1198–1209.

Cardinale BJ, Srivastava DS, Duffy JE, Wright JP, Downing AL, Sankaran M, and Jouseau C (2006) Effects of biodiversity on the functioning of trophic groups and ecosystems. *Nature*, **443**, 989–992.

Carey JR (1996) The future of the Mediterranean fruit fly *Ceratitis capitata* invasion of California: a predictive framework. *Biological Conservation*, **78**, 35–50.

Carlton JT (1985) Transoceanic and interoceanic dispersal of coastal marine organisms: the biology of ballast water. *Oceanography and Marine Biology Annual Review*, **23**, 313–374.

Carlton JT (1999) The scale and ecological consequences of biological invasions in the world's oceans. In OT Sandlund, PJ Schei, and A Viken, ed. *Invasive species and biodiversity management*, pp. 195–212. Kluwer Academic Publishers, Dordrecht, Netherlands.

Carlton JT (2002) Bioinvasion ecology: assessing invasion impact and scale. In E Leppäkoski, S Gollasch, and S Olenin, ed. *Invasive aquatic species in Europe. Distribution, impacts and management*, pp. 7–19. Kluwer Academic Publishers, Dordrecht.

Carlton JT and GM Ruiz. (2005) Vector science and integrated vector management in bioinvasion ecology: conceptual frameworks. In HA Mooney, RN Mack, JA McNeely, LE Neville, PJ Schei, and JK Waage, ed. *Invasive alien species: a new synthesis*, pp 36–58. Island Press, Washington DC.

Carrete M and Tella JL (2008) Wild-bird trade and exotic invasions: a new link of conservation concern? *Frontiers in Ecology and the Environment*, **6**, 207–211.

Carroll SP and H Dingle (1996) The biology of post-invasion events. *Biological Conservation*, **78**, 207–214.

Cassey P (2002) Life history and ecology influences establishment success of introduced land birds. *Biological Journal of the Linnean*, Society, **76**, 465–480.

Cassey P, Blackburn TM, Sol GD, Duncan RP, and Lockwood JL (2004) Global patterns of introduction effort and establishment success in birds. *Proceedings of the Royal Society B*, **271**, S405–S408.

Castilla JC, Lagos NA, Cerda M (2004) Marine ecosystem engineering by the alien ascidian *Pyura praeputialis* on a mid-intertidal rocky shore. *Marine Ecology Progress Series*, **268**, 119–130.

Castilla JC, Uribe M, Bahamonde N, Clarke M, Desqueyroux-Faundez R, Kong I, Moyano H, Rozbaczylo N, Santelices B, Valdovinos C, and Zavala P (2005) Down under the southeastern Pacific: marine non-indigenous species in Chile. *Biological Invasions*, **7**, 213–232.

Castle D (2005) Diversity and stability: theories, models, and data. In K. Cuddington and B Beisner, ed. *Ecological paradigms lost: routes of theory change*, pp. 201–209. Elsevier Academic Press, Amsterdam.

Cavieres LA, Quiroz CL, and Molina-Montenegro MA (2007) Facilitation of the non-native *Taraxacum officinale* by native nurse cushion species in the high Andes of central Chile: are there differences between nurses. *Functional Ecology*, **22**, 148–156.

Chapman MA and Burke JM (2006) Letting the gene out of the bottle: the population genetics of GM crops. *New Phytologist*, **170**, 429–443.

Chase JM and Leibold MA (2003) *Ecological niches.* University of Chicago Press, Chicago, IL.

Chen PF, Wiley EO, and McNyset KM (2007) Ecological niche modeling as a predictive tool: silver and bighead carps in North America. *Biological Invasions*, **9**, 43–51.

Cheptou P-O, Carrue O, Rouifed S, and Cantarel A (2008) Rapid evolution of seed dispersal in an urban environment in the weed *Crepis sancta. Proceedings of the National Academy of Sciences, USA*, **105**, 3796–3799.

Chesson P (1994) Multispecies competition in variable environments. *Theoretical Population Biology*, **45**, 227–276.

Chesson P (2000) Mechanisms of maintenance of species diversity. *Annual Review of Ecology and Systematics*, **31**, 343–366.

Chew MK (2006) *Ending with Elton: preludes to invasion biology*. Ph.D thesis, Arizona State University.

Chew MK and Laubichler MD (2003) Natural enemies: metaphor or misconception. *Science*, **301**, 52–53.

Choi J-K and Bowles S (2007) The coevolution of parochial altruism and war. *Science*, **318**, 636–640.

Christian JM and Wilson SD (1999) Long-term ecosystem impacts of an introduced grass in the northern great plains. *Ecology*, **80**, 2397–2407.

Christie WJ (1972) Lake Ontario: effects of exploitation, introductions and eutrophication on the salmonid community. *Journal of the Fisheries Research Board of Canada*, **29**, 913–929.

Christie WJ (1974) Changes in the fish species composition of the Great Lakes. *Journal of the Fisheries Research Board of Canada*, **31**, 827–854.

Chun YJ, Collyer ML, Moloney KA, and Nason JD (2007) Phenotypic plasticity of native vs invasive purple loosestrife: a two-state multivariate approach. *Ecology*, **88**, 1499–1512.

Chytrý M, Maskell LC, Pino J, Pyšek P, Vilà M, Font X, and Smart SM (2008) Habitat invasions by alien plants: a quantitative comparison among Mediterranean, subcontinental and oceanic regions of Europe. *Journal of Applied Ecology*, **45**, 448–458.

Clark JS (1991) Disturbance and tree life history on the shifting mosaic landscape. *Ecology*, **72**, 1102–1118.

Clarke M (2007) Flora and fauna guarantee—scientific advisory committee final recommendation on a nomination for listing (*Canis lupus* subsp. *dingo* (Meyer 1793)—dingo). Nomination no. 789, Item no. T1835. Department of Sustainability and Environment, Melbourne, AU.

Cleland EE, Smith MD, Andelman SJ, Bowles C, Carney KM, Horner-Devine MC, Drake JM, Emery SM, Gramling JM, and Vandermast DB (2004) Invasion in space and time: non-native species richness and relative abundance respond to interannual vari ation in productivity and diversity. *Ecology Letters*, **7**, 947–957.

Clergeau P and Nuñez MA (2006) The language of fighting invasive species. *Science*, **311**, 951.

Coates P (2006) *American perceptions of immigrant and invasive species*. University of California Press, Berkeley, CA.

Cody ML and Overton JM (1996) Short-term evolution of reduced dispersal in island plant populations. *The Journal of Ecology*, **84**, 53–61.

Colautti RI, Grigorovich, IA, and MacIsaac HJ (2006) Propagule pressure: a null model for biological invasions. *Biological Invasions*, **8**, 1023–1037.

Colautti RI and MacIsaac HJ (2004) A neutral terminology to define invasive species. *Diversity and Distributions*, **10**, 135–141.

Colautti RI, Ricciardi A, Grigorovich IA, and MacIsaac HJ (2004) Does the enemy release hypothesis predict invasion success? *Ecology Letters*, **7**, 721–733.

Conant JB (1947) *On understanding science*. Yale University Press, New Haven, Connecticut.

Condeso TE and Meentemeyer RK (2007) Effects of landscape heterogeneity on the emerging forest disease sudden oak death. *Journal of Ecology*, **95**, 364–375.

Connell JH (1978) Diversity in tropical rain forest and coral reefs. *Science*, **199**, 1302–1310.

Connor EF and Simberloff D (1979) The assembly of species communities: chance or competition? *Ecology*, **60**, 1132–1140.

Connor EF and Simberloff D (1983) Interspecific competition and species co-occurrence patterns: null models and the evaluation of evidence. *Oikos*, **41**, 455–465.

Cooper G (2001) Must there be a balance of nature? *Biology and Philosophy*, **16**, 481–506.

Cooper WS (1926) The fundamentals of vegetational change. *Ecology*, **7**, 391–413.

Cornell HV and Karlson R (1997) Local and regional processes as controls of species richness. In D Tilman and P Kareiva, ed. *Spatial ecology: the role of space in population dynamics and interspecific interactions*, pp. 250–268. Monographs in Population Biology 29. Princeton University Press, Princeton, NJ.

Correa SB, Winemiller KO, López-Fernández H, and Galetti M (2007) Evolutionary perspectives on seed consumption and dispersal by fishes. *Bioscience*, **57**, 748–756.

Costello C and McAusland C (2003) Protectionism, trade, and measures of damage from exotic species introductions. *American Journal of Agricultural Economics*, **85**, 964–975.

Costello CJ and Solow AR (2003) On the pattern of discovery of introduced species. *Proceedings of the National Academy of Sciences USA*, **100**, 3321–3323.

Courchamp F and Caut S (2005) Use of biological invasions and their control to study the dynamics of interacting populations. In MW Cadotte, SM McMahon, and T Fukami, ed. *Conceptual ecology and invasions biology: reciprocal approaches to nature*, pp. 235–279. Springer, London.

Courchamp F, Clutton-Brock T, and Grenfell B (1999) Inverse density dependence and the Allee effect. *Trends in Ecology and Evolution*, **14**, 405–410.

Courtenay WR Jr. and Robins CR (1975) Exotic organisms: an unsolved, complex problem. *Bioscience*, **25**, 306–313.

Couteron P and Kokou K (1997) Woody vegetation spatial patterns in a semi-arid savanna of Burkina Faso, West Africa. *Plant Ecology*, **132**, 211–227.

Cowie RH (2002) Invertebrate invasions on Pacific islands and the replacement of unique native faunas: a synthesis of the land and freshwater snails. *Biological Invasions*, **3**, 119–136.

Cowie RH and Robinson DG (2003) Pathways of introduction of nonindigenous land and freshwater snails and slugs. In GM Ruiz and JT Carlton, ed. *Invasive species: vectors and management strategies*, pp. 93–122. Island Press, Washington, DC.

Cox GW (1999) *Alien species in North America and Hawaii: impacts on natural ecosystems*. Island Press, Washington DC.

Cox GW (2004) *Alien species and evolution*. Island Press, Washington, DC.

Cox JG and Lima SL (2006) Naïveté and an aquatic-terrestrial dichotomy in the effects of introduced predators. *Trends in Ecology and Evolution*, **21**, 674–680.

Cox PA (1983) Extinction of the Hawaiian avifauna resulted in a change of pollinators for the ieie, *Freycinetia arborea*. *Oikos*, **41**, 195–199.

Cox-Foster DL, Conlan S, Holmes EC, Palacios G, Evans JD, Moran NA, Quan P-L, Briese T, Hornig M, Geiser DM, Martinson V, van Engelsdorp D, Kalkstein AL, Drysdale A, Hui J, Zhai J, Cui L, Hutchison SK, Simons JF, Egholm M, Pettis JS, and Lipkin WI (2007) A metagenomic survey of microbes in honey bee colony collapse disorder. *Science*, **318**, 283–287.

Crain CM and Bertness MD (2006) Ecosystem engineering across environmental gradients: implications for conservation and management. *Bioscience*, **56**, 211–218.

Craine JM (2005) Reconciling plant strategy theories of Grime and Tilman. *Journal of Ecology*, **93**, 1041–1052.

Craine JM (2007) Plant strategy theories: replies to Grime and Tilman. *Journal of Ecology*, **95**, 235–270.

Crall AW, Meyerson LA, Stohlgren TJ, Jarnevich CS, Newman GJ, and Graham J (2006) Show me the numbers: what data currently exist for non-native species in the USA? *Frontiers in Ecology and the Environment*, **4**, 414–418.

Cramer VA, Hobbs RJ, and Standish RJ (2008) What's new about old fields: land abandonment and ecosystem assembly. *Trends in Ecology and Evolution*, **23**, 104–112.

Crawley MJ (1987) What makes a community invasible? In AJ Gray, MJ Crawley, and PJ Edwards, ed. *Colonization, succession, and stability*, pp. 429–453. Blackwell, Oxford.

Croci S, Quilliec P Le, and Clergeau P (2007) Geographical range as predictor of spatial expansion of invading birds. *Biodiversity and Conservation*, **16**, 511–524.

Crooks JA (1998) Habitat alteration and community-level effects of an exotic mussel, *Musculista senhousia*. *Marine Ecology Progress Series*, **162**, 137–152.

Crooks JA (2002) Characterizing ecosystem-level consequences of biological invasions: the role of ecosystem engineers. *Oikos*, **97**, 153–166.

Crooks JA (2005) Lag times and exotic species: the ecology and management of biological invasions in slow-motion. *Ecography*, **12**, 316–329.

Crossland MR (2000) Direct and indirect effects of the introduced toad *Bufo marinus* (Anura: Bufonidae) on populations of native anuran larvae in Australia. *Ecography*, **23**, 283–290.

Crozier L and Dwyer G (2006) Combining population-dynamic and ecophysiological models to predict climate-induced insect range shifts. *American Naturalist*, **167**, 853–866.

Crutsinger GM, Souza L, and Sanders NJ (2008) Intraspecific diversity and dominant genotypes resist plant invasions. *Ecology Letters*, **11**, 16–23.

Cuddington K and Beisner B (2005) Kuhnian paradigms lost: embracing the pluralism of ecological theory. In K Cuddington and B Beisner, ed. *Ecological paradigms lost: routes to theory change*, pp. 419–426. Elsevier/Academic Press, Oxford.

Culliney TW (2005) Benefits of classical biological control for managing invasive plants. *Critical Reviews in Plant Sciences*, **24**, 131–150.

Curtin D, Selles F, Wang H, Zentner RP, and Campbell CA (2000) Restoring organic matter in a cultivated, semi-arid soil using crested wheatgrass. *Canadian Journal of Soil Science*, **80**, 429–435.

Cushman JH and Tierney TA, and Hinds JM (2004) Variable effects of feral pig disturbances on native and exotic plants in a California grassland. *Ecological Applications*, **14**, 1746–1756.

Daehler CC (2001a) Two ways to be an invader, but one is more suitable for ecology. *Bulletin of the Ecological Society of America*, **82**, 101–102.

Daehler CC (2001b) Darwin's naturalization hypothesis revisited. *American Naturalist*, **158**, 324–330.

Daehler CC (2003) Performance comparisons of co-occurring native and alien plants: implications for conservation and restoration. *Annual Review of Ecology, Evolution and Systematics*, **34**, 183–211.

Daehler CC (2005) Upper-montane plant invasions in the Hawaiian Islands: patterns and opportunities. *Perspectives in Plant Ecology, Evolution and Systematics*, **7**, 203–216.

Daehler CC (2006) Invasibility of tropical islands by introduced plants: partitioning the influence of isolation and propagule pressure. *Preslia*, **78**, 361–374.

Daehler CC and Carino DA (2000) Predicting invasive plants: Prospects for a general screening system based on current regional models. *Biological Invasions*, **2**, 92–103.

Daehler CC and Carino DA (2001) Hybridization between native and alien plants and its consequences. In JL Lockwood and ML. McKinney, ed. *Biotic homogenization: the loss of diversity through extinction and invasion*, pp. 81–102. Kluwer Academic Publishers, New York.

Daehler CC, Denslow JS, Ansari S, and Kuo H (2004) A risk assessment system for screening out invasive pest plants from Hawai'i and other Pacific Islands. *Conservation Biology*, **18**, 360–368.

Daehler CC and Strong DR (1993) Prediction and biological invasions. *Trends in Ecology and Evolution*, **8**, 380.

Daily G (1997) *Nature's services*. Island Press, Washington DC.

Dalton R (2006) Whitefly infestations: the Christmas invasion. *Nature*, **443**, 898–900.

Daniels RE and Sheil J (1999) Genetic pollution: concepts, concerns and transgenic crops. In Gene flow and agriculture: relevance to transgenic crops. In *Gene flow and agriculture: relevance for transgenic crops*, pp. 65–72. British Crop Protection Council, Farnham.

D'Antonio CM (2000) Fire, plant invasions and global changes. In H Mooney and R Hobbs, ed. *Invasive species in a changing world*, pp. 65–94. Island Press, Covelo.

D'Antonio CM and Corbin JD (2003) Effects of plant invaders on nutrient cycling: using models to explore the link between invasion and development of species effects. In CD Canham, JJ Cole, and WK Lauenroth, ed. *Models in ecosystem science*, pp. 363–384. Princeton University Press, Princeton NJ.

D'Antonio CM, Dudley T, and Mack M (1999) Disturbance and biological invasions. In L Walker, ed. *Ecosystems of disturbed ground*, pp. 429–468. Elsevier, Oxford.

D'Antonio CM and Hobbie S (2005) Plant species effects on ecosystem processes: insights from invasive species. In DF Sax, JJ Stachowicz, and SD Gaines, ed. *Species invasions. Insights into ecology, evolution, and biogeography*, pp. 65–84. Sinauer Associates, Sunderland, MA.

D'Antonio CM, Tunison JT, and Loh R (2000)Variation in impact of exotic grass fueled fires on species composition across an elevation gradient in Hawai'i. *Austral Ecology*, **25**, 507–522.

D'Antonio CM and Vitousek PM (1992) Biological invasions by exotic grasses, the grass-fire cycle, and global change. *Annual Review of Ecology and Systematics* **23**, 63–87.

Darwin C (1859) *On the origin of species*. Murray, London.

Dasmann RF (1988) Commentary: biosphere reserves, buffers, and boundaries. *BioScience*, **38**, pp. 487–489.

Daszak P, Cunningham AA, and Hyatt AD (2000) Emerging infectious diseases of wildlife: threats to biodiversity and human health. *Science*, **287**, 443–449.

Davies KF, Chesson P, Harrison S, Inouye BD, Melbourne BA, and Rice KJ (2005) Spatial heterogeneity explains the scale dependence of the native-exotic diversity relationship. *Ecology*, **86**, 1602–1610.

Davis MA (1986) Geographic patterns in the flight ability of a monophagous beetle. *Oecologia*, **69**, 407–412.

Davis MA (2003) Biotic globalization: does competition from introduced species threaten biodiversity? *Bioscience*, **53**, 481–489.

Davis MA (2006) Invasion biology 1958–2005: the pursuit of science and conservation. In MW Cadotte, SM McMahon, and T Fukami, ed. *Conceptual ecology and invasions biology: reciprocal approaches to nature*, pp. 35–64. Springer, London.

Davis MA, Grime JP, and Thompson K (2000) Fluctuating resources in plant communities: a general theory of invasibility. *Journal of Ecology*, **88**, 528–536.

Davis MA and Pelsor M (2001) Experimental support for a resource-based mechanistic model of invasibility. *Ecology Letters*, **4**, 421–428.

Davis MA and Thompson K (2000) Eight ways to be a colonizer; two ways to be an invader. *Bulletin of the Ecological Society of America*, **81**, 226–230.

Davis MA and Thompson K (2001) Invasion terminology: should ecologists define their terms differently than others? No, not if we want to be of any help! *Bulletin of the Ecological Society of America*, **82**, 206.

Davis MA and Thompson K (2002). 'Newcomers' invade the field of invasion ecology: question the field's future. *Bulletin of the Ecological Society of America*, **83**, 196–197.

Davis MA, Thompson K, and Grime JP (2001) Charles S. Elton and the dissociation of invasion ecology from the rest of ecology. *Diversity and Distributions*, **7**, 97–102.

Davis MA, Thompson K, Grime JP (2005a) Invasibility: the local mechanism driving community assembly and species diversity. *Ecography*, **28**, 696–704.

Davis MA, Curran C, Tietmeyer A, and Miller A (2005b) Dynamic tree aggregation patterns in a species-poor temperate woodland disturbed by fire. *Journal of Vegetation Science*, **16**, 167–174.

Davis MA, Pergl J, Truscott A, Kollmann J, Bakker JP, Domenech R, Prach K, Prieur-Richard A, Veeneklaas RM, Pyšek P, del Moral R, Hobbs RJ, Collins SL, Pickett STA, and Reich PB (2005c). Vegetation change: a reunifying concept in plant ecology. *Perspectives in Plant Ecology, Evolution and Systematics*, **7**, 69–76.

Davis MA, Bier L, Bushelle E, Diegel C, Johnson A, and Kujala B (2005d) Non-indigenous grasses impede woody succession. *Plant Ecology*, **178**, 249–264.

Davis MB, Woods KD, Webb SL, and Futyma RP (1986) Dispersal versus climate: expansion of Fagus and Tsuga into the Upper Great Lakes region. *Vegetatio* **67**, 93–103.

Davis HG, Taylor CM, Civille JC, and Strong DR (2004) An Allee effect at the frong on a plant invasion: *Spartina* in a Pacific estuary. *Journal of Ecology*, **92**, 321–327.

Daws MJ, Hall J, Flynn S, and Pritchard HW (2007) Do invasive species have bigger seeds? Evidence from intra- and inter-specific comparisons. *South African Journal of Botany*, **73**, 138–143.

Deckers B, Verheyen K, Hermy M, and Muys B (2005) Effects of landscape structure on the invasive spread of black cherry *Prunus serotina* in an agricultural landscape in Flanders, Belgium. *Ecography*, **28**, 99–109.

DeGasperis BG and Motzkin G (2007) Windows of opportunity: historical and ecological controls on *Berberis thunbergii* invasions. *Ecology*, **88**, 3115–3125.

de Laplante K (2004) Toward a more expansive conception of ecological science. *Biology and Philosphy*, **19**, 263–281.

de Poorter M, Browne M, Lowe S, and Clout M (2005) The ISSG global invasive species database and other aspects of an early warning system. In HA Mooney, RN Mack, JA McNeely, LE Neville, PJ Schei, and JK Waage, ed. *Invasive alien species: a new synthesis*, pp. 59–83. Island Press, Washington, DC.

de Rivera CE, Ruiz GM, Hines AH, and Jivoff P (2005) Biotic resistance to invasion: native predator limits abundance and distribution of an introduced crab. *Ecology*, **86**, 3364–3376.

Desprez-Loustau M-L, Robin C, Buée M, Courtecuisse R, Garbaye J, Suffert F, Sache I, and Rizzo DM (2007) The fungal dimension of biological invasions. *Trends in Ecology and Evolution*, **22**, 472–480.

Dethier MN and Hacker SD (2005) Physical factors vs. biotic resistance in controlling the invasion of an estuar ine marsh grass. *Ecological Applications*, **15**, 1273–1283.

Diamond J (1987) Reflections on goals and on the relationship between theory and practice. In WR Jordan III and ME Gilpin, ed. *Restoration ecology: a synthetic approach to ecological restoration*, pp. 329–326. Cambridge University Press, Cambridge, UK.

Didham RK, Tylianakis JM, Gemmell NJ, Rand TA, and Ewers RM (2007) Interactive effects of habitat modification and species invasion on native species decline. *Trends in Ecology and Evolution*, **22**, 489–496.

Dietz H and Edwards PJ (2006) Recognition of changing processes during plant invasions may help reconcile conflicting evidence of the causes. *Ecology*, **87**, 1359–1367.

Dietz H and Steinlein T (2004) Recent advances in understanding plant invasions. *Progress in Botany*, **65**, 539–573.

Ding J, Mack RN, Lu P, Ren M, and Huang H (2008) China's booming economiy is sparking and accelerating biological invasions. *Bioscience*, **58**, 317–324.

Doak DF, Estes JA, Halpern BS, Jacob U, Lindberg DR, Lovvorn J, Monson DH, Tinker MT, Williams TM, Wooton T, Carroll I, Emmerson M, Micheli F, and Novak M (2008) Understanding and predicting ecological dynamics: are major surprises inevitable? *Ecology*, **89**, 952–961.

Donald PF (2007) Adult sex ratios in wild bird populations. *Ibis*, **149**, 671–692.

Dornelas M, Connolly SR, and Hughes TP (2006) Coral reef diversity refutes the neutral theory of biodiversity. *Nature*, **440**, 80–82.

Douglas M (1966) *Purity and danger: an analysis of concepts of pollution and taboo.* Frederick A. Praeger, New York.

Dove ADM (1998) A silent tragedy: parasites and the exotic fishes of Australia. *Proceedings of the Royal Society of Queensland*, **107**, 109–113.

Drakare S, Lennon JJ, and Hillebrand H (2006) The imprint of the geographical, evolutionary and ecological context on species–area relationships. *Ecology Letters*, **9**, 215–227.

Drake JA, Mooney, HA, di Castri, F, Groves JH, Kruger FJ, Rejmánek M, and Williamson, M (1989). *Biological invasions: a global perspective*, SCOPE 37. John Wiley and Sons, Chichester, UK.

Drake JM (2004) Allee effects and the risk of biological invasion. *Risk Analysis*, **24**, 795–802.

Drake JM, Cleland EE, Horner-Devine MC, Fleishman E, Bowles C, Smith MD, Carney K, Emery, S, Gramling J, Vandermast DB, and Grace JB (2008) Do non-native plant species affect the shape of productivity-diversity relationships? *American Midland Naturalist*, **159**, 55–66.

Drake JM and Lodge DM (2006) Allee effects, propagule pressure and the probability of establishment: risk analysis for biological invasions. *Biological Invasions*, **8**, 365–375.

Drude O (1896) *Manuel de geographie botanique*. Librairie Des Sciences Naturelles, Paris.

Dukes JS (2000) Will the increasing atmospheric CO_2 concentration affect the success of invasive species? In HA Mooney and RJ Hobbs, ed. *Invasive species in a changing world*, pp. 95–113. Island Press, Washington DC.

Duncan RP (1997) The role of competition and introduction effort in the success of passeriform birds introduced into New Zealand. *American Naturalist*, **149**, 903–915.

Duncan RP, Blackburn TM, and Sol D (2003) The ecology of bird introductions. *Annual Review of Ecology and Systematics*, **34**, 71–98.

Duncan RP and Forsyth DM (2006) Competition and the assembly of introduced bird communities. In MW Cadotte, SM McMahon, and T Fukami, ed. *Conceptual ecology and invasions biology: reciprocal approaches to nature*, pp. 405–421. Springer, Dordrecht, The Netherlands.

Duncan RP and Williams PA (2002). Darwin's naturalization hypothesis challenged. *Nature*, **417**, 608–609.

Dunn E (1977) Predation by weasels (*Mustela nivalis*) on breeding tits (*Parus* spp.) in relation to the density of tits and rodents. *Journal of Animal Ecology*, **46**, 634–652.

Dunstan PK and Johnson CR (2004). Invasion rates increase with species richness in a marine epibenthic community by two mechanisms. *Oecologia*, **138**, 285–292.

Dunstan PK and Johnson CR (2006). Linking richness, community variability and invasion resistance with patch size in a model marine community. *Ecology*, **87**, 2842–2850.

Duponnois R, Colombet A, Hien V, and Thioulouse J (2005) The mychorrhizal fungus *Glomus intraradices* and rock phosphate amendment influence plant growth and microbial activity in the rhizosphere of *Acacia holosericea*. *Soil Biology and Biochemistry*, **37**, 1460–1468.

Dyck VA, Hendrichs J, and Robinson AS (2005) *Sterile insect techniques: principles and practice in area-wide integrated pest management*. Spring, Dordrecht, The Netherlands.

Dye PJ, Moses G, Vilakazi P, Ndlela R, and Royappen M (2001) Comparative water use of wattle thickets and indigenous plant communities at riparian sites in the Western Cape and KwaZulu-Natal. *Water SA*, **27**, 529–538.

Easter-Pilcher A (1996) Implementing the Endangered Species Act. *Bioscience*, **46**, 355–363.

Eby LA, Roach WJ, Crowder LB, and Stanford JA (2006) Effects of stocking-up freshwater food webs. *Trends in Ecology and Evolution*, **21**, 576–584.

Egler FE (1942) Indigene versus alien in the development of arid Hawaiian vegetation. *Ecology*, **23**, 14–23.

Egler FE (1951) A commentary on American plant ecology, based on the textbooks of 1947–1949. *Ecology*, **32**, 673–695.

Ehrenfeld JG (2003) Effects of plant invasions on soil nutrient cycling processes. *Ecosystems*, **6**, 503–523.

Ehrlén J, Münzbergova Z, Diekmann M, and Eriksson O (2006) Long-term assessment of seed limitation in plants: results from an 11-year experiment. *Journal of Ecology*, **94**, 1224–1232.

Ejrnæs R, Bruun HH and Graae BJ (2006) Community assembly in experimental grassland: suitable environment or timely arrival? *Ecology*, **87**, 1225–1233.

Elam DR, Ridley CE, Goodell K, and Ellstrand NC (2007) Population size and relatedness affect fitness of a self-incompatible invasive plant. *Proceedings of the National Academy of Sciences USA*, **104**, 549–552.

Ellis B (2005) Physical realism. *Ratio*, **18**, 371–384.

Ellstrand NC (2003) Going to 'great lengths' to prevent the escape of genes that produce specialty chemicals. *Plant Physiology*, **132**, 1770–1774.

Ellstrand NC and Schierenbeck K (2000) Hybridization as a stimulus for the evolution of invasiveness in plants? *Proceedings of the National Academy of Sciences USA*, **97**, 7043–7050.

Elser JJ, Chrzanowski TH, Sterner RW, and Mills KH (1998) Stoichiometric constraints on food-web dynamics: a whole lake experiment on the Canadian shield. *Ecosystem*, **1**, 120–136.

Elton C (1927) *Animal ecology*. Sidgwick and Jackson, London.

Elton C (1958) *The ecology of invasions by animals and plants*. Methuen, London.

Embree DG (1979) The ecology of colonizing species, with special emphasis on animal invaders. In DJ Horn, GR Stairs, and RD Mitchell, ed. *Analysis of ecological systems*, pp. 51–65. Ohio State University Press, Columbus, OH.

Emerson BC and Kolm N (2005) Species diversity can drive speciation. *Nature*, **434**, 1015–1017.

Emery SM and KL Gross (2007) Dominant species identity, not community evenness, regulates invasion in experimental plant communities. *Ecology*, **88**, 954–964.

Enquist BJ and Stark SC (2007). Follow Thompson to make biology a capital-S Science. *Nature*, **446**, 611.

Enserink M (2007) Chikungunya: no longer a third world disease. *Science*, **318**, 1860–1861.

Enserink M (2008) Exotic disease of farm animals tests Europe's responses. *Science*, **319**, 710–711.

Eppstein MJ and Molofsky J (2007) Invasiveness in plant communities with feedbacks. *Ecology Letters*, **10**, 253–263.

Eser U (1998) Assessment of plant invasions: theoretical and philosophical fundamentals. In U Starfinger, K Edwards, I Kowarik and M Williamson. *Plant invasions: ecological mechanisms and human responses*, pp 95–107. Backhuys Publishers, Leiden, The Netherlands.

Essington TE, Beaudreau AH, and Wiedenmann J (2006) Fishing through marine food webs. *Proceedings of the National Academy of Sciences USA*, **103**, 3171–3175.

Ewel JJ (1986) Invasibility lessons from South Florida. In HA Mooney and JA Drake, ed. *Ecology of biological invasions of North America and Hawaii*, pp. 214–230. Ecological Studies 58, Springer-Verlag.

Fa JE, Juste J, Burn RW, and Broad G (2002) Bushmeant consumption and preferences of two ethnic groups in Bioko Island, West Africa. *Human Ecology*, **30**, 397–416.

Facon B, Genton BJ, Shykoff J, Jarne P, Estoup A, and David P (2006) A general eco-evolutionary framework for understanding bioinvasions. *Trends in Ecology and Evolution*, **21**, 130–135.

Faden M (2008) Bye-bye to ballast water? *Frontiers in Ecology and Evolution*, **6**, 175.

Faliński JB (1968) Stadia neofityzmu I stosunek neofitow do innych komponentow zbiorowiska. [Stages of neophytism and the reaction of neophytes to other components of the community.] In JB Faliński, ed. Synantropizacja szaty roslinnej. I. Neofityzm i apofityzm w szacie roslinnej Polski. Materialy Sympozjum w Nowogrodzie. Mater. Zakl. *Fitosoc. Stos. UW*, **25**, 15–31.

Faliński JB (1969) Neofity I neofityzm. Dyskusje fitosocjologiczne (5). Neophytes et. neophytisme. [Discussion phytosociologiques (5)]. *Ekol. pol. B.*, **15**, 337–355.

Farber S, Costanza R, Childers DL, Erickson J, Gross K, Grove M, Hopkinson CS, Kahn J, Pincetl S, Troy, A, Warren P, and Wilson M (2006) Linking ecology and economics for ecosystem management. *Bioscience*, **56**, 121–133.

Fargione J, Brown CS, and Tilman D (2003) Community assembly and invasion: an experimental test of neutral versus niche processes. *Proceeding of the National Academy of Sciences USA*, **100**, 8916–8920.

Fargione J and Tilman D (2005) Diversity decreases invasion via both sampling and complementarity effects. *Ecology Letters*, **8**, 604–611.

Farji-Brener AG and Ghermandi L (2008) Leaf-cutting ant nests near roads increase fitness of exotic plant species in natural protected areas. *Proceedings of the Royal Society B*, **275**, 1431–1440.

Fauvergue X, Malausa JC, Giuge L, and Courchamp F (2007) Invading parasitoids suffer no Allee effect: a manipulative field experiment. *Ecology*, **88**, 2392–2403.

Ficetola GF, Miaud C, Pompanon F, and Taberlet P (2008) Species detection using environmental DNA from water samples. *Biology Letters*,**4**, 423–425.

Filchak KE, Roethele JB, and Feder JL (2000) Natural selection and sympatric divergence in the apple maggot *Rhagoletis pomonella*. *Nature*, **407**, 739–742.

Findlay DL, Vanni MJ, Paterson M, Mills KH, Kasian SEM, Findlay WJ, and Salki AG (2005) Dynamics of a boreal lake ecosystem during a long-term manipulation of top predators. *Ecosystems*, **8**, 603–618.

Fitch A (1861) Sixth report on the noxious and other insects of the state of New York. *New York State Agricultural Society Transactions*, **20**, 746–868.

Flanders AA, Kuvlesky Jr. WP, Tuthven III, DC, Zaiglan RE, Bingham RL, Fulbright TE, Hernandez F, and Brennan LA (2006) Effects of invasive exotic grasses on south Texas rangeland breeding birds. *Auk*, **123**, 171–182.

Fofonoff PW, Ruiz GM, Steves B, and Carlton JT (2003) In ships or on ships? Mechanisms of transfer and invasion for non-native species to the coasts of North America. In GM Ruiz and JT Carlton, ed. *Invasive species: vectors and management strategies*, pp. 152–182. Island Press, Washington, DC.

Forbes SA (1883) *Insects affecting corn*. Illinois State Entomology Office, Circulation 1–21.

Forbes SA (1886) *The clinch-bug in Illinois*. Illinois State Entomology Office, Circulation 8pp (September).

Forbes SA (1887) The lake as a microcosm. *Bulletin of the Scientific Association* (Peoria, IL), **1887**, 77–87.

Forbes SA (1898) The season's campaign against the San José and other scale insects in Illinois. *Transactions of the Illinois State Horticultural Society*, **31**, 105–119.

Ford PA, ed. (1892–99) *The writings of Thomas Jefferson*. GP Putnam's Sons, New York.

Forde SE, Thompson JN, and Bohannan BJM (2004) Adaptation varies through space and time in a coevolving host-parasitoid interaction. *Nature*, **431**, 841–844.

Foreman D (2004) *Rewilding North America: a vision for conservation in the 21st century*. Island Press, Washington DC.

Foster BL and Tilman D (2003) Seed limitation and the regulation of community structure in oak savanna grassland. *Journal of Ecology*, **91**, 999–1007.

Fowler AJ, Lodge DM, and Hsia JF (2007) Failure of the Lacey Act to protect US ecosystems against animal invasions. *Frontiers in Ecology and the Environment*, **5**, 353–359.

France KE and Duffy JE (2006) Consumer diversity mediates invasion dynamics at multiple trophic levels. *Oikos*, **113**, 515–529.

Francis AP and Currie DJ (2003) A globally consistent richness-climate relationship for angiosperms. *American Naturalist*, **161**, 523–536.

Freckleton RP, Dowling PM, and Dulvy NK (2006) Stochasticity, nonlinearity and instability in biological invasions. In MW Cadotte, SM McMahon, and T Fukami, ed. *Conceptual ecology and invasions biology: reciprocal approaches to nature*, pp. 125–146. Springer, Dordrecht, The Netherlands.

Freeman AS and Byers JE (2006) Divergent induced responses to an invasive predator in marine mussel populations. *Science*, **313**, 831–833.

Freestone A. and Inouye BD. (2006) Dispersal limitation and environmental heterogeneity shape scale-dependent diversity patterns in a plant community. *Ecology*, **87**, 2425–2432.

Frehlich LE, Hale CM, Scheu S, Holdsworth AR, Heneghan L, Bohlen PJ, and Reich PB (2006) Earthworm invasion into previously earthworm-free temperate and boreal forests. *Biological Invasions*, **8**, 1235–1245.

Frenot Y, Chown SL, Whinam J, Selkirk PM, Convey P, Skotnicki M, and Bergstrom DM (2005) Biological invasions in the Antarctic: extent, impacts and implications. *Biological Reviews of the Cambridge Philosophical Society*, **80**, 45–72.

Fréville H, McConway K, Dodd M, and Silvertown J (2007) Prediction of extinction in plants: interaction of extrinsic threats and life history traits. *Ecology*, **88**, 2662–2672.

Fridley JD, Brown RL, and Bruno JF (2004) Null models of exotic invasion and scale-dependent patterns of native and exotic species richness. *Ecology*, **85**, 3215–22.

Fridley JD, Stachowicz JJ, Naeem S, Sax DF, Seabloom EW, Smith MD, Stohlgren TJ, Tilman D, and von Holle B (2007) The invasion paradox: reconciling pattern and process in species invasions. *Ecology*, **88**, 3–17.

Frith, C (2007) *Making up the mind: how the brain creates our mental world*. Blackwell, Oxford.

Fritts TH and Rodda GH (1998) The role of introduced species in the degradation of island ecosystems: a case history of Guam. *Annual Review of Ecology and Systematics*, **29**, 113–140.

Fuhlendorf SD and Engle DM (2004) Application of the fire-grazing interaction to restore a shifting mosaic on tallgrass prairie. *Journal of Applied Ecology*, **41**, 604–614.

Fukami T, Wardle DA, Bellingham PJ, Mulder CPH, Towns DR, Yeates GW, Bonner KI, Durrett MS, Grant-Hoffman MN, and Williamson WM (2006) Above- and below-ground impacts of introduced predators in seabird-dominated island ecosystems. *Ecology Letters*, **9**, 1299–1307.

Fuller PL (2003) Freshwater aquatic vertebrate introductions in the United States: patterns and pathways. In GM Ruiz and JT Carlton, ed. *Invasive species: vectors and management strategies*, pp. 123–151. Island Press, Washington, DC.

Funk JL and Vitousek PM (2007) Resource use efficiency and plant invasion in low-resource systems. *Nature*, **446**, 1079–1081.

Galil B and Zenetos A (2002) A sea change: exotics in the eastern Mediterranean Sea. In E Leppäkoski, S Gollasch, and S Olenin, ed. *Invasive aquatic species in Europe. Distribution, impacts and management*, pp. 325–336. Kluwer Academic Publishers, Dordrecht.

Galil BS (2006) The marine caravan—the Suez Canal and the Erythrean invasion. In S Gollasch, BS Galil, and A Cohen, ed. *Bridging divides*. Kluwer Academic Publishers, Dordrecht.

García-Ramos G and Rodríguez D (2002) Evolutionary speed of species invasions. *Evolution*, **56**, 661–668.

Garrison VH, Shinn EA, Foreman WT, Griffin DW, Holmes CW, Kellogg CA, Majewski MS, Richardson LL, Ritchie KB, and Smith GW (2003) African and Asian dust: from desert soils to coral reefs. *Bioscience*, 53, 469–480.

Gaskin JF and Schaal BA (2002) Hybrid *Tamarix* widespread in US invasion and undetected in native Asian Range. *Proceedings of the National Academy of Sciences, USA*, **99**, 11256–11259.

Gassmann A and Louda SV (2001) *Rhinocyllus conicus*: initial evaluation and subsequent ecological impacts in North America. In E Wajnberg, JK Scott, and PC Quimby, ed. *Evaluating indirect ecological effects of biological control*, pp. 147–183. CABI Publishing, Wallingford, NZ.

Gause GF (1934) *The struggle for existence*. Williams and Wilkins, Baltimore, MD.

Gauthier-Clerc M, Lebarbenchon C, and Thomas F (2007) Recent expansion of highly pathogenic avian influenza H5N1: a critical review. *Ibis*, **149**, 202–214.

Geng Y-P, Pan X-Y, Xu C-Y, Zhang W-J, Li B, Chen J-K, Lu B-R, and Song Z-P (2007) Phenotypic plasticity rather than locally adapted ecotypes allows the invasive alligator weed to colonize a wide range of habitats. *Biological Invasions*, **9**, 245–256.

Gherardi F, ed. (2007) *Biological invaders in inland waters: profiles, distribution and threats*. Springer, Berlin.

Gherardi F and Angiolini C (2004) Eradication and control of invasive species. In F Gherardi, M Gualtieri, and C Corti, ed. *Biodiversity conservation and habitat management*. in *Encyclopedia of Life Support Systems (EOLSS)*. Developed under the Auspices of the UNESCO, Eolss Publishers, Oxford,UK.

Gianinazzi S and Vosátka M (2004) Inoculum of arbuscular mycorrhizal fungi for production systems: science meets business. *Canadian Journal of Botany*, **82**, 1264–1271.

Gido KB and Brown JH (1999) Invasion of alien fish species in North American drainages. *Freshwater Biology*, **42**,387–398.

Giere RN (2006) *Scientific perspectivism*. University of Chigago Press, Chicago.

Gilbert B and Lechowicz MJ (2005) Invasibility and abiotic gradients: the positive correlation between native and exotic plant diversity. *Ecology*, **86**, 1848–1855.

Gilbert GS and IM Parker (2006) Invasions and the regulation of plant populations by pathogens. In MW Cadotte, SM McMahon, and T Fukami, ed. *Conceptual ecology and invasions biology: reciprocal approaches to nature*, pp. 289–305. Springer, Dordrecht, The Netherlands.

Giles J (2008) Key biology databases go wiki. *Nature*, **445**, 691.

Gilpin M (1990) Ecological prediction. *Science*, **248**, 88–89.

Gitzen, RA (2007) The dangers of advocacy in science. *Science*, **317**, 748.

Gobster PA (2005) Invasive species as ecological threat: is restoration an alternative to fear-based resource management? *Ecological Restoration*, **23**, 261–270.

Goldenfeld N (2007) Biology's next revolution. *Nature*, **445**, 369.

Goldschmidt T, Witte F, and Wanink J (1993) Cascading effects of the introduced Nile perch on the detritiv orous/phytoplanktivorous species in the sublittoral areas of Lake Victoria. *Conservation Biology*, **7**, 686–700.

Goldsmith EW (2007) The seasonal impacts of urban development on roadside old field birds in east central Minnesota. Honors Thesis, Macalester College, St. Paul, MN (available at http://digitalcommons.macalester.edu/biology_honors/3/).

Goldstein JA. (2009) Aliens in the garden. *Colorado Law Review*, (in press).

Gollasch S and Leppäkoski E, ed. (1999) *Initial risk assessment of alien species in Nordic coastal waters*. Nordic Council of Ministers, Copenhagen.

Gollasch S, MacDonald E, Belson S, Botnen H, Christensen JT, Hamer JP, Houvenaghel G, Jelmert A, Lucas I, Masson D, McCollin T, Olenin S, Persson A, Wallentinus I, Wetseteyn LPMJ, Wittling T (2002) Life in ballast tanks. In E Leppäkoski, S Gollasch, and S Olenin, ed. *Invasive aquatic species in Europe. Distribution, impacts and management*, pp. 217–231. Kluwer Academic Publishers, Dordrecht.

Gonzalez A and Holt RD (2002). The inflationary effects of environmental fluctuations in source-sink systems. *Proceedings of the National Academy of Sciences, USA*, **99**, 14872–14877.

Goodman D (2005) Selection equilibrium for hatchery and wild spawning fitness in integrated breeding programs. *Canadian Journal of Fisheries and Aquatic Sciences*, **62**, 374–389.

Gorman J (2005) A frog brings cacophony to Hawaii's soundscape. *New York Times*, 25 January 2005.

Gosling LM (1989) Extinction to order. *New Scientist*, **4**, 44–49.

Gosling LM and Baker SJ (1989) The eradication of muskrats and coypus from Britain. *Biological Journal of the Linnean Society*, **38**, 39–51.

Gotelli NJ and McCabe DJ (2002) Species co-occurrence: a meta-analysis of J.M. Diamond's assembly rules model. *Ecology*, **83**, 2091–2096.

Gould SJ (1996) *Full house: spread of excellence from Plato to Darwin*. Three Rivers Press, New York.

Gould SJ (1998) An evolutionary perspective on strengths, fallacies, and confusions in the concept of native plants. *Arnoldia*, **58**, 2–10.

Gould SJ (2003) *The hedgehog, the fox, and the magister's pox: mending the gap between science and the humanities*. Harmony Books, New York.

Goulletquer P Bachelet G, Sauriau PG and Noel P. (2002) Open atlantic coast of europe: a century of introduced species into French waters. In E Leppäkoski, S Gollasch, and S Olenin, ed. *Invasive aquatic species in Europe. Distribution, impacts and management*, pp. 276–290. Kluwer Academic Publishers, Dordrecht.

Graham J, Simpson A, Crall A, Jarnevich C, Newman G, and Stohlgren TJ (2008) Vision of a cyberinfrastructure for nonnative, invasive species management. *Bioscience*, **58**, 263–268.

Graham MH and Dayton PK (2002) On the evolution of ecological ideas: paradigms and scientific progress. *Ecology*, **83**, 1481–1489.

Grant GS, Peitit TW, and Whittow EC 1981. Rat predation on Bonin petrel eggs on Midway Atoll. *Journal of Field Ornithology*, **52**, 336–338.

Graves SD and Shapiro AM (2003) Exotics as host plants of the California butterfly fauna. *Biological Conservation*, **110**, 413–433.

Gray A (1879) The pertinacity and predominance of weeds. *The American Journal of Science and Arts*, **118**, 161–167.

Green E and Galatowitsch SM (2002) Effects of *Phalaris arundinacea* and nitrate-N addition on wetland plant community establishment. *Journal of Applied Ecology*, **39**, 134–144.

Griffen BD and Delaney DG (2007) Species invasion shifts the importance of predator dependence. *Ecology*, **88**, 3012–3021.

Grigulis K, Lavorel S, Davies ID, Dossantos A, Llorets F, and Villa M (2005) Landscape-scale positive feedbacks between fire and expansion of the large tussock grass, *Ampelodesmos mauritanica*. Catalan shrublands. *Global Change Biology*, **11**, 1042–1053.

Grime JP (2007) Plant strategy theories: a comment on Craine (2005). *Journal of Ecology*, **95**, 227–230.

Grime JP, Hodgson JG, and Hunt R (1988) *Comparative plant ecology*. Unwin Hyman, London.

Grinnell J (1917) The niche-relationship of the California thrasher. *Auk*, **34**, 427–433.

Gröblacher S, Paterek T, Kaltenbaek R, Brukner Č, Żukowski M, Aspelmeyer M, and Zeilinger A (2007)

An experimental test of non-local realism. *Nature*, **446**, 871–875.

Grosholz ED, Ruiz GM, Dean CA, Shirely KA, Maron JL, and Connors PG (2000) The impacts of a nonindigenous marine predator in a California bay. *Ecology*, **81**, 1206–1224.

Gross CL (2001) The effect of introduced honeybees on native bee visitation and fruit-set in *Dillwynia juniperina* (Fabaceae) in a fragmented ecosystem. *Biological Conservation*, **102**, 89–95.

Gross CL and Mackay D (1998) Honeybees reduce fitness in the pioneer shrub *Melastoma affine* (Melastomataceae). *Biological Conservation*, **86**, 169–178.

Gross KL, Mittelbach GG, and Reynolds, HL (2005) Grassland invasibility and diversity: responses to nutrients, seed input, and disturbance. *Ecology*, **86**, 476–486.

Grotkopp E, Rejmánek M, and Rost TL (2002) Toward a causal explanation of plant invasiveness: seedling growth and life-history strategies of 29 pine (*Pinus*) species. *American Naturalist*, **159**, 396–419.

Groves RH and Burdon JJ, ed. (1986) *Ecology of biological invasions: an Australian perspective*. Australian Academy of Science, Canberra.

Guiler E (1985) *Thylacine: the tragedy of the tasmanian tiger*. Oxford University Press, Oxford, UK.

Guisan A and Thuiller W (2005) Predicting species distribution: offering more than simple habitat models. *Ecology Letters*, **8**, 993–1009.

Guisan A and Zimmermann NE (2000) Predictive habitat distribution models in ecology. *Ecological Modeling*, **135**, 147–186.

Guo J (2006) The Galápagos Islands kiss their goat problem goodbye. *Science*, **313**, 1567.

Guo Q, Qian, H, Ricklefs RE, and Xi W (2006) Distributions of exotic plants in eastern Asia and North America. *Ecology Letters*, **9**, 827–834.

Gurevitch J (2006) Commentary on Simberloff (2006) Meltdowns, snowballs and positive feedbacks. *Ecology Letters*, **9**, 919–921.

Gurevitch J and Padilla DK (2004) Are invasive species a major cause of extinctions? *Trends in Ecology and Evolution*, **19**, 470–474.

Gutiérrez-Yurrita PJ, Martínez JM, Bravo-Utrera MÁ., Montes C, Ilhéu M, and Bernardo JM (1999) The status of crayfish populations in Spain and Portugal. In F Gherardi and DM Holdich, ed. *Crayfish in Europe as alien species: how to make the best of a bad situation?* pp. 161–192. AA Balkema, Rotterdam.

Haas G, Brunke M, and Streit B (2002) Fast turnover in dominance of exotic species in the Rhine River determines biodiversity and ecosystem function: an affair between amphipods and mussels. In E Leppäkoski, S Gollasch, and S Olenin, ed. *Invasive aquatic species in Europe. Distribution, impacts and management*, pp. 426–432. Kluwer Academic Publishers, Dordrecht.

Hadfield MG, Miller SE, and Carwile AH (1993) The decimation of endemic Hawai'an tree snails by alien predators. *American Zoologist*, **33**, 610–622.

Hadwen WL, Small A, Kitching RL, and Drew RAI. (1998) Potential suitability of North Queensland rainforest sites as habitat for the Asian papaya fruit fly, *Bactrocera papayae* Drew and Hancock (Diptera: Tephritiade). *Australian Journal of Entomology*, **37**, 219–27.

Hager HA and McCoy KD (1998) The implications of accepting untested hypotheses: a review of the effects of purple loosestrive (*Lythrum salicaria*) in North America. *Biodiversity and Conservation*, **7**, 1069–1079.

Hairston NG, Ellner SP, Geber MA, Yoshida T, and Fox JA (2005) Rapid evolution and the convergence of ecological and evolutionary time. *Ecology Letters*, **8**, 1114–1127.

Hairston NG, Smith FE, and Slobodkin LB (1960) Community structure, population control, and competition. *American Naturalist*, **94**, 421–425.

Haldane JBS (1956) The relation between density regulation and natural selection. *Proceedings of the Royal Society of London B*, **145**, 306–308.

Hale CM, Frelich LE, and Reich PB (2006) Changes in hardwood forest understory plant communities in response to European earthworm invasions. *Ecology*, **87**, 1637–1649.

Halpern BS, Walbridge S, Selkoe KA, Kappel CV, Micheli F, D'Agrosa C, Bruno JF, Casey KS, Ebert C, Fox HE, Fujita R, Heinemann D, Lenihan HS, Madin EMP, Perry MT, Selig ER, Spalding M, Steneck R, and Watson R (2008) A global map of human impact on marine ecosystems. *Science*, **319**, 948–952.

Hamer A, Land S, and Mahoney M (2002) The role of introduced mosquitofish (*Gambusi holbrooki*) in excluding the native green and golden bell frog (*Litoria aurea*) from original habitats in south-eastern Australia. *Oecologia*, **132**, 445–452.

Hamilton MA, Murray BR, Cadotte MW, Hose GC, Baker AC, Harrris CJ, and Licari, D. (2005) Life-history correlates of plant invasiveness at regional and continental scales. *Ecology Letters*, **8**, 1066–1074.

Handley RJ, Steinger T, Treier UA, and Müller-Schärer H (2008) Testing the evolution of increased competitive ability (EICA) hypothesis in a novel framework. *Ecology*, **89**, 407–417.

Hansen DM Olesen JM, and Jones CG (2002) Trees, birds and bees in Mauritius: exploitative competition between introduced honey bees and endemic nectarivorous birds? *Journal of Biogeography*, **29**, 721–734 .

Harding JM (2003) Predation by blue crabs, *Callinectes sapidus*, on rapa whelks, *Rapana venosa*: possible natural controls for an invasive species? *Journal of Experimental Marine Biology and Ecology*, **297**, 161–177.

Harper JL (1965) Establishment, aggression and cohabitation in weedy species. In HG Baker and GL Stebbins, ed. *The genetics of colonizing* species, pp. 243–268. Academic Press, New York.

Harper JL (1977) *Population biology of plants*. Academic Press, London.

Harris DB and Macdonald DW (2007) Interference competition between introduced black rats and endemic Galápagos rice rats. *Ecology*, **88**, 2330–2344.

Harrison S (2008) Commentary on Stohlgren *et al.* (2008) the myth of plant species saturation. *Ecology Letters*, **11**, 322–324.

Harrison S, Grace JB, Davies KF, Safford HD, and Viers JH (2006) Invasion in a diversity hotspot: exotic cover and native richness in the Californian serpentine flora. *Ecology*, **87**, 695–703.

Hastings A (1980) Disturbance, coexistence, history, and competition for space. *Theoretical Population Biology*, **18**, 363–373.

Hastings A, Byers JE, Crooks JA, Cuddington K, Jones CG, Lambrinos JG, Talley TS, and Wilson WG (2007) Ecosystem engineering in space and time. *Ecology Letters*, **10**, 153–164.

Hastings A, Cuddington K, Davies KF, Dugaw CJ, Elmendorf S, Freestone A, Harrison S, Holland M, Lambrinos J, Malvadkar U, Melbourne BA, Moore K, Taylor C, and Thomson D (2005) The spatial spread of invasions: new developments in theory and evidence. *Ecology Letters*, **8**, 91–101.

Hastwell GT, Daniel AJ, and Vivian-Smith G (2008) Predicting invasiveness in exotic species: do subtropical native and invasive exotic plants differ in their growth responses to macronutrients? *Diversity and Distribution*, **14**, 243–251.

Hastwell GT and Panetta FD (2005) Can differential responses to nutrients explain the success of environmental weeds? *Journal of Vegetation Science*, **16**, 77–84.

Hattingh J (2001) Human dimensions of invasive alien species in philosophical perspective: towards an ethics of conceptual responsibility. In JA McNeely, ed. *The great reshuffling: human dimensions of alien invasive species*, pp. 183–194. IUCN, Gland, Switzerland.

Haubensak KA (2001) Invasion and impacts of nitrogen-fixing shrubs *Genista monspessulana* and *Cytisus scoparius* in grasslands of Washington and Coastal California. Ph.D thesis, University of California, Berkeley, CA.

Hauser MD (2006) *Moral minds: how nature designed our universal sense of right and wrong*. Harper Collins, New York.

Havel JE, Lee CE, and Vander Zanden MJ (2005a) Do reservoirs facilitate invasions into landscapes? *BioScience*, **55**, 518–252.

Havel JE and Medley KA (2006) Biological invasions across spatial scales: intercontinental, regional, and local dispersal of cladoceran zooplankton. *Biological Invasions*, **8**, 459–473.

Havel JE, Shurin JB, and Jones JR (2005b) Environmental limits to a rapidly spreading exotic cladoceran. *EcoScience*, **12**, 376–385.

Hawkes CV, Wren IF, Herman DJ, and Firestone MK (2005) Plant invasion alters nitrogen cycling bymodifying the soil nitrifying community. *Ecology Letters*, **8**, 976–985.

Hays WST and Conant S (2007) Impact of the small Indian mongoose (*Herpestes javanicus*) (Carnivora: Herpestidae) on native vertebrate populations in areas of introduction. *Pacific Science*, **61**, 3–16.

Hector A and Bagchi R (2007) Biodiversity and ecosystem multifunctionality. *Nature*, **448**, 188–190.

Hedge SG, Nason JD, Clegg JM, and Ellstrand NC (2006) The evolution of California's wild radish has resulted in the extinction of its progenitors. *Evolution*, **60**, 1187–1197.

Heeney JL (2006) Zoonotic viral diseases and the frontier of early diagnosis, control and prevention. *Journal of Internal Medicine*, **260**, 399–408.

Heger T and Trepl L (2003) Predicting biological invasions. *Biological Invasions*, **5**, 313–321.

Hejný S and Lhotská M (1964) Zu der Art der Ausbreitung von *Bidens frondosa* L. in die Teichgebiete der Tschechoslowakei. *Preslia*, **36**, 416–421.

Helford RM (2000) Constructing nature as constructing science: expertise, activist science, and public conflict in the Chicago wilderness. In PH Gobster and RB Hull, ed. *Restoring nature: perspectives from the social sciences and humanities*, pp. 119–142. Island Press, Washington DC.

Henderson S, Dawson TP, and Whittaker RJ (2006) Progress in invasive plants research. *Progress in Physical Geography*, **30**, 25–46.

Hendriks AJ, Pieters H, and de Boer J (1998) Accumulation of metals, polycyclic (halogenated) aromatic hydrocarbons, and biocides in zebra mussel and eel from the Rhine and Meuse Rivers. *Environmental Toxicology and Chemistry*, **17**, 1885–1898.

Heneghan L, Clay C, and Brundage C (2002) Observations on the initial decomposition rates and faunal colonization of native and exotic plant species in a urban forest fragment. *Ecological Restoration*, **20**, 108–111.

Herben T, Mandak B, Bimova K, and Muzbergova Z (2004) Invasibility and species richness of a community: a neutral model and a survey of published data. *Ecology*, **85**, 3223–3233.

Hierro JL, Vellarreal D, Eren O, Graham JM, and Callaway RM (2006) Disturbance facilitates invasion: the effects are stronger abroad than at home. *American Naturalist*, **168**, 144–156.

Hillebrand H, Watermann F, Karez R, and Berninger UG (2001) Differences in species richness patterns between unicellular and multicellular organisms. *Oecologia*, **126**, 114–124.

Hoagland P and Jin D (2006) Science and economics in the management of an invasive species. *BioScience*, **56**, 931–935.

Hobbs RJ, Arico S, Aronson J, Baron JS, Bridgewater P, Cramer VA, Epstein PR, Ewel JJ, Klink CA, Lugo AE, Norton D, Ojima D, Richardson DM, Sanderson EW, Valladares F, Vilà M, Zamora R, and Zobel M (2006) Novel ecosystems: theoretical and management aspects of the new ecological world order. *Global Ecology and Biogeography*, **15**, 1–7.

Hobbs RJ and Huenneke LF (1992) Disturbance, diversity and invasion: implications for conservation. *Conservation Ecology*, **6**, 324–337.

Hobbs RJ and Mooney HA (2005) Invasive species in a changing world: the interactions between global change and invasives. In HA Mooney, RN Mack, JA McNeely, LE Neville, PJ Schei, and JK Waage, ed. *Invasive alien species: a new synthesis*, pp. 310–331. Island Press, Washington DC.

Hoddle MS (2004) Restoring balance: using exotic species to control invasive exotic species. *Conservation Biology*, **18**, 38–49.

Hodges K (2008) Defining the problem: terminology and progress in ecology. *Frontiers in Ecology and the Environment*, **6**, 35–42.

Holden C (2007) Beetle battles. *Science*, **318**, 25.

Holdgate MW (1986) Summary and conclusions: characteristics and consequences of biological invasions. *Philosophical Transactions of the Royal Society B*, **314**, 733–742.

Holdsworth AR, Frelich LE, and Reich (2007) Effects of earthworm invasion on plant species richness in northern hardwood forests. *Conservation Biology*, **21**, 997–1008.

Holdsworth DK and Mark AF (1990) Water and nutrient input: output budgets: effects of plant cover at seven sites in upland snow tussock grasslands of Eastern and Central Otago, New Zealand. *Journal of the Royal Society of New Zealand*, **20**, 1–24.

Holland EM (2007) The risks and advantages of framing science (letter) *Science*, **317**, 1168.

Holling CS, ed. (1978) *Adaptive environmental assessment and management*. John Wiley and Sons, London.

Holmgren M (2002) Exotic herbivores as drivers of plant invasion and switch to ecosystem alternative states. *Biological Invasions*, **4**, 25–33.

Holt RD (2005) On the integration of community ecology and evolutionary biology: historical perspectives and current prospects. In K. Cuddington and B Beisner, ed. *Ecological paradigms lost: routes of theory change*, pp. 235–271. Elsevier Academic Press, Amsterdam.

Holt RD, Barfield M, and Gomulkiewicz R (2005) Theories of niche conservatism and evolution. In DF Sax, JJ Stachowicz, and SD Gaines, ed. *Species invasions. Insights into ecology, evolution, and biogeography*, pp. 259–290. Sinauer Associates, Sunderland, MA.

Holt RD and Keitt T (2000) Alternative causes for range limits: a metapopulation perspective. *Ecology Letters*, **3**, 41–47.

Holt RD and Keitt T (2005) Species borders: a unifying theme in ecology. *Oikos*, **108**, 3–6.

Holub J and Jirásek V (1967) Zur vereinheitlichung der terminologie in der phytogeographie. *Folia Geobotanica and Phytotaxonomica*, **2**, 69–113.

Holway DA and Suarez AV (1999) Animal behavior: an essential component of invasion biology. *Trends in Ecology and Evolution*, **14**, 328–330.

Holyoak M, Leibold MA, Mouquet NM, Holt RD, and Hoopes MF, ed. (2005) *Metacommunities: spatial dynamics and ecological communities*. University of Chicago Press, Chicago, IL.

Hoppe KN (2002) *Teredo navalis*—the cryptogenic shipworm. In E Leppäkoski, S Gollasch, and S Olenin, ed. *Invasive aquatic species in Europe. Distribution, impacts and management*, pp. 116–120. Kluwer Academic Publishers, Dordrecht.

Horton DR, Capinera JL, and Chapman PL (1988) Local differences in host use by two populations of the Colorado potato beetle. *Ecology*, **69**, 823–831.

Houlahan JE and Findlay CS (2004) Effect of invasive plant species on temperate wetland plant diversity. *Conservation Biology*, **18**, 1132–1138.

Howard LO (1893) *The fly weevil*. Address at the Farmers' Institute, Manassas, VA.

Howard LO (1897a) The spread of land species by the agency of Man: with especial reference to insects. *Proceedings of the American Association for the Advancement of Science*, **46**, 3–26.

Howard LO (1897b) Danger of importing insect pests. *Yearbook of Department of Agriculture*, 529–552.

Howarth DG and Baum DA (2005) Genealogical evidence of homoploid hybrid speciation in an adaptive radiation of *Scaevola* (Goodeniaceae) in the Hawaiian Islands. *Evolution*, **59**, 948–961.

Hubbell SP (2001) *The unified neutral theory of biodiversity and biogeography*. Princeton University Press, Princeton.

Huffaker CB (1951) The return of native perennial bunchgrass following the removal of Klamath weed (*Hypericum perforatum* L.) by imported beetles. Ecology, **32**, 443–458.

Hull BR and Robertson DP (2000) The language of nature matters: we need a more public ecology. In PH Gobster, R Bruce, ed. *Restoring nature: perspectives from the social sciences and humanities*, pp 97–118. Island Press, Washington, DC.

Hulme PE (2003) Biological invasions: winning the science battles but losing the conservation war? *Oryx*, **37**, 178–193.

Hulme P, Bacher S, Kenis M, Klotz S, Kühn I, Minchin D, Nentwig W, Olenin S, Panov V, Pergl J, Pyšek, Roques A, Sol D, Solarz W, and Vilà M (2008) Grasping at the routes of biological invasions: a framework for integrating pathways into policy. *Journal of Applied Ecology*, **45**, 403–414.

Hülsmann N and Galil BS (2002) Protists—a dominant component of the ballast-transported biota. In E Leppäkoski, S Gollasch, and S Olenin, ed. *Invasive aquatic species in Europe. Distribution, impacts and management*, pp. 20–26. Kluwer Academic Publishers, Dordrecht.

Humboldt A von (1850) *Views of nature: or contemplations on the sublime phenomena of creation*. Translated by EC Otté and HG Bohn. Henry G. Bohn, London.

Huston MA (1994) *Biological diversity*. Cambridge University Press, Cambridge.

Huston MA (1997) Hidden treatments in ecological experiments: re-evaluating the ecosystem function of biodiversity. *OecologiaI*, **108**, 449–460.

Huston MA (2004) Management strategies for plant invasions: manipulating productivity, disturbance, and competition. *Diversity and Distributions*, **10**, 167–178.

Huston MA and DeAngelis DL (1994) Competition and coexistence: the effects of resource transport and supply rates. *American Naturalist*, **144**, 954–977.

Hutchinson GE (1957). Concluding remarks. *Cold Spring Harbour Symposium on Quantitative Biology*, **22**, 415–427.

Hutchinson GE (1959) Homage to Santa Rosalia, or why are there so many kinds of animals? *The American Naturalist*, **93**, 145–159.

Inderjit (2005) *Invasive plants: ecological and agricultural aspects*. Birkhäuser Verlag, Basel, Switzerland.

Ingold T (2000) *The perceptions of the environment: essays in livelihood, dwelling and skill*. Routledge, London.

Innes J and Barker G (1999) Ecological consequences of toxin use for mammalian pest control in New Zealand: an overview. *New Zealand Journal of Ecology*, **23**, 111–127.

IUCN (2000) *IUCN guidelines for the prevention of biodiversity loss caused by alien invasive species*. Species Survival Commission, IUCN, Gland, Switzerland.

Ivors K, Garbelotto M, Vries IDe, Ruyter-Spira C, Hekkert BTe, Rosenzweig N, and Bonants P (2006) Microsatellite markers identify three lineages of *Phytophthora ramorum* in US nurseries, yet single lineages in US forest and European nursery populations. *Molecular Ecology*, **15**, 1493–1505.

James CS, Eaton JW, and Hardwick K (2006) Responses of three invasive aquatic macrophytes to nutrient enrichment do not explain their observed field displacements. *Aquatic Botany*, **84**, 347–353.

Janzen, DH (1985) On ecological fitting. *Oikos*, **45**, 308–310.

Jarnevich CS, Stohlgren TJ, Barnett D, and Kartesz J (2006) Filling in the gaps: modelling native species richness and invasions using spatially incomplete data. *Diversity and Distributions*, **12**, 511–515.

Jazdzewski K and Konopacka A (2002) Invasive Ponto-Caspian species in waters of the Vistual and Oder basins and the southern Baltic Sea. In E Leppäkoski, S Gollasch, and S Olenin, ed. *Invasive aquatic species in Europe. Distribution, impacts and management*, pp. 384–398. Kluwer Academic Publishers, Dordrecht.

Jehl JR and Everett WT (1985) History and status of the avifauna of Isla Guadalupe Mexico. *Transactions of the San Diego Society of Natural History*, **20**, 313–336.

Jehlík V and Slavík B (1968) Beitrag zum Erkennen des Verbreitungscharakters der Art *Bunias orientalis* L. in der Tschechoslowakei. *Preslia*, **40**, 274–293.

Jeschke JM and Strayer DL (2005) Invasion success of vertebrates in Europe and North America. *Proceedings of the National Academy of Sciences, USA*, **102**, 7198–7202.

Jewett EB, Hines AH, Ruiz GM (2005). Epifaunal disturbance by periodic low levels of dissolved oxygen: native vs invasive species response. *Marine Ecology*, **304**, 31–44.

Jewett T (2005) The agronomist of Monticello. *The Early America Review*, **6** (on-line journal: http://www.earlyamerica.com/review/2005_summer_fall/agronomist.htm).

Jiang L and Morin PJ (2004) Productivity gradients cause positive diversity-invasibility relationships in microbial communities. *Ecology Letters*, **7**, 1047–1057.

Jiguet F, Julliard R, Thomas CD, Dehorter O, Newson SE, and Couvet D (2006) Thermal range predicts bird population resilience to extreme high temperatures. *Ecology Letters*, **12**, 1321–1330.

John St MG, Wall DH, and Hunt HW (2006) Are soil mite assemblages structured by the identity of native and invasive alien grasses? *Ecology*, **87**, 1314–1324.

Johnson BE and Cushman JH (2007) Influence of a large herbivore reintroduction on plant invasions and community composition in a California grassland. *Conservation Biology*, **21**, 515–526.

Johnson CN, Isaac JL, and Fisher DO (2007) Rarity of top predator triggers continent-wide collapse of mammal prey: dingoes and marsupials in Australia. *Proceedings of the Royal Society B*, **274**, 341–346.

Johnson DM, Liebhold AM, Tobin PC, and Bjørnstad ON (2006) Allee effects and pulsed invasion by the gypsy moth. *Nature*, **444**, 361–3.

Johnstone IM (1986) Plant invasion windows: a time-based classification of invasion potential. *Biological Review*, **61**, 369–394.

Johnson LE and Padilla DK (1996) Geographic spread of exotic species: ecological lessons and opportunities from the invasion of the zebra mussel, *Dreissena polymorpha*. *Biological Conservation*, **78**, 23–33.

Johnson MTJ and Stinchcombe JR (2007) An emerging synthesis between community ecology and evolutionary biology. *Trends in Ecology and Evolution*, **22**, 250–257.

Jones CG and Callaway RM (2007) The third party. *Journal of Vegetation Science*, 18, 771–776.

Jones CG, Lawton JH, and Shachak M. (1994) Organisms as ecosystem engineers. *Oikos*, **69**, 373–386.

Jones KE, Patel NG, Levy MA, Storeygard A, Balk D,Gittleman JL, and Daszak P (2008) Global trends in emerging infectious diseases. *Nature*, **451**, 990–993.

Jørgensen RB, Andersen B, Hauser TP, Landbo L, Mikkelsen TR, and Østergård H (1998) Introgression of crop genes from oilseed rape (*Brassica napus*) to related wild species: an avenue for the escape of engineered genes. *Acta Horticulturae*, **459**, 211–217.

Juliano SA and Lounibos LP (2005) Ecology of invasive mosquitoes: effects on resident species and on human health. *Ecology Letters*, **8**, 558–574.

Julien MH, ed. (1992). *Biological control of weeds: a world catalogue of agents and their target weeds*, 3rd edition. CAB International, Oxon, UK.

Karatayev AY, Burlakova LE, and Padilla DK (2002) Impacts of zebra mussels on aquatic communities and their role as ecosystem engineers. In E Leppäkoski, S Gollasch, and S Olenin, ed. *Invasive aquatic species in Europe. Distribution, impacts and management*, pp. 433–446. Kluwer Academic Publishers, Dordrecht.

Kaufman L (1992) Catastrophic change in species-rich freshwater ecosystems: the lessons of Lake Victoria. *BioScience*, **42**, 846–858.

Keane RM and Crawley MJ (2002) Exotic plant invasions and the enemy release hypothesis. *Trends in Ecology and Evolution*, **17**, 164–170.

Keitt TH, Lewis MA, and Hold RD (2001) Allee effects, invasion pinning, and species' borders. *The American Naturalist*, **157**, 203–216.

Keitt BS, Wilcox CA, Teershy BR, Croll DA, and Donlan CJ (2002) The effect of feral cats on the population viability of black-vented shearwaters (*Puffinus opisthomelas*) on Natividad Island, Mexico. *Animal Conservation*, **5**, 217–223.

Keller EF (2007) A clash of two cultures. *Nature*, **445**, 603.

Keller RP and Lodge DM (2007) Species invasions from commerce in live aquatic organisms: problems and possible solutions. *BioScience*, **57**,428–436.

Keller RP, Lodge DM, and Finnoff DC (2007) Risk assessment for invasive species produces net bioeconomic benefits. *Proceedings of the National Academy of Sciences*, USA, **104**, 203–207.

Keller RP, Frang K, and Lodge DM (2008). Preventing the spread of invasive species: economic benefits of intervention guided by ecological predictions. *Conservation Biology*, **22**, 80–88.

Kellogg CA and Griffin DW (2006) Aerobiology and the global transport of desert dust. *Trends in Ecology and Evolution*, **21**, 638–644.

Kelly D, Ladley JJ, and Robertson AW (2007) Is the pollen-limited mistletoe *Peraxilla tetrapetala* (Loranthaceae) also seed limited? *Austral Ecology*, **32**, 850–857.

Kendall KC and Arno SA (1990) Whitebark pine: an important but endangered wildlife resource. In *Proceedings of the Whitebark Pine Symposium*, pp. 264–273. USDA Forest Service General Tech. Rep. INT-270.

Kepler CB (1967) Polynesian rat predation on nesting Laysan Albatrosses and other Pacific seabirds. *Auk*, **84**, 426–430.

Kessler CC (2001) Eradication of feral goats and pigs and consequences for other biota on Sarigan Island, Commonwealth of the Northern Mariana Islands. In CR Veitch and MN Clout, ed. *Turning the tide: the eradication of invasive species*, pp. 132–140. IUCN SSC Invasive Species Specialist Group. IUCN, Gland, Switzerland and Cambridge, UK.

Kideys AE (2002) The comb jelly *Mnemiopsis leidyi* in the Black Sea. In E Leppäkoski, S Gollasch, and S Olenin, ed. *Invasive aquatic species in Europe. Distribution, impacts and management*, pp. 56–61. Kluwer Academic Publishers, Dordrecht.

Kiesecker JM and Blaustein AR (1997) Population differences in responses of red-legged frogs (*Rana aurora*) to introduced bullfrogs. *Ecology*, **78**, 1752–1760.

Kilpatrick AM (2006) Facilitating the evolution of resistance to avian malaria (*Plasmodium relictum*) in Hawaiian birds. *Biological Conservation*, **128**, 475–485.

King C (1984) *Immigrant killers*. Oxford University Press, Auckland, NZ.

King C, ed. (1990) *The handbook of New Zealand mammals*. Oxford University Press, Aukland, NZ.

Kinlan BP and Gaines SD (2003) Propagule dispersal in marine. and terrestrial environments: a community perspective. *Ecology*, **84**, 2007–2020.

Kinlan BP and Hastings A (2005). Rates of population sperad and geographic range expansion: what exotic species tell us. In DF Sax, JJ Stachowicz, and SD Gaines, ed. *Species invasions. Insights into ecology, evolution, and biogeography*, pp. 381–419. Sinauer Associates, Sunderland, MA.

Kiritani K and Yamamura K (2003) Exotic insects and their pathways for invasion. In GM Ruiz and JT Carlton, ed. *Invasive species: vectors and management strategies*, pp. 44–67. Island Press, Washington, DC.

Klironomos JN (2002) Feedback with soil biota contributes to plant rarity and invasiveness in communities. *Nature*, **417**, 67–70.

Knick ST, Dobkin DS, Rotenberry JT, Schroeder MA,Vander Haegen WM, and Van Riper C III (2003) Teetering on the edge or too late? Conservation and research issues for avifauna of sagebrush habitats. *The Condor*, **105**, 611–634.

Knight KS, Kurylo JS, Endress AG, Stewart JR, and Reich PB (2007) Ecology and ecosystem impacts of common buckthorn (*Rhamnus cathartica*): a review. *Biological Invasions*, **9**, 925–937.

Knops JMH, Tilman D, Haddad NM, Naeem S, Mitchell CE, Haarstad J, Ritchie ME, Howe, KM, Reich PB, Siemann E, and Groth J (1999) Effects of plant species richness on invasion dynamics, disease outbreaks, insect abundances and diversity. *Ecology Letters*, **2**, 286–293.

Knorr-Cetina K (1999) *Epistemic cultures: how the sciences make knowledge*. Harvard University Press, Cambridge.

Kobayakawa K, Kobayakawa R, Matsumoto K, Oka Y, Imai T, Ikawa M, Okabe M, Ikeda T, Itohara S, Kikusui T, Mori K, and Sakano H. (2007) Innate versus learned odour processing in the mouse olfactory bulb. *Nature*, **450**, 503–508.

Kohler A. and Sukopp H (1964) Über die Gehölzentwicklung auf Berliner Trümmerstandorten. Zugleich ein Beitrag zum Studium neophytischer Holzarten. *Berichte der Deutschen Botanischen Gesellschaft*,**76**, 389–407.

Kolar CS and Lodge DM (2001) Progress in invasion biology: predicting invaders. *Trends in Ecology and Evolution*, **16**, 199–204.

Kolar CS and Lodge DM (2002) Ecological predictions and risk assessment for alien fishes in North America. *Science*, **298**, 1233–1236.

Kolb A and Alpert P (2003) Effects of nitrogen and salinity on growth and competition between a native grass and an invasive congener. *Biological Invasions*, **5**, 229–238.

Kolbe J, Glor RE, Schettino LR, and Chamizo A (2004) Genetic variation increases during biological invasion by a Cuban lizard. *Nature*, **431**, 177–184.

Kolbe JJ and Losos JB (2005) Hind-limb length plasticity in *Anolis carolinensis*. *Journal of Herpetology*, **39**, 674–678.

Köndgen S, Kühl H, N'Goran PK, Walsh PD, Schenk S, Ernst N, Biek R, Formenty P, Mätz-Rensing K, Schweiger B, Junglen S, Ellerbrok H, Nitsche A, Briese T, Lipkin WI, Pauli G, Boesch C, and Leendertz FH (2008) Pandemic human viruses cause decline of endangered great apes. *Current Biology*, **18**, 260–264.

Kondoh M (2006) Contact experience, alien-native interactions, and their community consequences: a theoretical consideration on the role of adaptation in biological invasion. In MW Cadotte, SM McMahon, and T Fukami, ed. *Conceptual ecology and invasions biology: reciprocal approaches to nature*, pp. 225–242. Springer, Dordrecht, The Netherlands.

Kornás J (1968b) Geograficzno-historyczna klasyfikacja roslin synantropijnych. A geographical-historical classification of synanthropic plants. In JB Faliński, ed. *Synantropizacja szaty roslinnej. I. Neofityzm i apofityzm w szacie roslinnej Polski. Materialy Sympozjum w Nowogrodzie. Mater. Zakl. Fitosoc. Stos. UW*, **25**, 33–41.

Kornberg H and MH Williamson, ed. (1987) *Quantitative aspects of the ecology of biological invasions*. The Royal Society, London.

Körner C (2007) The use of 'altitude' in ecological research. *Trends in Ecology and Evolution*, **22**, 569–574.

Koskinen MT, Haugen TO, and Primmer CR (2002) Contemporary fisherian life-history evolution in small salmonid populations. *Nature*, **419**, 826–830.

Kourtev PS, Ehrenfeld JG, and Haggblom M (2002) Exotic plant species alter the microbial community structure and function in the soil. *Ecology*, **83**, 3152–3166.

Kraus F (2003) Invasion pathways for terrestrial vertebrates. In GM Ruiz and JT Carlton, ed. *Invasive species: vectors and management strategies*, pp. 68–92. Island Press, Washington, DC.

Krieger MJB and Ross KG (2002) Identification of a major gene regulating complex social behavior. *Science*, **295**, 328–332.

Křivánek M and Pyšek P (2006) Predicting invasions by woody species in a temperate zone: a test of three risk assessment schemes in the Czech Republic (Central Europe). *Diversity and Distributions*, **12**, 319– 327.

Křivánek M, Pyšek P, and Jarošik V (2006) Planting history and propagule pressure as predictors of invasion by woody species in a temperate region. *Conservation Biology*, **20**, 1487–1498.

Kueffer C, Schumacher E, Fleischmann K, Edwards PJ, and Dietz H (2007) Strong below-ground competition shapes tree regeneration in invasive *Cinnamomum verum* forests. *Journal of Ecology*, **95**, 273–282.

Kulmatiski A (2006) Exotic plants establish persistent communities. *Plant Ecology*, **187**, 261–275.

Kumar S, Stohlgren TJ, and Chong GW (2006) Effects of spatial heterogeneity on native and non-native plant species richness. *Ecology*, **87**, 3186–3199.

Labra FA, Abades SR, and Marquet PA (2005) Distribution and abundance: scaling patterns in exotic and native bird species. In DF Sax, JJ Stachowicz, and SD Gaines, ed. *Species invasions. Insights into ecology, evolution, and biogeography*, pp. 421–446. Sinauer Associates, Sunderland, MA.

LaDeau SL, Kilpatrick AM, and Marra PP (2007) West Nile virus emergence and large-scale declines of North American bird populations. *Nature*, **447**, 710–713.

Lafferty KD, Smith KF, Torchin ME, Dobson AP, and Kuris AM (2005) The role of infectious disease in natural communities: what introduced species tell us. In DF Sax, JJ Stachowicz, and SD Gaines, ed. *Species invasions. Insights into ecology, evolution, and biogeography*, pp. 111–134. Sinauer Associates, Sunderland, MA.

Langerhans RB, Gifford ME, and Joseph EO (2007) Ecological speciation in *Gambusia* fishes. *Evolution*, **61**, 2056–2074.

Larson BMH (2005) The war of the roses: demilitarizing invasion biology. *Frontiers in Ecology and the Environment*, **3**, 495–500.

Larson BMH (2007) An alien approach to invasive species: objectivity and society in invasion biology. *Biological Invasions*, **9**, 947–956.

Lawton JH (1999) Are there general laws in ecology? *Oikos*, **84**, 177–192.

Lawton JH (2000) Community ecology in a changing world. In O Kinne, ed. *Excellence in ecology series*, Vol 11. International Ecology Institute, Oldendorf/Luhe, Germany.

Lawton JH (2007) Ecology, politics and policy. *Journal of Applied Ecology*, **44**, 465–474.

Lee CE (2002) Evolutionary genetics of invasive species. *Trends in Ecology and Evolution*, **17**, 386–391.

Lehtonen H (2002) Alien freshwater fishes of Europe. In E Leppäkoski, S Gollasch, and S Olenin, ed. *Invasive aquatic species in Europe. Distribution, impacts and management*, pp. 153–161. Kluwer Academic Publishers, Dordrecht.

Leibold MA, Holyoak M, Mouquet N, Amarasekare P, Chase JM, Hoopes MF, Holt RD, Shurin JB, Law R, Tilman D, Loreau M, and Gonzalez A (2004) The metacommunity concept: a framework for multi-scale community ecology. *Ecology Letters*, **7**, 601–613.

Leimu R and Koricheva J (2005) What determines the citation frequency of ecological papers? *Trends in Ecology and Evolution*, **20**, 28–32.

Leishman MR, Haslehurst T, Ares A, and Baruch Z (2007) Leaf trait relationships of native and invasive plants: community- and global-scale comparisons. *New Phytologist*, **176**, 635–643.

Leishman MR and Thomson VP (2005) Experimental evidence for the effects additional water, nutrients and physical disturbance on invasive plants in low fertility Hawkesbury Sandstone soil, Sydney, Australia. *Journal of Ecology*, **93**, 38–49.

Lennon JT, Smith VH, and Dzialowski AR (2003) Invasibility of plankton food webs along a trophic state gradient. *Oikos*, **103**, 191–203.

Lenormand T (2002) Gene flow and the limits to natural selection. *Trends in Ecology and Evolution*, **17**, 183–189.

Lepori F and Hjerdt N (2006) Disturbance and aquatic biodiversity: reconciling contrasting views. *Bioscience*, **56**, 809–818.

Leppäkoski E and Olenin, S (2000a). Xenodiversity of the European brackish water seas: the North American contribution. In J Pederson, J, ed. *Marine bioinvasions: Proceedings of the First National Conference*, 24–27 January 1999, pp. 107–119. Massachusetts Institute of Technology, Cambridge, MA, USA.

Leppäkoski E and Olenin S (2000b) Non-native species and rates of spread: lessons from the brackish Baltic Sea. *Biological Invasions*, **2**, 151–163.

Leppäkoski E and Olenin, S, and Gollasch S (2002) The Baltic Sea: a field laboratory for invasion biology. In E Leppäkoski, S Gollasch, and S Olenin, ed. *Invasive aquatic species in Europe. Distribution, impacts and management*, pp. 253–259. Kluwer Academic Publishers, Dordrecht.

Lertzman KP, Sutherland GD, Inselberg A, and Saunders SC (1996) Canopy gaps and the landscape mosaic in a coastal temperate rain forest. *Ecology*, **77**, 1254–1270.

Lester SE, Ruttenberg BI, Gaines SD, and Kinlan BP (2007) The relationship between dispersal ability and geographic range size. *Ecology Letters*, **10**, 745–758.

Leung B, Drake JM, and Lodge DM (2004) Predicting invasions: propagule pressure and the gravity of Allee effects. *Ecology*, **85**, 1651–1660.

Leung B, Lodge DM, Finnoff DA, Shogren A, Lewis MA, and Lamberti G (2002) An ounce of prevention or a pound of cure: bioeconomic risk analysis of invasive species. *Proceedings of the Royal Society B*, **269**, 2407–2413.

Leveque C (1997) Introductions of exotic fish species in tropical freshwaters: purposes and consequences. *Bulletin Francais De La Peche Et De La Pisciculture*, **344–345**, 79–91.

Levin LA, Neira C, and Grosholz ED (2006) Invasive cordgrass modifies wetland trophic function. *Ecology*, **87**, 419–432.

Levine JM (2000) Species diversity and biological invasions: relating local process to community pattern. *Science*, **288**, 852–854.

Levine JM, Adler PB, and Yelenik SG (2004) A meta-analysis of biotic resistance to exotic plant invasions. *Ecology Letters*, **7**, 975–989.

Levine JM and D'Antonio CM (1999) Elton revisited: a review of evidence linking diversity and invasibility. *Oikos*, **87**, 15–26.

Levine JM and D'Antonio CM (2003) Forecasting biological invasions with.increasing international trade. *Conservation Biology*, **17**, 322–326.

Levins R (1966) The strategy of model building in population biology. *American Scientist*, **54**, 421–431.

Leyval C, Joner E J, del Val C, and Haselwandter K (2002) Potential of arbuscular mycorrhizal fungi for bioremediation. In S Gianinazzi, H Schüepp, JM Barea, and K Haselwandter, ed. *Mycorrhiza technology in agriculture: from genes to bioproducts*, pp. 175–186. Birkhäuser Verlag, Basel.

Lindqvist OV and Huner JV (1999) Life history characteristics of crayfish: What makes some of them good colonizers? In F Gherardi F and DM Holdich, ed. *Crayfish in Europe as alien species. How to make the best of a bad situation?* pp. 23–30. AA Balkema, Rotterdam.

Lippert KA, Gunn JM, and Morgan GE (2007) Effects of colonizing predators on yellow perch (*Perca flavescens*) populations in lakes recovering from acidification and metal stress. *Canadian Journal of Fisheries and Aquatic Sciences*, **64**, 1413–1428.

Litton AF, Creighton M, Sandquist DR, and Cordell S (2006) The effects of non-native grass invasion on aboveground carbon pools and tree population structure in a dry tropical forest of Hawaii. *Forest Ecology and Management*, **231**, 105–113.

Liu H, Stiling P, and Pemberton RW (2007a) Does enemy release matter for invasive plants? Evidence from a comparison of insect herbivore damage among invasive, non-invasive and native congeners. *Biological Invasions*, **9**, 773–781.

Liu S-S, De Barro PJ, Xu J, Luan J-B, Zang L-S, Ruan Y-M, and Wan F-H (2007b) Asymmetric mating interactions drive widespread invasion and displacement in a whitefly. *Science*, **318**, 1769–1772.

Lockwood JL, Cassey P, and Blackburn TM. (2005) The role of propagule pressure in explaining species invasions. *Trends in Ecology and Evolution*, **20**, *223–228.*

Lockwood JL, Hoopes MF, and Marchetti MP (2007) *Invasion ecology.* Blackwell, Oxford.

Lockwood JL and McKinney M, ed. (2001) *Biotic homogenization.* Kluwer Academic/Plenum Publishers, New York.

Lodge DM, Williams S, MacIsaac HJ, Hayes KR, Leung B, Reichard S, Mack RN, Moyle PB, Smith M, Andow DA, Carlton JT, and McMichael A (2006) Biological invasions: recommendations for US policy and management. *Ecological Applications*, **16**, 2035–2054.

Lohrer AM, Chiaroni LD, Hewitt JE, and Thrush SF (2008) Biogenic disturbance determines invasion success in a subtidal soft-sediment system. *Ecology*, **89**, 1299–1307.

Longino HE (1990) *Science as social knowledge: values and objectivity in scientific inquiry*. Princeton University Press, Princeton.

Lonsdale WM (1999) Global patterns of plant invasions and the concept of invasibility. *Ecology*, **80**, 1522–1536.

Loope LL, Hamann O, and Stone CP (1988) Comparative conservation biology of oceanic archipelagoes. *BioScience*, **38**, 272–282.

Losos JB, Warheit KL, and Schoener TW (1997) Adaptive differentiation following experimental island colonization in *Anolis* lizards. *Nature*, **387**, 70–73.

Losos, JB, Creer DA, Glossip D, Goellner R, Hampton A, Roberts G, Haskell N, Taylor P, and Ettling J (2000) Evolutionary implications of phenotypic plasticity in the hindlimb of the lizard *Anolis sagrei*. *Evolution*, **54**, 301–305.

Losos JB, Schoener TW, and Spiller DA (2004) Predator-induced behaviour shifts and natural selection in field-experimental lizard populations. *Nature*, **432**, 505–508.

Lotze, HK, Lenihan HS, Bourque BJ, Bradbury RH, Cooke RG, Kay MC, Kidwell SM, Kirby MX, Peterson CH, and Jackson JBC (2006) Depletion, degradation, and recovery potential of estuaries and coastal seas. *Science*, **312**, 1806–1809.

Louda SM, Arnett AE, Rand TA, and Russell FL (2003) Invasiveness of some biological control insects and adequacy of their ecological risk assessment and regulation. *Conservation Biology*, **17**, 73–82.

Louda SM, Kendall D, Connor J, and Simberloff D (1997) Ecological effects of an insect introduced for the biological control of weeds. *Science*, **277**, 1088–1090.

Lovett GM, Canham CD, Arthur MA, Weathers KC, and Fitzhugh RD (2006) Forest ecosystem responses to exotic pests and pathogens in eastern North America. *Bioscience*, **56**, 395–405.

Lugo AE (1994) Maintaining an open mind on exotic species. In Meffe GK and Caroll R, ed. *Principles of conservation biology*, pp. 218–220. Sinauer Associates, Sunderland Massachusetts.

Lugo AE (1997) The apparent paradox of re-establishing species richness on degraded lands with tree monocultures. *Forest Ecology and Management*, **99**, 9–19.

Lugo AE (2004) The outcome of alien tree invasions in Puerto Rico. *Frontiers in Ecology and the Environment*, **2**, 265–273.

Lynch EA and Saltonstall K (2002) Paleological and genetic analyses provide evidence for recent colonization of *Phragmites australis* populations in a Lake Superior wetland. *Wetlands*, **22**, 637–646.

Macdonald DW and Harrington LA (2003) The American mink: the triumph and tragedy of adaptation out of context. *New Zealand Journal of Zoology*, **30**, 421–441.

Macdonald IAW and Jarman ML, ed. (1984) *Invasive alien organisms in the terrestrial ecosystems of the fynbos biome, South Africa*. South African National Scientific Programmes Report No. 85, CSIRO, Pretoria.

Macdonald IAW, Kruger FJ, and Ferrar AA (1986) *The ecology and management of biological invasions in southern Africa*. Oxford University Press, Cape Town.

Macdonald IAW, Loope LL, Usher MB, and Hamann O (1989) Wildlife conservation and the invasion of nature reserves by introduced species: a global perspective. In JA Drake, HA Mooney, F di Castri, RH Groves, F J Kruger, M Rejmanek, and M Williamson, ed. *Biological invasions: a global perspective*, pp. 215–255. John Wiley and Sons, New York.

MacDougall AS (2005) Response of diversity and invasibility to burning in a northern oak savanna. *Ecology*, **86**, 3354–3363.

MacDougall AS and Turkington R (2005) Are invasive species the drivers or passengers of ecological change in highly disturbed plant communities? *Ecology*, **86**, 42–55.

MacDougall AS and Turkington R (2006) Dispersal, competition, and shifting patterns of diversity in a degraded oak savanna. *Ecology*, **87**, 1831–1843.

MacDougall AS and Wilson SD (2007) Herbivory limits recruitment in an old-field seed addition experiment. *Ecology*, **88**, 1105–1111.

Mack RN (1981) Invasion of Bromus tectorum into western North America : an ecological chronicle. *Agroecosystems*, **7**, 145–65.

Mack RN (1996) Predicting the identity and fate of plant invaders: emergent and emerging approaches. *Biological conservation*, **78**, 107–121.

Mack RN (2001) Motivations and consequences of the human dispersal of plants. In JA McNeely, ed. *The great reshuffling: human dimensions of alien invasive species*, pp. 23–34. IUCN, Gland, Switzerland.

Mack RN (2003) Global plant dispersal, naturalization and invasion: pathways, modes and circumstances. In GM Ruiz and JT Carlton, ed. *Invasive species: vectors and management strategies*, pp. 3–30. Island Press, Washington, DC.

Mack MC and D'Antonio CM (1998) Impacts of biological invasions on disturbance regimes. *Trends in Ecology and Evolution*, **13**, 195–198.

Mack RN, Simberloff D, Lonsdale WM, Evans H, Clout M, and Bazzaz FA (2000) Biotic invasions: causes, epidemiology, global consequences and control. *Ecological Applications*, **10**, 689–710.

Mack RN, Von Holle B, and Meyerson L (2007) Assessing the impacts of invasive alien species across multiple spatial scales: the need to work globally and locally. *Frontiers in Ecology and the Environment*, **5**, 217–220.

Maesako Y (1999) Impacts of streaked shearwater (*Calonectris leucomelas*) on tree seedling regeneration in a warm-temperate evergreen forest on Kanmurijima Island, Japan. *Plant Ecology*, **145**, 183–190.

Magurran AE (2007) Species abundance distributions over time. *Ecology Letters*, **10**, 347–354.

Mallet J (2007) Hybrid speciation. *Nature*, **446**, 279–283.

Malmstrom CM, McCullough AJ, Johnson HA, Newton LA, and Borer ET (2005) Invasive annual grasses indirectly increase virus incidence in California native perennial bunchgrasses. *Oecologia*, **145**, 153–164.

Maran T and Henttonen H (1995) Why is the European mink (*Mustela lutreola*) disappearing? A review of the process and hypotheses. *Annales Zoologici Fennici*, **32**, 47–54.

Maran T, Macdonald DW, Kruuk H, Sidorovich V, and Rozhnov VV (1998) The continuing decline of the European mink *Mustela lutreola*: evidence for the intraguild aggression hypothesis. In N Dunstone and M Gorman, ed. *Behaviour and ecology of riparian mammals*, pp. 297–324. Cambridge University Press, Cambridge, UK.

Marchetti MP, Light T, Feliciano J, Armstrong T, Hogan Z, Viers J, and Moyle PB (2001) Homogenization of California's fish fauna through abiotic change. In JL Lockwood and ML McKinney, ed. *Biotic homogenization*, pp. 259–278. Kluwer Academic Publishers, New York.

Marchetti MP, Moyle PB, and Levine R (2004) Invasive species profiling: exploring the characteristics of exotic fishes across invasion stages in California. *Freshwater Biology*, **49**,646–661.

Margolis M, Shogren J, and Fischer C (2005) How trade politics affect invasive species control, *Ecological Economics*, **52**, 305–313.

Mark AF and Dickinson KJM (2008) Maximizing water yield with indigenous non-forest vegetation: a New Zealand perspective. *Frontiers in Ecology and the Environment*, **6**, 25–34.

Maron JL, Elmendorf S, and Vilà M (2007) Contrasting plant physiological adaptation to climate in the native and introduced range. *Evolution*, **61**, 1912–1924.

Maron JL and Marler M (2007) Native plant diversity resists invasion at both low and high resource levels. *Ecology*, **88**, 2651–2661.

Maron JL and Vilà M. (2007) Exotic plants in an altered enemy landscape: effects on enemy resistance In J Kelley and J Tilmon, ed. *Specialization, speciation and radiation: the evolutionary biology of herbivorous insects*, pp. 280–285. University of California Press, Berkeley.

Maron JL, Vilà M, Bommarco R, Elmendorf S, and Beardsley P (2004) Rapid evolution of an invasive plant. *Ecological Monographs*, **74**, 261–280.

Marris E (2008) Bagged and boxed: it's a frog's life. *Nature*, **452**, 394–395.

Martí R and del Moral JC, ed. (2003) *Atlas de las aves reproductoras de España*. Dirección General de Conservación de la Naturaleza. Madrid.

Martin FN and Tooley PW (2003) Phylogenetic relationships of *Phytophthora ramorum* and a *Phytophthora ilicis*-like species associated with sudden oak death in California. *Mycological Research*, **107**, 1379–1391.

Maskell LC, Bullock JM, Smart SM, Thompson K and Hulme PE (2006) *The distribution and vegetation associations of non-native plant species in urban riparian habitat. Journal of Vegetation Science*, **17**, 499–508.

Matarczyk JA, Willis AJ, Vranjic JA, and Ash JE (2002) Herbicides, weeds and endangered species: management of bitou bush (*Chrysanthemoides monilifera* ssp. *rotundata*) with glyphosate and impacts on the endangered shrub, *Pimelea spicata. Biological Conservation*, **108**, 133–141.

Matthews DP and Gonzalez A (2007) The inflationary effects of environmental fluctuations ensure the persistence of sink metapopulations. *Ecology*, **88**, 2848–2856.

Matthews S (2007) *Directed evolution of native species: a new approach to managing invasive species and their impacts.* Senior paper, Department of Biology, Macalester College, Saint Paul, MN.

Mavarez J, Salazar CA, Bermingham E, Salcedo C, Jiggins CD, and Linares M (2006) Speciation by hybridization in *Heliconius* butterflies. *Nature*, **441**, 868–871.

Mayr E (1939) The sex ratio in wild birds. *American Naturalist*, **73**, 156–179.

Mayr E (1963) *Animal species and evolution*. Harvard University Press, Cambridge, MA.

Mazak EJ, MacIsaac HJ, Servos MR, and Hesslein RH (1997) Influence of feeding habits on organochlorine contaminant accumulation in waterfowl on the Great Lakes. *Ecological Applications*, **7**, 1133–1143.

McChesney GJ and Tershy BR (1998) History and status of introduced mammals and impacts to breeding seabirds on the California Channel and Northwestern Baja California Islands. *Colonial Waterbirds*, **21**, 335–347.

McClure MS and Cheah CAS-J (1999) Reshaping the ecology of invading populations of hemlock woolly Adelgid, *Adelges tsugae* (Homoptera: Adelgidae) in eastern North America. *Biological Invasions*, **1**, 247–254.

McDonald R and Urban D (2006) Edge effects on species composition and exotic species abundance in the North Carolina Piedmont. *Biological Invasions*, **8**, 1049–1060.

McDougall KL, Morgan JW, Walsh NG, and Williams RJ (2005) Plant invasions in treeless vegetation of the Australian Alps. *Perspectives in Plant Ecology, Evolution, and Systematics*, **7**, 159–171.

McElreath R, Boyd R, and Richerson PJ (2003) Shared norms can lead to the evolution of ethnic markers. *Current Anthropology*, **44**, 122–129.

McFadyen REC (1998) Biological control of weeds. *Annual Review of Entomology*, **43**, 369–393.

McGill BJ, Enquist BJ, Weiher E, and Westoby M (2006) Rebuilding community ecology from functional traits. *Trends in Ecology and Evolution*, **21**, 178–184.

McIntosh RP (1987) Pluralism in ecology. *Annual Review of Ecology and Systematics*, **18**, 321–341.

McKinney ML and Lockwood JL (2001) Biotic homogenization: a sequential and selective process. In JL Lockwood and ML McKinney, eds. *Biotic Homogenization*, pp. 1–18. Kluwer Academic/Plenum Press, New York.

McKinney ML and JL Lockwood (2005) Community composition and homogenization: evenness and abundance of native and exotic species. In DF Sax, JJ Stachowicz, and SD Gaines, ed. *Species invasions. Insights into ecology, evolution, and biogeography*, pp. 365–381. Sinauer Associates, Sunderland, MA.

McKnight BN (1993) *Biological pollution*. Indiana Academy of Science, Indianopolis.

McLain DK, Moulton MP, and Sanderson JG (1999) Sexual selection and extinction: the fate of plumage-dimorphic and plumage-monomorphic birds introduced onto islands. *Evolutionary Ecology Research*, **1**, 549–565.

McMichael AJ and Bouma A (2000) Global changes, invasive species and human health. In HA Mooney and JR Hobbs, ed. *Invasive species in a changing world*, pp. 191–210. Island Press, Washington DC.

McNeely JA (2005) Human dimensions of alien invasive species. In HA Mooney, RN Mack, JA McNeely, LE Neville, PJ Schei, and JK Waage, ed. *Invasive alien species: a new synthesis*, pp. 285–309. Island Press, Washington, DC.

McNeely JA, Mooney HA, Neville LE, Schei P, and Waage JK, ed. (2001) *A global strategy on invasive alien species.* IUCN Gland, Switzerland and Cambridge, UK.

Meiners SJ (2007) Native and exotic plant species exhibit similar population dynamics during succession. *Ecology*, **88**, 1098–1104.

Meinesz A (1999) *Killer algae, the true tale of biological invasion*. Translated by D Simberloff. The University of Chicago Press, Chicago.

Melbourne BA, Cornell HV, Davies KF, Dugaw CJ, Elmendorf S, Freestone AL, Hall RJ, Harrison S, Hastings A, Holland M, Holyoak M, Lambrinos J, Moore K, and Yokomizo H (2007) Invasion in a heterogeneous world: resistance, coexistence or hostile takeover? *Ecology Letters*, **10**, 77–94.

Mercer KL, Andow DA, Wyse DL, and Shaw RG (2007) Stress and domestication traits increase the relative fitness of crop-wild hybrids in sunflower. *Ecology Letters*, **10**, 383–393.

Mercer KL, Wyse DL, and Shaw RG (2006) Effects of competition on the fitness of wild and crop-wild hybrid sunflower from a diversity of wild populations and crop lines. *Evolution*, **60**, 2044–2055.

Mergeay J, Verschuren D, and de Meester L (2006) Invasion of an asexual American water flea clone throughout Africa and rapid displacement of a native sibling species. *Proceedings of the Royal Society B*, **273**, 2839–2844.

Meyer JR and Kassen R (2007) The effects of competition and predation on diversification in a model adaptive radiation. *Nature*, **446**, 432–435.

Meyerson LA and Mooney HA (2007) Invasive alien species in an era of globalization. *Frontiers in Ecology and the Environment*, **5**, 199–208.

Midgley GFL, Millar HD, Thuiller W, and Booth A (2003) Developing regional and species-level assessments of climate change impacts on biodiversity in the Cape Floristic Region. *Biological Conservation*, **112**, 1–2.

Miller C (2007) Australia's wild dogs under threat. *Frontiers in Ecology and the Environment*, **5**, 401.

Miller RM and Jastrow JD (1992) The application of va mycorrhizae to ecosystem restoration and reclamation. In MF Allen, ed. *Mycorrhizal functioning: an integrative plant-fungal process*, pp. 438–467. Chapman and Hall, New York.

Milton SJ, Wilson JRU, Richardson DM, Seymour CL, Dean WRJ, Iponga DM, and Procheş Ş (2007) Invasive alien plants infiltrate bird-mediated shrub nucleation processes in arid savanna. *Journal of Ecology*, **95**, 648–66.

Minchinton TE (2002) Precipitation during El Niño correlates with increasing spread of *Phragmites australis* in New England, USA, coastal marshes. *Marine Ecology Progress Series*, **242**, 305–309.

Mitchell CE, Agrawal AA, Bever JD, Gilbert GS, Hufbauer RA, Klironomos JN, Maron JL, Morris WF, Parker IM, Power AG, Seabloom EW, Torchin ME, and Vázquez DP (2006) Biotic interactions and plant invasions. *Ecology Letters*, **9**, 726–740.

Mitchell CE and Power AG (2003) Release of invasive plants from fungal and viral pathogens. *Nature*, **421**, 625–627.

MN DNR (Minnesota Department of Natural Resources) (2008). *Invasive species of aquatic plants and wild animals in Minnesota: annual report summary for 2007.* Minnesota Department of Natural Resources, Saint Paul, MN.

Møller AP and Nielson JT (2007) Malaria and risk of predation: a comparative study of birds. *Ecology*, **88**, 871–881.

Mooney HA and Drake JA, ed. (1986) *Ecology of biological invasions of North America and Hawaii*. Springer-Verlag, New York.

Mooney HA, Mack RN, McNeely JA, Neville LE, Schei PJ, and Waage JK, ed. (2005) *Invasive alien species: a new synthesis*. Island Press, Washington DC.

Morrison SA, Macdonald N, Walker K, Lozier L, and Shaw MR (2007) Facing the dilemma at eradication's end: uncertainty of absence and the Lazarus effect. *Frontiers in Ecology and the Environment*, **5**, 271–276.

Moulton MP and Pimm SL (1983) The introduced Hawaiian avifauna: biogeographic evidence for competition. *American Naturalist*, **121**, 669–690.

Moyle PB (1973) Effects of introduced bullfrogs, *Rana catesbeiana*, on the native frogs of the San Joaquin Valley, California. *Copeia*, **1973**, 18–22.

Moyle PB (1986) Fish introductions into North America: Patterns and ecological impact. In HA Mooney and JA Drake, ed. *Ecology of biological invasions of North America and Hawaii*, pp. 27–43. Springer-Verlag, New York.

Moyle PB and Marchetti MP (2006) Predicting invasion success: freshwater fishes in California as a model. *Bioscience*, **56**, 515–524.

Mueller-Dombois D and Spatz G (1975) The influence of feral goats on the lowland vegetation in Hawaii Volcanoes National Park. *Phytocoenologia*, **3**, 1–29.

Muirhead JR and MacIsaac HJ (2005) Development of inland lakes as hubs in an invasion network. *Journal of Applied Ecology*, **42**, 80–90.

Murdoch WW (1969) Switching in general predators: experiments on predator specificity and stability of prey populations. *Ecological Monographs*, **39**, 335–354.

Murphy E and Bradfield P (1992) Change in diet of stoats following poisoning of rats in a New Zealand forest. *New Zealand Journal of Ecology*, **16**, 137–140.

Murray C and Marmorek D (2003) Adaptive management and ecological restoration. In P Freiderici, ed. *Ecological restoration of southwestern ponderosa pine forests*, pp. 417–428. Island Press, Washington DC.

Murray-Rust P (2008) Chemistry for everyone. *Nature*, 451, 648–651.

Muth NZ and Pigliucci M (2007) Implementation of a novel framework for assessing species plasticity in biological invasions: responses of *Centaurea* and *Crepis* to phosphorus and water availability. *Journal of Ecology*, **95**, 1001–1013.

Naeem S, Knops JMH, Tilman D, Howe KM, Kennedy T, and Gale S (2000) Plant neighborhood diversity increases resistance to invasion in experimental grassland plots. *Oikos*, **91**, 97–108.

Nehrbass N, Winkler E, Müllerová, Pergl J, Pyšek P, and Perglová (2007) A simulation model of plant invasion: long-distance dispersal determines the pattern of spread. *Biological Invasions*, **9**, 383–395.

Nekola JC and Brown JH (2007) The wealth of species: ecological communities, complex systems, and the legacy of Frank Preston. *Ecology Letters*, **10**, 188–196.

Nellis DW and Small V (1983) Mongoose predation on sea turtle eggs and nests. *Biotropica*, **15**, 159–160.

Nentwig W, ed. (2007) *Biological Invasions*. Springer, Berlin.

Nicholls H (2007) Thre royal raccoon from Swedesboro. *Nature*, **446**, 255–256.

Nisbet MC and Mooney C (2007a) Science and society: framing science. *Science*, **316**, 56.

Nisbet MC and Mooney C (2007b) The risks and advantages of framing science, response. *Science*, **317**, 1169–1170.

Norden N, Chave J, Caubère A, Châtelet P, Ferroni N, Forget P, and Thébaud C (2007) Is temporal variation of seedling communities determined by environment or by seed arrival? A test in a neotropical forest. *Journal of Ecology*, **95**, 507–516.

Normile D (2008) Driven to extinction. *Science*, **319**, 1606–1609.

Nowacki GJ and Abrams MD (2008) The demise of fire and 'mesophication' of forests in the eastern United States. *Bioscience*, **58**, 123–138.

Nummi P (2002) Introduced semiaquatic mammals and birds in Europe. In E Leppäkoski, S Gollasch, and S Olenin, ed. *Invasive aquatic species in Europe. Distribution, impacts and management*, pp. 162–172. Kluwer Academic Publishers, Dordrecht.

Nyberg B (1999) *An introductory guide to adaptive management for project leaders and participants*. British Columbia Forest Service, Victoria, BC.

O'Brien W (2006) Exotic invasions, nativism, and ecological restoration: on the persistence of a contentious debate. *Ethics, Place and Environment*, **9**, 63–77.

Occhipinti-Ambrogi A and Galil BS (2004) A uniform terminology on bioinvasions: a chimera or an operative tool? *Marine Pollution Bulletin*, **49**, 688–694.

O'Connor MI, Bruno JF, Gaines SD, Halpern BS, Lester SE, Kinlan BP, and Weiss JM (2006) Temperature control of larval dispersal and the implications for marine ecology, evolution, and conservation. *Proceedings of the National Academy of Science USA*, **104**, 1266–1271.

Odum EP (1997) *Ecology: a bridge between science and society*. Sinauer Associates, Sunderland, MA.

Ogada MO (2005) Effects of the Louisiana crayfish invasion on the food and territorial ecology of the African clawless otter in the Ewaso Ng'iro ecosystem. Ph.D thesis, Kenyatta University.

Ogutu-Ohwayo R (1990) The decline of the native fishes of lakes Victoria and Kyoga (East Africa) and the impact of introduced species, especially the Nile perch, *Lates niloticus* and the Nile tilapia, *Oreochroms niloticus*. *Environmental Biology of Fishes*, **27**, 81–96.

Ohlemuller R, Walker S, and Wilson JB (2006) Local vs regional factors as determinants of the invasibility of indigenous forest fragments by alien plant species. *Oikos*, **112**, 493–501.

Ojaveer, Leppäkoski E, Olenin S, and Ricciardi A (2002) Ecological impact of Ponto-Caspian invaders in the Baltic Sea, European inland waters and the Great Lakes: an inter-ecosystem comparison. In E Leppäkoski, S Gollasch, and S Olenin, ed. *Invasive aquatic species in Europe. Distribution, impacts and management*, pp. 412–425. Kluwer Academic Publishers, Dordrecht.

Olden JD and Poff NL (2003) Toward a mechanistic understanding and prediction of biotic homogenization. *American Naturalist*, **162**, 442–460.

Olden JD and Poff NL (2004) Ecological processes driving biotic homogenization: testing a mechanistic model using fish faunas. *Ecology*, **85**, 1867–1875.

Olden JD, Poff NL, and Bestgen KR (2006) Life-history strategies predict fish invasions and extirpations in the Colorado River basin. *Ecological Monographs*, **76**, 25–40.

Olden JD, Poff NL, Douglas MR, Douglas ME, and Fausch KD (2004) Ecological and evolutionary consequences of biotic homogenization. *Trends in Ecology and Evolution*, **19**, 18–24.

Olyarnik SV, Bracken MES, Byrnes JE, Hughes AR, Hultgren KM, Stachowicz JJ (2008) Ecological factors affecting community invasibility. In G Rilov and JA Crooks, ed. *Biological invasions of marine ecosystems: ecological, management, and geographic perspectives*. Springer, Heidelberg, Germany.

O'Riordan T and Cameron J, ed. (1994) *Interpreting the precautionary principle.* Earthscan, London.

Orrock JL, Witter MS, and Reichman OJ (2008) Apparent competition with an exotic plant reduces native plant establishment. *Ecology*, **89**, 1168–1174.

Owens S (2005) Making a difference? Some perspectives on environmental research and policy. *Transactions of the Institute of British Geographers*, **30**, 287–292.

Owre OT (1973) A consideration of the exotic avifauna of southeastern Florida. *Wilson Bulletin*, **85**, 495.

Ozinga WA, Schaminée JHJ, Bekker RM, Bonn S, Poschlod P, Tackenberg O, Bakker J, and van Groenendael JM (2005) Predictability of plant species composition from environmental conditions is constrained by dispersal limitation. *Oikos*, **108**, 555–561.

Paavola M, Olenin S, and Leppäkoski E (2005) Are invasive species most successful in habitats of low native species richness across European brackish water seas? *Estuarine, Coastal, and Shelf Science*, **64**, 738–750.

Padilla FM and Pugnaire FI (2006) The role of nurse plants in the restoration of degraded environments. *Frontiers in Ecology and the Environment*, **4**, 196–202.

Page SE (2007) *The difference: how the power of diversity creates better groups, firms, schools, and societies.* Princeton University Press, Princeton, NJ.

Paine RT (1966) Food web complexity and species diversity. *American Naturalist*, **100**, 65–75.

Pala C (2008) Invasion biologist suck it up. *Frontiers in Ecology and the Environment*, **6**, 63.

Palsbøll, PJ, Bérubé M, and Allendorf FW (2007) Identification of management units using population genetic data. *Trends in Ecology and Evolution*, **22**, 11–16.

Panov VE and Berezina NA (2002) Invasion history, biology and impacts of the Baikalian amphipod *Gmelinoides fasciatus.* In E Leppäkoski, S Gollasch, and S Olenin, ed. *Invasive aquatic species in Europe. Distribution, impacts and management*, pp. 96–103. Kluwer Academic Publishers, Dordrecht.

Parker IM and Gilbert GS (2004) The evolutionary ecology of novel plant-pathogen interactions. *Annual Review of Ecology and Systematics*, **35**, 675–700.

Parker IM and Gilbert GS (2007) When there is no escape: the effects of natural enemies on native, invasive, and noninvasive plants. *Ecology*, **88**, 1210–1224.

Parker IM, Simberloff D, Lonsdale WM, Goodell K, Wonham M, Kareiva PM, Williamson MH, Von Holle B, Moyle PB, Byers JE, and Goldwasser L (1999) Impact: toward a framework for understanding the ecological effects of invaders. *Biological Invasions*, **1**, 3–19.

Parker JD, Burkepile DE, and Hay ME (2006) Opposing effects of native and exotic herbivores on plant invasions. *Science*, **311**, 1459–1461.

Parker JD and Hay ME (2005) Biotic resistance to plant invasions? Native herbivores prefer non-native plants. *Ecology Letters*, **8**, 959–967.

Parmesan C (2006) Ecological and evolutionary responses to recent climate change. *Annual Review of Ecology, Evolution, and* Systematics, **37**, 637–69.

Parmesan, C, Ryrholm N, Stefanescu C, Hill JK, Thomas CD, Descimon H, Huntley B, Kaila L, Kulberg J, Tammaru T, Tennent WJ, Thomas JA, and Warren M (1999) Poleward shifts in geographical ranges of butterfly species associated with regional warming. *Nature*, **399**, 579–583.

Passi P (2006) A $3.2 million exotic species research program that gets the private sector involved in announced today. *Duluth Tribune*, 12 July 2006.

Pauchard A. Alaback P, and Edlund E (2003) Plant invasions in protected areas at multiple scales: *Linaria vulgaris* (Scrophulariaceae) in the West Yellowstone area. *Western North American Naturalist*, **63**, 416–428.

Pauchard A and Shea K (2006) Integrating the study of non-native plant invasions across spatial scales. *Biological Invasions*, **8**, 399–413.

Paul MJ and Meyer JL (2001) Streams in the urban landscape. *Annual Review of Ecological Systems*, **32**, 333–365.

Pauly D, Christensen V, Dalsgaard J, Froese R, and Torres Jr. F (1998) Fishing down marine food webs. *Science*, **279**, 860–863.

Pauly PJ (2008) *Fruits and plains: the horticultural transformation of America.* Harvard University Press, Cambridge, MA.

Pearson DE and Callaway RM (2006) Biological control agents elevate hantavirus by subsidizing deer mouse populations. *Ecology Letters*, **9**, 443–450.

Pedersen AB and Fenton A (2007) Emphasizing the ecology in parasite community ecology. *Trends in Ecology and Evolution*, **22**, 133–139.

Pemberton RW and Wheeler GS (2006) Orchid bees don't need orchids: evidence from the naturalization of an orchid bee in Florida. *Ecology*, **87**, 1995–2001.

Perelman SB, Chaneton EJ, Batista WB, Burkart SE, and Leon RJC (2007) Habitat stress, species pool size and biotic resistance influence exotic plant richness in the Flooding Pampas grasslands. *Journal of Ecology*, **95**, 662–673.

Perrings C, Dalmazzone S, and Williamson M (2005) The economics of biological invasions. In HA Mooney, RN Mack, JA McNeely, LE Neville, PJ Schei, and JK Waage, ed. *Invasive alien species: a new synthesis*, pp. 16–35. Island Press, Washington, DC.

Perrings C, Williamson M, and Dalmazzone S (2000) *The economics of biological invasions.* Edward Elgar, Cheltenham, UK.

Perry AL, Low PJ, Ellis JR, and Reynolds JD (2005) Climate change and distribution shifts in marine fishes. *Science*, **308**, 1912–1915.

Perry LG, Galatowitsch SM, and Rosen CJ (2004) Competitive control of invasive vegetation: a native wetland sedge suppresses *Phalaris arundinacea* in carbon-enriched soil. *Journal of Applied Ecology*, **41**, 151–162.

Perry WL, Feder JL, and Lodge DM (2001) Implications of hybridization between introduced and resident *Orconectes* crayfishes. *Conservation Biology*, **15**, 1656–1666.

Peterson AT, Ortega-Huerta MA, Bartley J, Sanchez-Cordero V, Soberon J, Buddemeier RH, and Stockwell DRB (2002) Future projections for Mexican faunas under global climate change scenarios. *Nature*, **416**, 626–629.

Peterson AT and Vieglais DA (2001) Predicting species invasions using ecological niche modeling. *Bioscience*, **51**, 363–371.

Peterson GD (2005) Ecological management control, uncertainty, and understanding. In K Cuddington and B Beisner, ed. *Ecological paradigms lost: routes of theory change*, pp.371–395. Elsevier Academic Press, Amsterdam.

Petit C (2007) In the Rockies, pines die and bears feel it. *New York Times*, 30 January (http://www.nytimes.com/2007/01/30/science/30bear.html).

Petrie SA and Knapton RW (1999) Rapid increase and subsequent decline in zebra and quagga mussels in Long Point Bay, Lake Erie: possible influence of waterfowl predation. *Journal of Great Lakes Research*, **25**, 772–782.

Pheloung PC, Williams PA, and Halloy SR (1999) A weed risk assessment model for use as a biosecurity tool evaluating plant introductions. *Journal of Environmental Management*, **57**, 239–251.

Phillips BL, Brown GP, Webb JK, and Shine R (2006) Invasion and the evolution of speed in toads. *Nature*, **439**, 803.

Pianka ER and Horn HS (2005) Ecology's legacy from Robert MacArthur. In K. Cuddington and B Beisner, ed. *Ecological paradigms lost: routes of theory change*, pp. 213–232. Elsevier Academic Press, Amsterdam.

Pickering J and Norris CA (1996) New evidence concerning the extinction of the endemic murid *Rattus macleari* Thomas 1887 from Christmas Island, Indian Ocean. *Australian Mammalogy*, **19**, 35–41.

Pickett STA (2007) The paper trail, W. S. Cooper's 'fundamentals of vegetation change' and a fluent mode of thought for ecology. *Bulletin of the Ecological Society of America*, **88**, 98–102.

Pickett STA, Kolasa J, and Jones CG (1994) *Ecological understanding: the nature of theory and the theory of nature*. Academic Press, San Diego.

Pickett STA, Kolasa J, and Jones CG (2007) *Ecological understanding: the nature of theory and the theory of nature*, 2nd edn. Elsevier, Amsterdam.

Pickett STA and White PS (1985) *The Ecology of natural disturbance and patch dynamics*. Academic Press, New York.

Pierce S, Luzzaro A, Caccianiga M, Ceriani RM, and Cerabolini B (2007) Disturbance is the principal σ-scale filter determining niche differentiation, coexistence and biodiversity in an alpine community. *Journal of Ecology*, **95**, 698–706.

Pilkey OH and Pilkey-Jarvis L (2007) *Useless arithmetic: why environmental scientists can't predict the future*. Columbia University Press, New York.

Pimentel D, Lach L, Zuniga R, and Morrison D (2000) Environmental and economic costs of nonindigenous species in the United States. *BioScience*, **50**, 53–65.

Pimentel D, Zuniga R, and Morrison D (2005) Update on the environmental and economic costs associated with alien-invasive species in the United States. *Ecological Economics*, **52**, 273–288.

Pleasant A (2007) The risks and advantages of framing science (letter) *Science*, **317**, 1168.

Plowright W (1982) The effects of rinderpest and rinderpest control on wildlife in Africa. *Symposia of the Zoological Society of London*, **50**, 1–28.

Poff NL, Olden JD, Merritt D, and Pepin D (2007) Homogenization of regional river dynamics by dams and global biodiversity implications. *Proceedings of the National Academcy of Sciences, USA*, **104**, 5732–5737.

Pörtner HO and Knust R (2007) Climate change affects marine fishes through the oxygen limitation of thermal tolerance. *Science*, **315**, 95–97.

Potts BM, Barbour RC, Hingston AB, and Vaillancourt RE (2003) Corrigendum to: Turner Review No. 6 Genetic pollution of native eucalypt gene pools: identifying the risks. *Australian Journal of Botany*, **51**, 333–333.

Potts MD (2003) Drought in a Bornean everwet rain forest. *Journal of Ecology*, **91**, 467–474.

Prach K and Rehounkova K (2006) Restoration ecology: the new frontier. *Restoration Ecology*, **14**, 323–324.

Pressey RL, Cabeza M, Watts ME, Cowling RM, and Wilson KA (2007) Conservation planning in a changing world. *Trends in Ecology and Evolution*, **22**, 583–592.

Preston FW (1960) Time and space and the variation of species. *Ecology*, **41**, 611–627.

Priddel DM (2007) Compensatory mitigation. *Frontiers in Ecology and the Environment*, **5**, 407–408.

Priddel DM (2008) Compensatory mitigation. *Frontiers in Ecology and the Environment*, **6**, 68.

Priddel DM, Carlile N, Fullagar P, Hutton I, and O'Neill L (2006) Decline in the distribution and abundance of flesh-footed shearwaters (*Puffinus carneipes*) on Lord Howe Island, Australia. *Biological Conservation*, **128**, 412–24.

Pueyo S, Fangliang H, and Tommaso Z (2007) The maximum entropy formalism and the idiosyncratic theory of biodiversity. *Ecology Letters*, **10**, 1017–1028.

Pyšek P and Hulme PE (2005) Spatio-temporal dynamics of plant invasions: Linking pattern to process. *Ecoscience*, **12**, 302–315.

Pyšek P, Müllerová J, and Jarošík V (2007) Historical dynamics of *Heracleum mantegazzianum* invasion at regional and local scales. In P Pyšek, MJW Cock, W Nentwig, and HP Ravn, ed. *Ecology and management of giant hogweed (Heracleum mantegazzianum)*, pp. 42–54. CAB International, Wallingford, UK.

Pyšek P and Prach K (1993) Plant invasions and the role of riparian habitats: a comparison of four species alien to central Europe. *Journal of Biogeography*, **20**, 413–420.

Pyšek P and Prach K (1995) Invasion dynamics of *Impatiens glandulifera*—a century of spreading reconstructed. *Biological Conservation*, **74**, 41–48.

Pyšek P and Richardson DM (2006) The biogeography of naturalization in alien plants. *Journal of Biogeography*, **33**, 2040–2050.

Pyšek P and Richardson DM (2007) Traits associated with invasiveness in alien plants: Where do we stand? In W Nentwig, ed. *Biological invasions, ecological studies 193*, pp 97–126. Springer-Verlag, Berlin.

Pyšek P, Richardson DM, and Jarošík V (2006) Who cites who in the invasion zoo: insights from an analysis of the most highly cited papers in invasion ecology. *Preslia*, **78**, 437–468.

Pyšek P, Richardson DM, Pergl J, Vojtěch J, Zuzana S, and Weber E (2008) Geographical and taxonomical biases in invasion biology. *Trends in Ecology and Evolution*, **23**, 237–244.

Pyšek P, Sádlo J, and Mandák B (2002) Catalogue of alien plants of the Czech Republic. *Preslia*, **74**, 97–186.

Qian H and Ricklefs RE (2006) The role of exotic species in homogenizing the North American flora. *Ecology Letters*, **9**, 1293–1298.

Quinos PM, Insausti P, and Soriano A (1998) Facilitative effect of *Lotus tenuis* on *Paspalum dilatatum* in a lowland grassland of Argentina. *Oecologia*, **114**, 427–431.

Radomski PJ and Goeman TJ (1995) The homogenizing of Minnesota lake fish assemblages. *Fisheries*, **20**, 20–23.

Raghu S, Anderson RC, Daehler CC, Davis AS, Wiedenmann RN, Simberloff D, and Mack RN (2006) Adding biofuels to the invasive species fire? *Science*, 313, 1742.

Rahel FJ (2000) Homogenization of fish faunas across the United States. *Science*, **288**, 854–56.

Rahel FJ (2002) Homogenization of freshwater faunas. *Annual Review of Ecology and Systematics*, 33, 291–315.

Rahel FJ (2007) Biogeographic barriers, connectivity and homogenization of freshwater faunas: it's a small world after all. *Freshwater Biology*, **52**, 696–710.

Reader J (1999) *Africa: a biography of the continent.* Vintage, New York.

Reaser JK (2001) Human dimensions of resource management. In JA McNeely, ed. *The great reshuffling: human dimensions of alien invasive species*, pp. 89–104. IUCN, Gland, Switzerland.

Reaser JK, Gutierrez AT, and Meyerson LA (2003) Biological invasions: do the costs outweigh the benefits? *Bioscience*, **53**, 598–599.

Rebertus AJ, Williamson GB, and Moser EB (1989) Fire induced changes in *Quercus laevis* spatial pattern in Florida sandhills. *Journal of Ecology*, **77**, 638–650.

Regan TJ, McCarthy MA, Baxter PWJ, Dane Panetta F, and Possingham HP (2006) Optimal eradication: when to stop looking for an invasive plant. *Ecology Letters*, **10**, 759–766.

Regnier EE, Harrison SK, and Schmoll JT (2006) Impact of seed caching by the earthworm, Lumbricus terrestris, on giant ragweed (*Ambrosia trifida*) establishment. *Abstracts of the Weed Science Society of America*, **46**, 247.

Rehage JS, Barnett BK, and Sih A (2005) Foraging behaviour and invasiveness: do invasive *Gambusia* exhibit higher feeding rates and broader diets than their non-invasive relatives? *Ecology of Freshwater Fish*, **14**, 352–360.

Reichard SH and White P (2001) Horticulture as a pathway of invasive plant introductions in the United States. *Bioscience*, **51**, 103–113.

Reinhart KO (2006) Invasive plants. *Science*, **311**, 1865.

Reise K, Gollasch S, and Wolff WJ (2002) Introduced marine species of the north sea coasts. In E Leppäkoski, S Gollasch, and S Olenin, ed. *Invasive aquatic species in Europe. Distribution, impacts and management*, pp. 260–266. Kluwer Academic Publishers, Dordrecht.

Reise K., Olenin S, and Thieltges DW (2006) Are aliens threatening aquatic coastal ecosystems? *Helgoland Marine Research*, **60**, 77–83.

Rejmánek M (1989) Invasibility of plant communities. In JA Drake, HA Mooney, F di Castri, RH Groves, FJ Kruger, M Rejmánek, and M Williamson, ed. *Biological invasions: a global perspective*, pp. 369–388. John Wiley & Sons, Brisbane, Australia.

Rejmánek M (1996) A theory of seed plant invasiveness: the first sketch. *Biological Conservation*, **78**, 171–180.

Rejmánek M (1998) Invasive plant species and invadible ecosystems. In OT Sandlund, PJ Schei, and A Vilken, ed. *Invasive species and biodiversity management*, pp. 79–102. Kluwer, Dordrecht.

Rejmánek M and Pitcairn MJ (2002) When is eradication of exotic plant pests a realistic goal? In CR Veitch and MN Clout, ed. *Turning the tide: the eradication of invasive species*, pp. 249–253. IUCN, Gland, Switzerland.

Rejmánek M and Richardson DM (1996) What attributes make some plant species more invasive? *Ecology*, **77**, 1655–166.

Rejmánek M, Richardson DM, Barbour MG, Crawley MJ, Hrusa GF, Moyle PB, Randall JM, Simberloff D, and Williamson M (2002) Biological invasions: politics and the discontinuity of ecological terminology. *Bulletin of the Ecological Society of America*, **83**, 131–133.

Rejmánek M, Richardson DM, Higgins SI, Pitcairn MJ, and Grotkopp E (2005a) Ecology of invasive plants: state of the art. In HA Mooney, RN Mack, JA McNeely, LE Neville, PJ Schei, and JK Waage, ed. *Invasive alien species: a new synthesis*, pp. 104–161. Island Press, Washington, DC.

Rejmánek M, Richardson DM, and Pyšek P (2005b) Plant invasions and invasibility of plant communities. In E Van der Maarel, ed. *Vegetation ecology*, pp. 332–355. Blackwell, Oxford.

Ren M-X, Zhang Q-G, Zhang D-Y (2005) RAPD markers reveal low genetic variation and monodominance of one genotype in *Eichhornia crassipes* populations throughout China. *Weed Research*, **45**, 236–244.

Renne IJ, Tracy BF, and Colonna IA (2006) Shifts in grassland invasibility: effects of soil resources, disturbance, composition, and invader size. *Ecology*, **87**, 2264–2277.

Renner S (2004) Plant dispersal across the tropical Atlantic by wind and sea currents. *International Journal of Plant Sciences*, **165**(4 Suppl.), S23–S33.

Renwick JAA, Zhang W, Haribal M, Attygalle AB, and Lopez KD (2001) Dual chemical barriers protect a plant against different larval stages of an insect. *Journal of Chemical Ecology*, **27**, 1575–1583.

Reynolds JD (1988) Crayfish extinctions and crayfish plague in central Ireland. *Biological Conservation*, **45**, 279–285.

Ricciardi A (2005) Facilitation and synergistic interactions among introduced aquatic species. In HA Mooney and RJ Hobbs, ed. *Invasive species in a changing world*, pp. 162–178. Island Press, Washington DC.

Ricciardi A (2007) Are modern biological invasions an unprecedented form of global change? *Conservation Biology*, **21**, 329–336.

Ricciardi A and Cohen J (2007) The invasiveness of an introduced species does not predict its impact. *Biological Invasions*, **9**, 309–315.

Ricciardi A and MacIsaac HJ (2008) The book that began invasion ecology. *Nature*, **452**, 34.

Ricciardi A and Mottiar M (2006) Does Darwin's naturalization hypothesis explain fish invasions? *Biological Invasions*, **8**, 1403–1407.

Ricciardi A and Rasmussen JB (1998) Predicting the identity and impact of future biological invaders: a priority for aquatic resource management. *Canadian Journal of Fisheries and Aquatic Sciences*, **55**, 1759–1765.

Ricciardi A, Whoriskey FG, and Rasmussen JB (1997) The role of the zebra mussel (*Dreissena polymorpha*) in structuring macroinvertebrate communities on hard substrata. *Canadian Journal of Fisheries and Aquatic Sciences*, **54**, 2596–2608.

Richards, CL, Bossdorf O, Muth, NZ, Gurevitch J, and Pigliucci M (2006) Jack of all trades, master of some? On the role of phenotypic plasticity in plant invasions. *Ecology Letters*, **9**, 981–993.

Richardson DM (2006) *Pinus*: a model group for unlocking the secrets of alien plant invasions? *Preslia*, **78**, 375–388.

Richardson DM, Allsopp N, D'Antonio CM, Milton SJ, and Rejmánek M (2000b) Plant invasions: The role of mutualisms. *Biological Reviews*, **75**, 65–93.

Richardson DM, Holmes PM, Esler KJ, Galatowitsch SM, Stromberg JC, Kirkman SP, Pyšek P, and Hobbs RJ (2007) Riparian vegetation: degradation, alien plant invasions, and restoration prospects. *Diversity and Distributions*, **13**, 126–139.

Richardson DM and Pyšek P (2006) Plant invasions: merging the concepts of species invasiveness and community invasibility. *Progress in Physical Geography*, **30**, 409–431.

Richardson DM and Pyšek P (2007) Elton, C.S. 1958: the ecology of invasions by animals and plants. London, Methuen. *Progress in Physical Geography*, **31**, 659–666.

Richardson DM and Pyšek P (2008) Fifty years of invasion ecology: the legacy of Charles Elton. *Diversity and Distributions*, **14**, 161–168.

Richardson DM, Pyšek P, Rejmánek, M., Barbour, MG, Panetta ED, and West CJ (2000a) Naturalization and invasion of alien plants: concepts and definitions. *Diversity and Distributions*, **6**, 93–107.

Ricklefs RE (1987) Community diversity: relative roles of local and regional processes. *Science*, **235**, 167–171.

Ricklefs RE (2004) A comprehensive framework for global pattern in biodiversity. *Ecology Letters*, **7**, 1–15.

Ricklefs RE (2005) Taxon cycles: insights from invasive species. In DF Sax, JJ Stachowicz, and SD Gaines, ed. *Species invasions. Insights into ecology, evolution, and biogeography*, pp. 165–199. Sinauer Associates, Sunderland, MA.

Ricklefs RE (2006) Evolutionary diversification and the origin of the diversity-environment relationship. *Ecology*, **87**, S3-S13.

Ricklefs RE and Bermingham E (2002) The concept of the taxon cycle in biogeography. *Global Ecology and Biogeography*, **11**, 353–361.

Ricklefs RE and Schluter D, ed. (1993) *Species diversity: historical and geographical perspectives*. University of Chicago Press, Chicago, IL.

Ridenour WM and Callaway RM (2001) The relative importance of allelopathy in interference: the effects of an invasive week on a native bunchgrass. *Oecologia*, **126**, 444–450.

Rizzo DM, Garbelotto M, Davidson JM, Slaughter GW, and Koike ST (2002) *Phytophthora ramorum* as the cause of extensive mortality of *Quercus* spp. and *Lithocarpus densiflorus* in California. *Plant Disease*, **86**, 205–214.

Robertson GP and Swinton SM (2005) Reconciling agricultural productivity and environmental integrity: a grand challenge for agriculture. *Frontiers in Ecology and the Environment*, **3**, 38–46.

Robbins J (2008) In a warmer Yellowstone Park, a shifting environmental balance. *New York Times*, 18 March 2008.

Robinson GR, Quinn JF, and Stanton ML (1995) Invasibility of experimental habitat islands in a California winter annual grassland. *Ecology*, **76**, 786–794.

Roderick GK and Navajas M (2003) Genotypes in new environments: genetics and evolution in biological control. *Nature Reviews Genetics*, **4**, 889–899.

Robinson JV and Edgemon MA (1988) An experimental evaluation of the effect of invasion history on community structure. *Ecology*, **69**, 1410–1417.

Rodriguez, LF (2006) Can invasive species facilitate native species? Evidence of how, when, and why these impacts. *Biological Invasions*, **8**, 927–939.

Rodgers VL, Stinson KA, and Finzi AC (2008) Ready or not, garlic mustard is moving in: *Alliaria petiolata* as a member of eastern North American forests. *Bioscience*, **58**, 426–436.

Roe GH and Baker MB (2007) Why is climate sensitivity so unpredictable? *Science*, **318**, 629–632.

Roff DA and Roff RJ (2003) Of rats and Maoris: a novel method for the analysis of patterns of extinction in the New Zealand avifauna before human contact. *Evolutionary Ecology Research*, **5**, 759–779.

Roman J and Darling JA (2007) Paradox lost: genetic diversity and the success of aquatic invasions. *Trends in Ecology and Evolution*, **22**, 454–464.

Romero S (2007) War in the Pacific: it's hell, especially if you're a goat. *New York Times*, 1 May.

Rood SB, Gourley CR., Ammon, EM, et al. (2003). Flows for floodplain forests: a successful riparian restoration. *BioScience*, **53**, 647–656.

Rosenfield JA, Nolasco S, Lindauer S, Sandoval C, and Kodric-Brown A (2004) The role of hybrid vigor in the replacement of Pecos pupfish by its hybrids with sheeps head minnow. *Conservation Biology*, **18**, 1589–1598.

Rosenzweig ML (1995) *Species diversity in space and time*. Cambridge University Press, Cambridge, UK.

Rosenzweig ML (2001) The four questions: what does the introduction of exotic species do to diversity? *Evolutionary Ecology Research*, **3**, 361–367.

Rossiter NA, Setterfield SA, Douglas MM, and Hutley LB (2003) Testing the grass-fire cycle: alien grass invasion in the tropical savannas of northern Australia. *Diversity and Distributions*, **9**, 169–176.

Rouget M and Richardson DM (2003). Inferring process from pattern in alien plant invasions: a semi-mechanistic model incorporating propagule pressure and environmental factors. *American Naturalist*, **162**, 713–724.

Rudgers JA, Hola J, Orr SP, and Clay K (2007) Forest succession suppressed by an introduced plant-fungal symbiosis. *Ecology* **88**, 18–25.

Ruiz GM and Carlton JT, ed. (2003a) *Invasive species: vectors and management strategies*. Island Press, Washington, DC.

Ruiz GM and Carlton JT (2003b) Preface. In GM Ruiz and JT Carlton, ed. *Invasive species: vectors and management strategies*, pp. ix-xii. Island Press, Washington, DC.

Ruiz GM and Carlton JT (2003c) Invasion vectors: a conceptual framework for management. In GM Ruiz and JT Carlton, ed. *Invasive species: vectors and management strategies*, pp. 459–504. Island Press, Washington, DC.

Ruiz GM, Rawlings TK, Dobbs FC, Drake LA, Mullady T, Hug A, and Colwell RR (2000) Global spread of microorganisms by ships. *Nature*, **408**, 49–50.

Russell CA, Jones TC, Barr IG, Cox NJ, Garten RJ, Gregory V, Gust ID, Hampson AW, Hay AJ, Hurt AC, de Jong JC, Kelso A, Klimov AI, Kageyama T, Komadina N, Lapedes AS, Lin YP, Mosterin A, Obuchi M, Odagiri T, Osterhaus ADME, Rimmelzwaan GF, Shaw MW, Skepner E, Stohr K, Tashiro M, Fouchier RAM, and Smith DJ (2008) The global circulation of seasonal influenza A (H3N2) viruses. *Science*, **320**, 340–346.

Russell FL, Louda SM, Rand TA, and Kachman SD (2007) Variation in herbivore-mediated indirect effects of an invasive plant on a native plant. *Ecology*, **88**, 413–423.

Russell JC, Towns DR, Anderson SH, and Clout MN (2005) Intercepting the first rat ashore. *Nature*, **437**, 1107.

Ryan RL (2000) A people-centered approach to designing and managing restoration projects: insights from understanding attachment to urban natural areas. In *PH Gobster and R Bruce, ed. Restoring nature: perspectives from the social sciences and humanities*, pp. 209–228. Island Press, Washington, DC.

Sagarin RD and Gaines SD (2002) The 'abundant centre' distribution: to what extent is it a biogeographical rule? *Ecology Letters*, **5**, 137–147.

Sagoff M (1999) What's wrong with alien species? *Report of the Institute for Philosophy and Public Policy*. University of Maryland, College Park, Maryland (available at http://www.puaf.umd.edu/IPPP/fal11999/exotic_species.htm).

Sale PF (1977) Maintenance of high diversity in coral reef fish communities. *American Naturalist*, **111**,337–359.

Salo P, Korpimä E, Banks PB, and Nordtröm M (2007) Alien predators are more dangerous than native predators to prey populations. *Proceedings of the Royal Society B*, **274**, 1237–1243.

Sanford E. Holzman SB, Haney RA, Rand DM, and Bertness MD (2006) Larval tolerance, gene flow and the northern geographic range limit of fiddler crabs. *Ecology*, **87**, 2882–2894.

Savidge JA (1987) Extinction of an island forest avifauna by an introduced snake. Ecology **68**, 660–668.

Savolainen P, Leitner T, Wilton AN, Matisoo-Smith E, and Lundeberg J (2004) A detailed picture of the origin of the Australian dingo, obtained from the study of mitochondrial DNA. *Proceedings of the National Academy of Sciences, USA*, **101**, 12387–12390.

Sax DF (2001) Latitudenal gradients and geographic ranges of exotic species: implications for biogeography. *Journal of Biogeography*, **28**, 139–150.

Sax DF (2002) Native and naturalized plant diversity are positively correlated in scrub communities of California and Chile. *Diversity and Distributions*, **8**, 193–210.

Sax DF and Brown JN (2000) The paradox of invasion. *Global Ecology and Biogeography*, **9**, 363–372.

Sax DF and Gaines SD (2003) Species diversity: from global decreases to local increases. *Trends in Ecology and Evolution*, **18**, 561–566.

Sax DF and Gaines SD (2008) Species invasions and extinctions: the future of native biodiversity on islands. *Proceedings of the National Academy of Sciences USA*, **105**, 11490–11497.

Sax DF, Gaines SD, and Brown JH (2002) Species invasions exceed extinctions on islands worldwide: a comparative study of plants and birds. *American Naturalist*, **160**, 766–783.

Sax DF, Stachowicz JJ, and Gaines SD, ed. (2005a) *Species invasions: insights into ecology, evolution and biogeography*. Sinauer Associates, Sunderland, MA.

Sax D, Gaines SD, and Stachowicz JJ (2005b) Introduction. In DF Sax, JJ Stachowicz, and SD Gaines, ed. *Species invasions: insights into ecology, evolution and biogeography*, pp. 1–7. Sinauer Associates, Sunderland, MA.

Sax DF, Stachowicz JJ, and Gaines SD (2005c) Capstone: Where do we go from here? In DF Sax, JJ Stachowicz, and SD Gaines, ed. *Species invasions: insights into ecology, evolution and biogeography*, pp. 457–480. Sinauer Associates, Sunderland, MA.

Sax DF, Kinlan P, and Smith KF (2005d) A conceptual framework for comparing species assemblages in native and exotic habitats. *Oikos*, **108**, 457–464.

Sax DF, Brown JH, White EP, and Gaines SD (2005e) Insights into the mechanisms that limit species diversity. In DF Sax, JJ Stachowicz, and SD Gaines, ed. *Species invasions: insights into ecology, evolution and biogeography*, pp. 447–465. Sinauer Associates, Sunderland, MA.

Sax DF, Stachowicz JJ, Brown JH, Bruno JF, Dawson MN, Gaines SD, Grosberg RK, Hastings A, Holt RD, Mayfield MM, O'Connor MI, and Rice WR. (2007) Ecological and evolutionary insights from species invasions. *Trends in Ecology and Evolution*, **22**, 465–471.

Scheiner SM (2008) Humans, disasters, and human disasters. *Bioscience*, **58**, 79–80.

Scheu S and Parkinson D (1994) Effects of earthworms on nutrient dynamics, carbon turnover and microorganisms in soils from cool temperate forests of the Canadian Rocky Mountains: laboratory studies. *Applied Soil Ecology*, **1**, 113–125.

Schierenbeck KA and Aïnouche ML (2006) The role of evolutionary genetics in studies of plant invasions. In MW Cadotte, SM McMahon, and T Fukami, ed. *Conceptual ecology and invasions biology: reciprocal approaches to nature*, pp. 193–221. Springer, Dordrecht, The Netherlands.

Schierenbeck KA, Mack RN, and Sharitz RR (1994) Effects of herbivory on growth and biomass allocation in native and introduced species of *Lonicera*. *Ecology*, **75**, 1661–1672.

Schindler DE, Kitchell JF, He S, Carpenter SR, Hodgson JR, and Cottingham KL (1993) Food-web structure and phosphorus cycling in lakes. *Transactions of the American Fisheries Society*, **122**, 756–772.

Schlaepfer MA, Sherman PW, Blossey B, and Runge MC (2005) Introduced species as evolutionary traps. *Ecology Letters*, **8**, 241–246.

Schmidt KA and Whelan CJ (1999) Effects of exotic *Lonicera* and *Rhamnus* on songbird nest predation. *Conservation Biology*, **13**, 1502–1506.

Schoen DJ, Reichman JR, and Ellstrand NC (2008) Transgene escape monitoring, population genetics, and the law. *Bioscience*, **58**, 71–77.

Schroeder HW (2000) The restoration experience: volunteers' motives, values, and concepts of nature. *In PH Gobster, R Bruce, ed. Restoring nature: perspectives from the social sciences and humanities*, pp. 247–264. Island Press, Washington, DC.

Schwartz MW (2006) How conservation scientists can help develop social capital for biodiversity. *Conservation Biology*, **20**, 1550–1552.

Schwartz MW, Hoeksema JD, Gehring CA, Johnson NC, Klironomos JN, Abbott LK, and Pringle A (2006). The promise and the potential consequences of the global transport of mycorrhizal fungal inoculum. *Ecology Letters*, **9**, 501–515.

Schwartz MK, Luikart G, and Waples RS (2007) Genetic monitoring as a promising tool for conservation and management. *Trends in Ecology and Evolution*, **22**, 25–33.

Schwindt E and Iribarne OO (2000) Settlement sites, survival and effects on benthos of an introduced reef-building polychaete in a SW Atlantic coastal lagoon. *Bulletin of Marine Science*, **67**, 73–82.

Sclater, PL (1858) On the general geographical distribution of the members of the Class Aves. *Journal of the Proceedings of the Linnean Society: Zoology*, **2**, 130–145.

Scott JJ and Kirkpatrick JB (2008) Rabbits, landslips and vegetation change on the coastal slopes of subantarctic Macquarie Island, 1980–2007: implications for management. *Polar Biology*, **31**, 409–419.

Scott PE, DeVault TL, Bajema RA, and Lima SL (2002) Grassland vegetation and bird abundances on reclaimed midwestern coal mines. *Wildlife Society Bulletin*, **30**, 1006–1014.

Seabloom E, Bjørnstad O, Bolker B, and Reichman OJ (2005) The spatial signature of environmental heterogeneity, dispersal, and competition in successional grasslands. *Ecological Monographs*, **75**, 199–214.

Seabloom EW, Harpole WS, Reichman OJ, and Tilman D (2003) Invasion, competitive dominance, and resource use by exotic and native California grassland species. *Proceedings of the National Academy of Sciences, USA*, **100**, 13384–13389.

Seaman GA and Randall JE (1962) The mongoose as a predator in the Virgin Islands. *Journal of Mammalogy*, **43**, 544–546.

Searcy KB, Pucko C, and McClelland D (2006) The distribution and habitat preferences of introduced species in the Mount Holyoke Range, Hampshire Co., MA. *Rhodora*, **108**, 43–61.

Seastedt TR (2005) Soil biology and the emergence of adventive grassland ecosystems. In SC Jaravis, PJ Murray, and JA Roker, ed. *Optimization of nutrient cycling and soil quality for sustainable grasslands*, pp. 15–24. Wageningen Academic Publishers, The Netherlands.

Semmens B, Buhle E, Salomon A, and Pattengill-Semmens C (2004) A hotspot of non-native marine fishes: evidence for the aquarium trade as an invasion pathway. *Marine Ecology Progress Series*, **266**, 239–244.

Shapiro A. (2002) The Californian urban butterfly fauna is dependent on alien plants. *Diversite and Distributions*, **8**, 31–40.

Sharov AA, Liebhold AM, and Ravlin FW (1995) Prediction of gypsy moth (Lepidoptera, Lymantriidae) mating success from pheromone trap counts. *Environmental Entomology*, **24**, 1239–1244.

Shea K and Chesson P (2002) Community ecology theory as a framework for biological invasions. *Trends in Ecology and Evolution*, **17**, 170–176.

Sheldon SP and Jones KN (2001) Restricted gene flow according to host plant in an herbivore feeding on native and exotic watermilfoils (*Myriophyllum*: Haloragaceae). *International Journal of Plant Sciences*, **162**, 793–799.

Shigesada N and Kawasaki K (1997) *Biological invasions. theory and practice*. Oxford University Press, Oxford.

Shine C, Williams N, and Burhenne-Guilmin F (2005) Legal and institutional frameworks for invasive-alien species. In HA Mooney, RN Mack, JA McNeely, LE Neville, PJ Schei, and JK Waage, ed. *Invasive alien species: a new synthesis*, pp. 233–284. Island Press, Washington, DC.

Shipley, B, Vile D, and Garnier E (2006) From plant traits to plant communities: a statistical mechanistic approach to biodiversity. *Science*, **314**, 812–814.

Shurin JB (2000) Dispersal limitation, invasion resistance, and the structure of pond zooplankton communities. *Ecology*, **81**, 3074–3086

Silvertown, JW (2005) *Demons in Eden, the paradox of plant diversity*. University of Chicago Press, Chicago.

Simberloff D (1981) Community effects of introduced species. In H Nitecki, ed. *Biotic crises in ecological and evolutionary time*, pp. 53–81. Academic Press, New York.

Simberloff D (1996) Impacts of introduced species in the United States. *Consequences: the nature and implications of environmental change*, **2**, pp.1–13. US Global Change Research Information Office (available at: http://www.gcrio.org/CONSEQUENCES/vol2no2/article2.html).

Simberloff D (2003) Confronting introduced species: a form of xenophobia? *Biological Invasions*, **5**, 179–192.

Simberloff D (2005) The politics of assessing risk for biological invasions: the USA as a case study. *Trends in Ecology and Evolution*, **20**, 216–222.

Simberloff D and Gibbons L (2004) Now you see them, now you don't! Population crashes of established introduced species. *Biological Invasions*, **6**, 161–172.

Simberloff D, Schmitz DC, and Brown TC (1997) *Strangers in paradise, impact and management of nonindigenous species in Florida.* Island Press, Washington, DC.

Simberloff D and Stiling P (1996) How risky is biological control? *Ecology,* **77,** 1065–1074.

Simberloff D and Von Holle B (1999) Positive inter actions of nonindigenous species: Invasional meltdown? *Biological Invasions,* **1,** 21–32.

Simpson GG, Miller Jr F, Nagel E, King EJ, and Bremer J (1961) *Notes on the nature of science.* Harcourt, Brace and World, Inc, New York.

Slobodkin LB (2001) The good, the bad and the reified. *Evolutionary Ecology Research,* **3,** 1–13.

Smart SM, Thompson K, Marrs RH, Le Duc MG, Maskell LC, and Firbank LG (2006) Biotic homogenization and changes in species diversity across human-mediated ecosystems. *Proceedings of the Royal Society B,* **273,** 2659–2665.

Smith JM (1974) *Models in ecology.* Cambridge University Press, London.

Smith KF, Sax DF, Gaines SD, Guernier V, and Guégan J-F (2007) Globalization of human infectious disease. *Ecology,* **88,** 1903–1910.

Smith RG, Maxwell BD, Menalled FD, and Rew LJ (2006) Lessons from agriculture may improve the management of invasive plants in wildland systems. *Frontiers in Ecology and the Environment,* **4,** 428–434.

Smith SA and Shurin JB (2006) Room for one more? Evidence for invasibility and saturation in ecological communities. In MW Cadotte, SM McMahon, and T Fukami, ed. *Conceptual ecology and invasions biology: reciprocal approaches to nature,* pp. 423–447. Springer, Dordrecht, The Netherlands.

Soberón J (2007) Grinellian and Eltonian niches and geographic distributions of species. *Ecology Letters,* **10,** 1115–1123.

Sol D, Duncan RP, Blackburn TM, Cassey P, and Lefebvre L (2005) Big brains, enhanced cognition, and response of birds to novel environments. *Proceedings of the National Academy of Sciences USA,* **102,** 5460–5465.

Sol D and Lefebvre L (2000) Behavioural flexibility predicts invasion success in birds introduced to New Zealand. *Oikos,* **90,** 599–605.

Sol, D, Timmermans S, and Lefebvre L (2002) Behavioral flexibility and invasion success in birds. *Animal Behavior,* **63,** 495–502.

Solow AR and Costello CJ (2004) Estimating the rate of species introductions from the discovery record. *Ecology,* **85,** 1822–1825.

Spalding VM (1909) *Distribution and movements of desert plants.* Carnegie Institute of Washington, Washington, DC.

Spinage CA (2003) *Cattle plague: a history.* Kluwer, New York.

Stachowicz JJ, Bruno JF, and Duffy JE (2007) Understanding the effects of marine biodiversity on community and ecosystem processes. *Annual Review of Ecology, Evolution, and Systematics,* **38,** 739–766.

Stachowicz JJ and Byrnes JE (2006) Species diversity, invasion success and ecosystem functioning: disentangling the influence of resource competition, facilitation and extrinsic factors. *Marine Ecology Progress Series,* **311,** 251–262.

Stachowicz JJ, Fried H, Osman RW, and Whitlatch RB (2002) Biodiversity, invasion resistance, and marine ecosystem function: reconciling pattern and process. *Ecology,* **83,** 2575–2590.

Stachowicz JJ and Tilman D (2005) What species invasions tell us about the relationship between community saturation, species diversity and ecosystem functioning. In, D Sax, J Stachowicz, and S Gaines, ed. *Species invasions: insights into ecology, evolution and biogeography,* pp. 41–64. Sinauer, Sunderland, MA.

Stachowicz JJ and Whitlatch RB (2005) Multiple mutualists provide complementary benefits to their seaweed host. *Ecology,* **86,** 2418–2427.

Stadler B, Muller T, and Orwig D (2006) The ecology of energy and nutrient fluxes in hemlock forests invaded by hemlock woolly adelgid. *Ecology,* **87,** 1792–1804.

Steadman DW (2006) *Extinction and biogeography of tropical Pacific birds.* University of Chicago Press, Chicago.

Stewart G and Hull AC (1949) Cheatgrass (*Bromus tectorum* L.): an ecological intruder in southern Idaho. *Ecology,* **30,** 58–74.

Stinson KA, Campbell SA, Powell JR, Wolfe BE, Callaway RM, Thelan GC, Hallett SG, Prati D, and Klironomos JN (2006) Invasive plant suppresses the growth of native tree seedlings by disrupting belowground mutualisms. *PLoS Biology,* **4,** e140 doi:10.1371/journal.pbio.0040140.

Stinson KA, Kaufman SK, Durbin L, and Lowenstein F (2007) Impact of garlic mustard invasion on a forest understory community. *Northeastern Naturalist,* **14,** 73–88.

Stockwell CA, Hendry AP, and Kinnison MT (2003) Contemporary evolution meets conservation biology. *Trends in Ecology and Evolution,* **18,** 94–101.

Stohlgren TJ, Barnett D, Flather C, Kartesz J, and Peterjohn B (2005) Plant species invasions along the latitudinal gradient in the United States. *Ecology,* **86,** 2298–2309.

Stohlgren TJ, Barnett D, and Kartesz J (2003) The rich get richer: patterns of plant invasions in the United States. *Frontiers in Ecology and the Environment,* **1,** 11–14.

Stohlgren TJ, Barnett D, Flather C, Fuller P, Peterjohn B, Kartesz J, and Master LL (2006a) Species richness and patterns of invasion in plants, birds, and fishes in the United States. *Biological Invasions*, **8**, 427–447.

Stohlgren TJ, Jarnevich C, Chong GW, and Evangelista PH (2006b) Scale and plant invasions: a theory of biotic acceptance. *Preslia*, **78**, 405–426.

Stohlgren TJ, Barnett D, Jarnevich CS, Flather C, and Kartesz J (2008a) The myth of plant species saturation. *Ecology Letters*, **11**, 313–322.

Stohlgren TJ, Binkley D, Chong GW, Kalkhan MA, Schell LD, Bull KA, Otsuki Y, Newman G, Bashkin M, and Son Y (1999) Exotic plant species invade hot spots of native plant diversity. *Ecological Monographs*, **69**, 25–46.

Stohlgren TJ, Flather C, Jarnevich CS, Barnett DT, and Kartesz J (2008b) Rejoinder to Harrison (2008): the myth of plant species saturation. *Ecology Letters*, **11**, 322–324.

Stokes VL, Pech R, Banks PB, and Arthur A (2004) Foraging behaviour and habitat use by *Antechinus flavipes* and *Sminthopsis murina* (Marsupialia: Dasyuridae) in response to predation risk in eucalypt woodland. *Biological Conservation*, **117**, 331–342.

Stokstad E (2007) Feared quagga mussel turns up in western United States. *Science*, **26**, 453.

Storch D, Davies RG, Zajicek S, Orme CDL, Olson V, Thomas GH, Ding TS, Rasmussen PC, Ridgely RS, Bennett, PM, Blackburn TM, Owens IPF, and Gaston KJ (2006) Energy, range dynamics and global species richness patterns: reconciling mid-domain effects and environmental determinants of avian diversity. *Ecology Letters*, **9**, 1308–1320.

Storey AA, Ramirez JM, Quiroz D, Burley DV, Addison DJ, Walter R, Anderson AJ, Hunt TL, Athens S, Huynen L, and Matisoo-Smith EA (2007) Radiocarbon and DNA evidence for a pre-Columbian introduction of Polynesian chickens to Chile. *Proceedings of the National Academy of Sciences, USA*, **104**, 10335–10339.

Storfer A, Alfaro ME, Ridenhour BJ, Jancovich KK, Mech SG, Parris MJ, and Collins JP (2007) Phylogenetic concordance analysis shows an emerging pathogen is novel and endemic. *Ecology Letters*, **10**, 1075–1083.

Strauss SY, Campbell CO, and Salamin N (2006a) Exotic taxa less related to native species are more invasive. *Proceedings of the National Academy of Sciences, USA*, **103**, 5841–5845.

Strauss SY, Lau JA, and Carroll SP (2006b) Evolutionary responses of natives to introduced species: what do introductions tell us about natural communities? *Ecology Letters*, **9**, 357–374.

Strayer DL (1999) Effects of alien species on fresh water mollusks in North America. *Journal of the North American Benthological Society*, **18**, 74–98.

Strayer DL, Caraco NF, Cole JJ, Findlay S, and Pace ML (1999) Transformation of freshwater ecosystems by bivalves: a case study of zebra mussels in the Hudson River. *Bioscience*, **49**, 19–27.

Strayer DL and Malcom HM (2007) Effects of zebra mussels (*Dreissena polymorpha*) on native bivalves: the beginning of the end or the end of the beginning? *Journal of the North American Benthological Society*, **26**, 111–122.

Sukopp H (1962) Neophyten in naturlichen Pflanzengesellschaften Mitteleuropas. *Berichte der Deutschen Botanischen Gesellschaft*, **75**, 193–205.

Sultan SE (2007) Development in contexts: the timely emergence of eco-devo. *Trends in Ecology and Evolution*, **22**, 575–582.

Surowiecki J (2004) *The wisdom of crowds: why the many are smarter than the few and how*. Doubleday, New York.

Svecar T (2003) Applying ecological principles to wildland weed management. *Weed Science*, **51**, 266–270.

Svenning J-C and Skov F (2004) Limited filling of the potential range in European tree species. *Ecological Letters*, **7**, 565–573.

Sweitzer RA and Van Vuren D (2002) Rooting and foraging effects of wild pigs on tree regeneration and acorn survival in California's oak woodland ecosystems. *Proceedings of the 5th symposium on oak woodlands: oaks in California 's changing landscape*, pp. 218–238. Pacific Southwest Research Station, Forest Service, U.S. Department of Agriculture. General Technical Report PSW-GTR-184, Albany, CA.

Sykes K (2007) The quality of public dialogue. *Science*, **318**, 1349.

Symstad AJ (2000) A test of the effects of functional group richness and composition on grassland invasibility. *Ecology* **81**, 99–109.

Syrett P, Briese DT, and Hoffmann JH (2000) Success in biological control of terrestrial weeds by arthropods. In G Gurr and S Wratten, ed. *Biological control: measures of success*, pp 189–230. Kluwer Academic Publishers, Dordrecht, The Netherlands.

Tabashnik B (1983) Host range evolution: Shift from native legume host to alfalfa by the butterfly, *Colias philodice eryphile*. *Evolution*, **37**, 150–162.

Tapper S (1979) The effect of fluctuating vole numbers (*Microtus agrestis*) on a population of weasels (*Mustela nivalis*) on farmland. Journal *of Animal Ecology*, **48**, 603–617.

Tavares M and De Melo GAS (2004) Discovery of the first known benthic invasive species in the Southern Ocean:

the North Atlantic spider crab *Hyas araneus* found in the Antarctic Peninsula. *Antarctic Science*, **16**, 129–131.

Taylor CM and Hastings A (2005) Allee effects in biological invasions. *Ecology Letters*, **8**, 895–908.

Taylor DJ and Hebert PDN (1993) Cryptic intercontinental hybridization in *Daphnia* (Crustacea): the ghost of introductions past. *Proceedingsof the Royal Society of London B*, **254**, 163–168.

Taylor LH, Latham SM, and Woolhouse ME (2001) Risk factors for human disease emergence. *Philosophical Transactions of the Royal Society of London Series B*, **356**, 983–989.

Taylor PJ (2005) *Unruly complexity: ecology, interpretation, engagement*. University of Chicago Press, Chicago.

Telesh IV, Bolshagin PV, and Panov VE (2001) Quantitative estimation of the impact of alien species *Cercopagis pengoi* (Crustacea: Onychopoda) on the structure and functioning of plankton community in the Gulf of Finland, Baltic Sea. *Doklady Biological Sciences*, **377**, 157–159.

Telesh IV and Ojaveer H (2002) The predatory water flea *Cercopagis pengoi* in the Baltic sea: invasion history, distribution and implications to ecosystem dynamics. In E Leppäkoski, S Gollasch, and S Olenin, ed. *Invasive aquatic species in Europe. Distribution, impacts and management*, pp. 62–65. Kluwer Academic Publishers, Dordrecht.

Tella JL and Carrete M (2008) Broadening the role of parasites in biological invasions. *Frontiers in Ecology and the Environment*, **6**, 11–12.

Temple SA (1992) Exotic birds: a growing problem with no easy solution. *The Auk*, **109**, 395–397.

Templeton AR (2002) Out of Africa again and again. *Nature*, **416**, 45–51.

Terborgh J (1992) Maintenance of diversity in tropical forests. *Biotropica*, **24**, 283–292.

Thébaud C and Debussche M (1991) Rapid invasion of *Fraxinus ornus* L. along the Herault River system in southern France: the importance of seed dispersal by water. *Journal of Biogeography*, **18**, 7–12.

Theoharides KA and Dukes JS (2007) Plant invasion across space and time: factors affecting nonindigenous species success during four stages of invasion. *New Phytologist*, **276**, 256–273.

Thomas CD and Kunin WE (1999) The spatial structure of populations. *Journal of Animal Ecology*, **68**, 647–657.

Thomas MB and Reid AM (2007) Are exotic natural enemies an effective means for controlling invasive plants? *Trends in Ecology and Evolution*, **22**, 447–453.

Thompson JK (2005a) One estuary, one invasion, two responses–phytoplankton and benthic community dynamics determine the effect of an estuarine invasive suspension-feeder. In RF Dame and S Olenin, ed. *The comparative roles of suspension-feeders in ecosystems*, pp. 291–316, Springer, The Netherlands.

Thompson JN (1994) *The coevolutionary process*. University of Chicago Press, Chicago.

Thompson JN (1998) Rapid evolution as an ecological process. *Trends in Ecology and Evolution*, **13**, 329–332.

Thompson JN (1999) Specific hypotheses on the geographic mosaic of coevolution. *American Naturalist*, **153**, S1–S14.

Thompson JN (2005b) *The geographic mosaic of coevolution*. Univesity of Chicago Press, Chicago.

Thompson K, Hodgson JG, Grime JP, and Burke MJW (2001) Plant traits and temporal scale: evidence from a 5-year invasion experiment using native species. *Journal of Ecology*, **89**, 1054–1060.

Thompson K, Hodgson JG, and Rich TCG (1995) Native and alien invasive plants: more of the same? *Ecography*, **18**, 390–402.

Thomson DM (2004) Competitive effects of the invasive European honey bee on the reproductive success of a native bumble bee. *Ecology*, **85**, 458–470.

Thomson GM (1922) *The naturalisation of animals and plants in New Zealand*. Cambridge University Press, Cambridge.

Thorpe AS and Callaway RM (2006) Interactions between invasive plants and soil ecosystems: positive feedbacks and their potential to persist. In MW adotte, SM McMahon, and T Fukami, ed. *Conceptual ecology and invasions biology: reciprocal approaches to nature*, pp. 323–341. Springer, Dordrecht, The Netherlands.

Thuiller W, Richardson DM, Rouget M, Procheş Ş, and Wilson JR (2006) Interactions between environment, species traits, and human uses describe patterns of plant invasions. *Ecology*, **87**, 1755–1769.

Tilman D (1982) *Resource competition and community structure*. Princeton University Press, Princeton, NJ.

Tilman D (1994) Competition and biodiversity in spatially structured habitats. *Ecology*, **75**, 2–16.

Tilman D (1996) Biodiversity: population versus ecosystem stability. *Ecology*, **77**, 350–363.

Tilman D (2004) Niche tradeoffs, neutrality, and community structure: a stochastic theory of resource competition, invasion, and community assembly. *Proceedings of the National Academy of Sciences, USA*, **101**, 10854–10861.

Tobin PC, Whitmire SL, Johnson DM, Bjørnstad ON, and Liebhold AM (2007) Invasion speed is affected by geographical variation in the strength of Allee effects. *Ecology Letters*, **10**, 36–43.

Todd K (2002) *Tinkering with Eden: a natural history of exotic species in America*. WW Norton and Co., New York.

Toivonen H and Meriläinen J (1980) Impact of the muskrat (*Ondatra zibethica*) on aquatic vegetation in small Finnish lakes. *Developments in Hydrobiology*, **3**, 131–138.

Tompkins DM, White AR, and Boots M (2003) Ecological replacement of native red squirrels by invasive greys driven by disease. *Ecology Letters*, **6**, 189–196.

Torchin ME, Lafferty KD, Dobson AP, McKenzie VJ, and Kuris AM (2003) Introduced species and their missing parasites. *Nature*, **421**, 628–630.

Torchin ME and Mitchell CE (2004) Parasites, pathogens, and invasions by plants and animals. *Frontiers in Ecology and the Environment*, **2**, 183–190.

Townsend CR (2003) Individual, population, community, and ecosystem consequences of a fish invader in New Zealand streams. *Conservation Biology*, **17**, 38–47.

Traveset A and Richardson DM (2006) Biological invasions as disruptors of plant reproductive mutualisms. *Trends in Ecology and Evolution*, **21**, 208–216.

Trombulak SC and Frissell CA. (2000) Review of ecological effects of roads on terrestrial and aquatic communities. *Conservation Biology*, **14**, 18–30.

Truscott AM, Soulsby C, Palmer SCF, Newell L, and Hume PE (2006) The dispersal characteristics of the invasive plant *Mimulus guttatus* and the ecological significance of increased occurrence of high-flow events. *Journal of Ecology*, **94**, 1080–1091.

Tsutsui ND and Suarez AV (2003) The colony structure and population biology of invasive ants. *Conservation Biology*, **17**, 48–58.

Tsutsui ND Suarez AV, Holway DA, and Case TJ (2000) Reduced genetic variation and the success of an invasive species. *Proceedings of the National Academy of Sciences, USA*, **97**, 5948–5953.

Turnbull LA, Rahm S, Baudois O, Eichenberger-Glinz S, Wacker L, and Schmid B (2005) Experimental invasion by legumes reveals non-random assembly rules in grassland communities. *Journal of Ecology*, **93**, 1062–1070.

Turner MG (2005) Landscape ecology in North America: past, present and future. *Ecology*, **86**, 1967–1974.

Urban MC, Phillips BL, Skelly DK, and Shine R (2007) The cane toad's (*Chaunus* [*Bufo*] *marinus*) increasing ability to invade Australia is revealed by a dynamically updated range model. *Proceedings of the Royal Society B*, **274**, 1413–1419.

Valiente-Banuet A and Verdu M (2007) Facilitation can increase the phylogenetic diversity of plant communities. *Ecology Letters*, **10**, 1029–1036.

Vanacker V, von Blanckenburg F, Govers G, Molina A, Poesen J, Deckers J, and Kubik P (2007) Restoring dense vegetation can slow mountain erosion to near natural benchmark levels. *Geology*, **35**, 303–306.

van den Belt H (2003) Debating the precautionary principle: 'guilty until proven innocent' or 'innocent until proven guilty'? *Plant Physiology*, **132**, 1122–1126.

van der Putten WH, Klironomous JN, and Wardle DA (2007b) Microbial ecology of biological invasions. *International Society for Microbial Ecology Journal*, **1**, 28–37.

van der Putten WH, Kowalchuk GA, Brinkman EP, Doodeman GTA, van der Kaaij RM, Kamp, AFD, Menting FBJ, and Veenendaal EM (2007a) Soil feedback of exotic savanna grass relates to pathogen absence and mycorrhizal selectivity. *Ecology*, **88**, 978–988.

van der Velde G, Nagelkerken I, Rajagopal S, and Bij de Vaate A (2002) Invasions by alien species in inland freshwater bodies in western Europe: the Rhine delta. In E Leppäkoski, S Gollasch, and S Olenin, ed. *Invasive aquatic species in Europe. Distribution, impacts and management*, pp. 360–372. Kluwer Academic Publishers, Dordrecht.

Vander Zanden MJ, Casselman JM, and Rasmussen JB (1999) Stable isotope evidence for the food web consequences of species invasions in lakes. *Nature*, **401**, 464–467.

Van Driesche J and Van Driesche R (2000) *Nature out of place: biological invasions in the global age*. Island Press, Washington, DC.

Van Driesche RS, Lyon S, Blossey B, Hoddle M, and Reardon R (2003) *Biological control of invasive plants in the eastern United States*. USDA Forest Service. FHTET-2002–04.

van Kleunen M and Johnson SD (2007) South African Iridaceae with rapid and profuse seedling emergence are more likely to become naturalized in other regions. *Journal of Ecology*, **95**, 674–681.

van Kleunen M and Richardson, DM (2007) Invasion biology and conservation biology: time to join forces to explore the links between species traits and extinction risk and invasiveness. *Progress in Physical Geography*, **31**, 447–50.

van Riper C III, van Riper SG, Goff ML, and Laird M (1986) The epizootiology and ecological significance of malaria in Hawaiian land birds. *Ecological Monographs*, **56**, 327–344.

van Riper C III, van Riper SG, and Hansen WR (2002) Epizootiology and effect of avian pox on Hawaiian forest birds. *The Auk*, **119**, 929–942.

Van Vuren D and Coblentz BE (1987) Some ecological effects of feral sheep on Santa Cruz Island, California, USA. *Biological Conservation*, **41**, 253–268.

van Wilgen BW (2004) Guest Editorial: scientific challenges in the field of invasive alien plant management. *South African Journal of Science*, **100**, 19–20.

Vellend M (2002) A pest and an invader: white-tailed deer (*Odocoileus virginianus* Zimm.) as a seed dispersal agent for honeysuckle shrubs (*Lonicera* L.). *Natural Areas Journal*, **22**, 230–234.

Vellend M, Harmon LJ, Lockwood JL, Mayfield MM, Hughes AR, Wares JP, and Sax DF (2007) Effects of exotic species on evolutionary diversification. *Trends in Ecology and Evolution*, **22**, 481–488.

Vellend M, Hughes AR, Grosberg RK, and Holt RD (2005) Introduction: insights into evolution. In DF Sax, JJ Stachowicz, and SD Gaines, ed. *Species invasions: insights into ecology, evolution and biogeography*, pp. 135–137. Sinauer Associates, Inc., Sunderland, MA.

Venter O, Brodeur NN, Nemiroff L, Belland B, Dolinsek IJ, and Grant JWA (2006) Threats to endangered species in Canada. *Bioscience*, **56**, 903–910.

Verlaque M (2001) Checklist of the Thau Lagoon, a hot-spot of marine species introduction in Europe. *Oceanologica Acta*, **24**, 29–49.

Verling E, Ruiz GM, Smith LD, Galil B, Miller AW, and Murphy KR (2005) Supply-side invasion ecology: characterizing propagule pressure in coastal ecosystems. *Proceedings of the Royal Society B*, **272**, 1249–1257.

Vermeij GJ (1991) Anatomy of an invasion: the Trans-Arctic interchange. *Paleobiology*, **17**, 218–307.

Vermeij GJ (2005) Invasion as expectation: a historical fact of life. In DF Sax, JJ Stachowicz, and SD Gaines, ed. *Species invasions: insights into ecology, evolution and biogeography*, pp. 315–339. Sinauer Associates, Inc., Sunderland, MA.

Vilà M and D'Antonio CM (1998) Fruit choice and seed dispersal of invasive vs noninvasive *Carpobrotus* (Aizoaceae) in coastal California. *Ecology*, **79**, 1053–1060.

Vilà M, and Gimeno I (2007) Does invasion by an alien plant species affect the soil seed bank? *Journal of Vegetation Science*, **18**, 423–430.

Vilà M, Pino J, and Font X (2007) Regional assessment of plant invasions across different habitat types. *Journal of Vegetation Science*, **18**, 35–42.

Vitousek PM (1990) Biological invasions and ecosystem process: towards an integration of population biology and ecosystem studies. *Oikos*, **57**, 7–13.

Vitousek PM, Walker LR, Whiteaker LD, Mueller-Dombois D, and Matson PA (1987) Biological invasion by *Myrica faya* alters ecosystem development in Hawaii. *Science*, **238**, 802–804.

Vogel G (2008) Proposed frog ban makes a splash. *Science*, **319**, 1472.

Waddington CH (1965) Introduction to the symposium. In HG Baker and GL Stebbins, ed. *The genetics of colon izing species*, pp. 1–7. Academic Press, London.

Wake HM (2008) Integrative biology: science for the 21st century. *Bioscience*, **58**, 349–353.

Walker S, Eilson JB, and Lee WG (2005) Does fluctuating resource availability increase invasibility? Evidence from field experiments in New Zealand short tussock grassland. *Biological Invasions*, **7**, 195–211.

Wardle DA (2001) Experimental demonstration that plant diversity reduces invisibility—evidence of a biological mechanism or a consequence of sampling effect. *Oikos*, **95**, 161–170.

Wares JP, Hughes AR, and Grosberg RK (2005) Mechanisms that drive evolutionary change: insights from species introductions and invasions. In DF Sax, JJ Stachowicz, and SD Gaines, ed. *Species Invasions. Insights into ecology, evolution, and biogeography*, pp. 229–257. Sinauer Associates, Sunderland, MA.

Warming E (1909) *Oecology of plants: an introduction to the study of plant communities*. Clarendon Press, Oxford, UK.

Warner RE (1968) The role of introduced diseases in the extinction of the endemic Hawaiian avifauna. *The Condor*, **70**, 101–120.

Warren CR (2007) Perspectives on the 'alien' versus 'native' species debate: a critique of concepts, language and practice. *Progress in Human Geography*, **31**, 427–446.

Warren PH, Law R, and AJ Weatherby (2006) Invasion biology as a community process: messages from microbial microcosms. In MW Cadotte, SM McMahon, and T Fukami, ed. *Conceptual ecology and invasions biology: reciprocal approaches to nature*, pp. 343–367. Springer, Dordrecht, The Netherlands.

Watari Y, Takatsuki S, and Miyashita T (2008) Effects of exotic mongoose (*Herpestes javanicus*) on the native fauna of Amami-Oshima Island, southern Japan, estimated by distribution patterns along the historical gradient of mongoose invasion. *Biological Invasions*, **10**, 7–17.

Watson HC (1847) *Cybele Britannica* Vol. I. Longman and Co., London.

Watson HC (1859) *Cybele Britannica* Vol. IV. Longman and Co., London.

Weatherby AJ (2000) Species coexistence and community assembly in protest microcosms. Ph.D thesis, University of Sheffield, UK.

Weber E and Li B (2008) Plant invasions in China: what is to be expected in the wake of economic development? *Bioscience*, **58**, 437–444.

Webb CO, Ackerly DD, McPeek MA, and Donoghue MJ (2002) Phylogenies and community ecology. *Annual Review of Ecology and Systematics*, **33**, 475–505.

Weiss SB (1999) Cars, cows, and checkerspot butterflies: nitrogen deposition and management of nutrient-poor grasslands for a threatened species. *Conservation Biology*, **13**, 1476–1486.

Weldon C, du Preez LH, Hyatt AD, Muller R, and Speare R. (2004) Origin of the amphibian chytrid fungus. *Emerging Infectious Diseases* [serial on the Internet]. Available from http://www.cdc.gov/ncidod/EID/vol10no12/03–0804.htm

Wescott, DA, Setter M, Bradford MG, McKeown A, and Setter S (2008) Cassowary dispersal of the invasive pond apple in a tropical rainforest: the contribution of subordinate dispersal modes in invasion. *Biodiversity and Distributions*, **14**, 432–439.

Westbrooks R, Maden J, and Brown R (2006) *Detection and reporting of cactus moth in the United States. Fact Sheet.* Mississippi State University, Starkville, MS.

Westley F, Carpenter SR, Brock WA, Holling CS, and Gunderson LH (2002) Why systems of people and nature are not just social and ecological systems. In LH Gunderson and CS Holling, ed. *Panarchy: understanding transformations in human and natural systems*, pp. 103–119. Island Press, Washington, DC.

Westman K (2002) Alien crayfish in Europe: negative and positive impacts and interactions with native crayfish. In E Leppäkoski, S Gollasch, and S Olenin, ed. *Invasive aquatic species in Europe. Distribution, impacts and management*, pp. 79–65. Kluwer Academic Publishers, Dordrecht.

White, EM Wilson JC, and Clarke AR (2006) Biotic indirect effects: a neglected concept in invasion biology. *Diversity and Distributions*, **12**, 443–455.

White PCL and King CM (2006) Predation on native birds in New Zealand beech forests: the reole of functional relationships between stoats *Mustela erminea* and rodents. *Ibis*, **148**, 765–771.

White PS and Pickett STA (1985) Natural disturbance and patch dynamics: an introduction. In STA Pickett and PS White, ed. *The ecology of natural disturbance and patch dynamics*, pp. 3–13. Academic Press, Orlando FL.

Whittier TR, Ringold PL, Herlihy AT, and Pierson SM (2008) A calcium-based invasion risk assessment for zebra and quagga mussels (*Dreissena* spp.). *Frontiers in Ecology and the Environment*, **6**, 180–184.

Wilkinson DM (2004) The parable of Green Mountain: Ascension Island, ecosystem construction and ecological fitting. *Journal of Biogeography*, **31**, 1–4.

Wilcove DS, Rothstein D, Dubow J, Phillips A, and Losos E (1998) Quantifying threats to imperiled species in the United States. *BioScience*, **48**, 607–615.

Wilcox C and Donlan CJ (2007) Resolving economic inefficiencies: compensatory mitigation as a solution to fisheries bycatch–biodiversity conservation conflicts. *Frontiers in Ecology and the Environment*, **5**, 325–331.

Williams JL and Crone EE (2006) The impact of invasive grasses on the population growth of *Anemone patens*, a long-lived native forb. *Ecology*, **87**, 3200–3208.

Williams JW and Jackson ST (2007) Novel climates, no-analog plant communities, and ecological surprises: past and future. *Frontiers in Ecology and Evolution*, **5**, 475–482.

Williams JW, Jackson ST, and Kutzbach JE (2007) Projected distributions of novel and disappearing climates by 2100 AD. *Proceedings of the National Academy of Sciences USA*, **104**, 5738–5742.

Williams JW, Shuman BN, and Webb III T (2001) No-analog conditions and rates of change in the climate and vegetation of eastern North America. *Ecology*, **82**, 3346–3362.

Williams PA (1981) Aspects of the ecology of broom (*Cytisus scoparius*) in Canterbury, New Zealand. *New Zealand Journal of Botany*, **19**, 31–43.

Williams PA (1992) *Hakea sericea*: seed production and role in succession in Golden Bay, Nelson. *Journal of the Royal Society of New Zealand*, **22**, 307–320.

Williams SL and Smith JE (2007) A global review of the distribution, taxonomy, and impacts of introduced seaweeds. *Annual Review of Ecology and Systematics*, **38**, 327–359.

Williamson M (1996) *Biological invasions.* Chapman & Hall, London.

Williamson M and Brown KC (1986) The analysis and modeling of British invasions. *Philosophical Transactions of the Royal Society B*, **314**, 505–522.

Williamson M and Fitter A (1996) The varying success of invaders. *Ecology*, **77**, 1661–1666.

Willis KJ and Birks HJB (2006) What is natural? The need for a long-term perspective in biodiversity conservation. *Science*, **314**, 1261–1265.

Wilson JRU, Richardson DM, Rouget M, Procheş Ş, Armis MA, Henderson L, and Thuiller W (2007) Residence time and potential range: crucial considerations in modelling plant invasions. *Diversity and Distributions*, **13**, 11–22.

Winsome T, Epstein L, Hendrix PF, and Horwath WR (2006) Competitive interactions between native and exotic earthworm species as influenced by habitat quality in a California grassland. *Applied Soil Ecology*, **32**, 38–53.

Wiser SK and Allen RB (2006) What controls invasion of indigenous forests by alien plants? *In* RB Allen and WG Lee, ed. *Biological invasions in New Zealand*, pp 195–209. Springer, Berlin.

Wiser SK, Allen RB, Clinton PW, and Platt KH (1998) Community structure and forest invasion by an exotic herb over 23 years. *Ecology*, **79**, 2071–2081.

Wittenberg R and Cock MJW (2005) Best practices for the prevention and management of invasive alien species. In In HA Mooney, RN Mack, JA McNeely, LE Neville, PJ Schei, and JK Waage, ed. *SCOPE 63—Invasive alien species: a new synthesis*, pp. 209–232. Island Press, Washington, DC.

Witkowski ETF (1991) Effects of invasive alien acacias on nutrient cycling in the coastal lowlands of the Cape fynbos. *Journal of Applied Ecology*, **28**, 1–15.

Wolfe BE and Kliornomos JN (2005) Breaking new ground: soil communities and exotic plan invasion. *Bioscience*, **55**, 477–487.

Wolfe ND, Dunavan CP, and Diamond J (2007) Origins of major human infectious diseases. *Nature*, 447, 279–283.

Wolff WJ (2005) Non-indigenous marine and estuarine species in The Netherlands. *Zoologische Mededelingen (Leiden)*, **79**, 1–116.

Wolff WJ and Reise K (2002) Oyster imports as a vector for the introduction of alien species into northern and western European waters. In E Leppäkoski, S Gollasch, and S Olenin, ed. *Invasive aquatic species in Europe. Distribution, impacts and management*, pp. 193–205. Kluwer Academic Publishers, Dordrecht.

Wonham M and Carlton J (2005) Trends in marine invasions at local and regional scales: the northeast Pacific Ocean as a model system. *Biological Invasions*, **7**, 369–392.

Wonham M and Pachepsky E. (2006) Accumulation of introduced species: a null model of temporal patterns. *Ecology Letters*, **9**, 663–672.

Wonham MJ, Carlton JT, Ruiz GM, and Smith LD (2000) Fish and ships: relating dispersal frequency to success in biological invasions. *Marine Biology*, **136**, 1111—1121.

Wonham MJ, O'Connor M, and Harley CDG (2005) Positive effects of a dominant invader on introduced and native mudflat species. *Marine Ecology Progress Series*, **289**, 109–116.

Woolfrey AR and Ladd PG (2001) Habitat preference and reproductive traits of a major Australian riparian tree species (*Casuarina cunninghamiana*). *Australian Journal of Botany*, **49**, 705–715.

Woolhouse MEJ and Gowtage-Sequeira S (2005) Host range and emerging and reemerging pathogens. *Emerging Infectious Disease*, **11**, 1842–1847.

Wootton LS, Halsey SD, Bevaart K, McGough A, Ondreicka J, and Patel P (2005) When invasive species have benefits as well as costs: managing *Carex kobomugi* (Asiatic sand wedge) in New Jersey's coastal dunes. *Biological Invasions*, **7**, 1027–1017.

WRI (World Resources Institute) (2005) *Millenium ecosystem assessment: living beyond our means—natural assets and human well-being*. World Resources Institute, Washington DC.

Wyatt T and Carlton JT (2002) Phytoplankton introductions in European coastal waters: why are so few invasions reported? In CIESM (Commission Internationale pour l'Exploration Scientifique de la mer Mediterranee) *Workshop Monographs no. 20*, pp. 41–46. CIESM, Monaco.

Xu K, Ye W, Cao H, Deng X, Yang Q, Zhang Y. (2004) The role of diversity and functional traits of species in community invasibility. *Botanical Bulletin of Academia Sinica* **45**, 149–157.

Yang LH, Bastow JL, Spence KO, and Wright AN (2008) What can we learn from resource pulses? *Ecology*, **89**, 621–634.

Yoshida T, Ellner SP, Jones LE, Bohannan BJM, Lenski RE, and Hairston Jr. NG (2007) Cryptic population dynamics: rapid evolution masks trophic interactions. *PloS Biology*, **5**: e235. doi:10.1371/journal.pbio.0050235.

Zaiko A, Olenin S, Daunys D, and Nalepa T (2007) Vulnerability of benthic habitats to the aquatic invasive species. *Biological Invasions*, **9**, 703–714.

Zamith R (2007) Fight against buckthorn rages on. *St. Paul Pioneer Press*, 12 October.

Zangerl AR and Berenbaum MR (2005) Increase in toxicity of an invasive weed after reassociation with its coevolved herbivore. *Proceedings of the National Academy of Sciences, USA*, **102**, 15529–15132.

Zavaleta ES, Hobbs RJ, and Mooney HA (2001) Viewing invasive species removal in a whole-ecosystem context. *Trends in Ecology and Evolution*, **16**, 454–459.

Zeisset I and Beebee TJC (2003) Population genetics of a successful invader: the marsh frog *Rana ridibunda* in Britain. *Molecular Ecology*, **12**, 639–646.

Zeiter M, Stampfli A, and Newbery DM (2006) Recruitment limitation constrains local species richness and productivity in dry grassland. *Ecology*, **87**, 942–951.

Zuk M, Rotenberry JT, and Simmons LW (1998) Calling songs of field crickets (*Teleogryllus oceanicus*) with and without phonotactic parasitoid infection. *Evolution*, **52**, 166–171.

Zuk M, Rotenberry JT, and Tinghitella RM (2006) Silent night: Adaptive disappearance of a sexual signal in a parasitized population of field crickets. *Biology Letters*, **2**, 521–524.

Geographic index

This index is not hierarchical in organization. Page numbers provided are only for pages containing the actual geographic entry listed. For example, page 103 is listed for New York City but not for North America, since 'New York City', but not 'North America', was referred to on this page.

Taxonomic index

This index is not hierarchical in organization. Page numbers provided are only for pages containing the actual taxonomic entry listed. For example, page 114 is listed for voles but not for mammals, since 'voles', but not 'mammals', were referred to on this page.

Subject index